A STRATEGY FOR RESEARCH IN SPACE BIOLOGY AND MEDICINE IN THE NEW CENTURY

Committee on Space Biology and Medicine
Space Studies Board
Commission on Physical Sciences, Mathematics, and Applications
National Research Council

NATIONAL ACADEMY PRESS
Washington, D.C. 1998

...ct of this report was approved by the Governing Board of the National Research Council, whose members are drawn from the councils of the National Academy of Sciences, the National Academy of Engineering, and the Institute of Medicine. The members of the committee responsible for the report were chosen for their special competences and with regard for appropriate balance.

The National Academy of Sciences is a private, nonprofit, self-perpetuating society of distinguished scholars engaged in scientific and engineering research, dedicated to the furtherance of science and technology and to their use for the general welfare. Upon the authority of the charter granted to it by the Congress in 1863, the Academy has a mandate that requires it to advise the federal government on scientific and technical matters. Dr. Bruce Alberts is president of the National Academy of Sciences.

The National Academy of Engineering was established in 1964, under the charter of the National Academy of Sciences, as a parallel organization of outstanding engineers. It is autonomous in its administration and in the selection of its members, sharing with the National Academy of Sciences the responsibility for advising the federal government. The National Academy of Engineering also sponsors engineering programs aimed at meeting national needs, encourages education and research, and recognizes the superior achievements of engineers. Dr. William A. Wulf is president of the National Academy of Engineering.

The Institute of Medicine was established in 1970 by the National Academy of Sciences to secure the services of eminent members of appropriate professions in the examination of policy matters pertaining to the health of the public. The Institute acts under the responsibility given to the National Academy of Sciences by its congressional charter to be an adviser to the federal government and, upon its own initiative, to identify issues of medical care, research, and education. Dr. Kenneth I. Shine is president of the Institute of Medicine.

The National Research Council was organized by the National Academy of Sciences in 1916 to associate the broad community of science and technology with the Academy's purposes of furthering knowledge and advising the federal government. Functioning in accordance with general policies determined by the Academy, the Council has become the principal operating agency of both the National Academy of Sciences and the National Academy of Engineering in providing services to the government, the public, and the scientific and engineering communities. The Council is administered jointly by both Academies and the Institute of Medicine. Dr. Bruce Alberts and Dr. William A. Wulf are chairman and vice chairman, respectively, of the National Research Council.

Support for this project was provided by Contract NASW 96013 between the National Academy of Sciences and the National Aeronautics and Space Administration. Any opinions, findings, conclusions, or recommendations expressed in this publication are those of the author(s) and do not necessarily reflect the view of the organizations or agencies that provided support for this project.

The cover was designed by Penny Margolskee.

Library of Congress Catalog Card Number 98-86544
International Standard Book Number 0-309-06047-8

Additional copies of this report are available from:

National Academy Press
2101 Constitution Ave., NW
Box 285
Washington, DC 20055
800-624-6242
202-334-3313 (in the Washington metropolitan area)
http://www.nap.edu

COMMITTEE ON SPACE BIOLOGY AND MEDICINE

Foreword

The space life sciences occupy a unique niche in the nation's extensive biomedical research enterprise. Only in space is it possible to explore fully the role of gravity on biological systems. In the case of the most complex systems, namely humans, the possible effects of long-term exposure to zero gravity is of more than academic interest. Astronauts have been spending increasing amounts of time in low Earth orbit, extended sojourns in the International Space Station will become routine, and someday humans will likely return to the moon and venture farther. In studies of fundamental biological processes at the cellular or organismic level, the ability to fully manipulate the gravity vector enables a range of studies that cannot be performed in terrestrial laboratories.

The cost and complexity of doing any experiment in space demand that careful priorities be set for research. This was done by the National Research Council for space biology and medicine over a decade ago. The present strategy is a complete reformulation of research agendas in the context of current scientific understanding and current or projected opportunities for conducting investigations in space. It is particularly timely given the nation's decision to make a large investment in an orbiting laboratory on the space station.

Biological research is a relative newcomer to NASA and still occupies a relatively modest portion of the agency's resources. But there is a growing appreciation of the importance of life sciences within NASA. Outside NASA, space research has often been seen by bench biologists as far from the mainstream of their discipline. However, successful life sciences missions on the space shuttle, joint programs with the National Institutes of Health, and effective peer review have enhanced perceptions about the program. In preparing this report, the Space Studies Board's Committee on Space Biology and Medicine, which itself includes many biologists with little or no connection to space research, convened workshops involving participants drawn widely from the relevant disciplines. The product should help to reinforce the positive trends in both the reality and perceptions about space biology and medicine by providing a science-based assessment of the most important topics to pursue for the decade to come.

Claude R. Canizares, *Chair*
Space Studies Board

Preface

In 1987, the Committee on Space Biology and Medicine (CSBM) produced a research strategy, *A Strategy for Space Biology and Medical Science for the 1980s and 1990s.*[1] In 1991, the committee's *Assessment of Programs in Space Biology and Medicine 1991*[2] examined the National Aeronautics and Space Adminstration's (NASA's) progress in implementing the 1987 strategy. Since publication of these reports there have been major changes in the direction and status of NASA's life sciences program. The unprecedented amount of biological and medical data gathered from Spacelab missions since 1987 has allowed NASA investigators to move from experiments of an exploratory nature to those that address more fundamental questions. This development has been accompanied by a program shift away from human physiology, the area of major emphasis in the 1987 CSBM report, to more diverse plant and animal studies.

As a consequence of these and numerous programmatic changes at NASA, the committee believed that a new strategy, which builds on the current scientific understanding of space biology questions and issues, was needed. After a series of discussions with NASA's Life Sciences Division, the committee agreed to undertake a comprehensive review of the status of research in the various fields of space life sciences and to develop a science strategy that could guide NASA in its long-term research and mission planning. This study was carried out over a 3-year period, and its objectives remained the same as those outlined in the 1987 report: "(1) to identify and describe those areas of fundamental scientific investigation in space biology and medicine that are both exciting and important to pursue and (2) to develop

---

[1]Space Science Board, National Research Council. 1987. A Strategy for Space Biology and Medical Science for the 1980s and 1990s. National Academy Press, Washington, D.C.

[2]Space Studies Board, National Research Council. 1991. Assessment of Programs in Space Biology and Medicine 1991. National Academy Press, Washington, D.C.

the foundation of knowledge and understanding that will make long-term manned space habitation and/ or exploration feasible."[3]

Specifically, the committee attempted to provide the following in this report:

• A review of the disciplines of biology and medicine that can usefully be studied in the space environment, including sciences that study plant, animal, and human systems at the molecular, cellular, system, and whole-organism levels;
    • Discussion of the fundamental research issues and questions within these disciplines;
    • Identification of the most promising experimental challenges and opportunities within each discipline;
    • Evaluation of the potential for space research to provide advances within each discipline; and
    • Prioritization of research topics to the extent feasible.

In addition to numerous expert speakers from NASA and academia, who were invited to give presentations at regular committee meetings, the CSBM used a variety of approaches to gather information for its task. Three workshops were organized by the committee, each focusing on a broad life sciences discipline, and both NASA and non-NASA investigators were invited to participate. The committee also sent delegates to several international life sciences workshops organized by NASA and its international partners. Each workshop was directed at reviewing progress in a specific discipline and included participation by space life sciences investigators from around the world. Of course, the committee also reviewed both NASA source materials and the relevant literature, published and online, on flight- and ground-based research.

Separate discipline panels, each chaired by a member of the CSBM, were developed to review and discuss the areas of space radiation and human behavioral studies. These two groups were given responsibility for drafting the sections of this report representing their disciplines, although the final report is the responsibility of the committee as a whole. As orginally planned, the recommendations and analysis developed by the Task Group on the Biological Effects of Space Radiation and published separately in 1996[4] form the basis of Chapter 11, "Radiation Hazards," in CSBM's new strategy for research.

---

[3]Space Science Board, 1987, A Strategy for Space Biology and Medical Science for the 1980s and 1990s, p. xi.
[4]Space Studies Board. 1996. Radiation Hazards to Crews of Interplanetary Missions: Biological Issues and Research Strategies. National Academy Press, Washington, D.C.

Acknowledgment of Reviewers

This report has been reviewed by individuals chosen for their diverse perspectives and technical expertise, in accordance with procedures approved by the National Research Council's (NRC's) Report Review Committee. The purpose of this independent review is to provide candid and critical comments that will assist the authors and the NRC in making the published report as sound as possible and to ensure that the report meets institutional standards for objectivity, evidence, and responsiveness to the study charge. The contents of the review comments and draft manuscript remain confidential to protect the integrity of the deliberative process. We wish to thank the following individuals for their participation in the review of this report:

S. James Adelstein, Harvard Medical School,
Robert M. Berne, University of Virginia,
Joseph V. Brady, Johns Hopkins University,
Robert R. Burris, University of Wisconsin–Madison,
Robert A. Frosch, Harvard University,
Sally K. Frost-Mason, University of Kansas,
Ursula W. Goodenough, Washington University,
J. Richard Hackman, Harvard University,
Jack P. Landolt, Defence and Civil Institute of Environmental Medicine, Ontario, Canada,
Philip Osdoby, Washington University,
Robert O. Scow, National Institute of Diabetes and Digestive and Kidney Diseases, and
Frank A. Witzman, Indiana University Purdue University-Columbus.

Although the individuals listed above have provided many constructive comments and suggestions, responsibility for the final content of this report rests solely with the authoring committee and the NRC.

Contents

A Strategy for Research in Space Biology and Medicine in the New Century

Executive Summary

The core of the National Aeronautics and Space Administration's (NASA's) life sciences research lies in understanding the effects of the space environment on human physiology and on biology in plants and animals. The strategy for achieving that goal as originally enunciated in the 1987 Goldberg report, *A Strategy for Space Biology and Medical Science for the 1980s and 1990s*,[1] remains generally valid today. However, during the past decade there has been an explosion of new scientific understanding catalyzed by advances in molecular and cell biology and genetics, a substantially increased amount of information from flight experiments, and the approach of new opportunities for long-term space-based research on the International Space Station. A reevaluation of opportunities and priorities for NASA-supported research in the biological and biomedical sciences is therefore desirable.

The strategy outlined in the Goldberg report had two main purposes: "(1) to identify and describe those areas of fundamental scientific investigation in space biology and medicine that are both exciting and important to pursue and (2) to develop the foundation of knowledge and understanding that will make long-term manned space habitation and/or exploration feasible."[2] To achieve these purposes, the Goldberg report identified four major goals of space life sciences:

"1. To describe and understand human adaptation to the space environment and readaptation upon return to earth.

"2. To use the knowledge so obtained to devise procedures that will improve the health, safety, comfort, and performance of the astronauts.

"3. To understand the role that gravity plays in the biological processes of both plants and animals.

"4. To determine if any biological phenomenon that arises in an individual organism or small group of organisms is better studied in space than on earth."[3]

These goals remain valid and form the basis of the present report.

Both the Goldberg report and the 1991 follow-up assessment, *Assessment of Programs in Space Biology and Medicine 1991*,[4] emphasized basic research and the importance of vigorous ground-based programs aimed at addressing the fundamental mechanisms that underlie observed effects of the space environment on human physiology and other biological processes. The present report strongly reemphasizes that strategy and calls for an integrated, multidisciplinary approach that encompasses all levels of biological organization—the molecule, the cell, the organ system, and the whole organism—and employs the full range of modern experimental approaches from molecular and cellular biology to organismic physiology.

The sections that follow summarize the Committee on Space Biology and Medicine's priorities for NASA-supported research, its recommendations for high-priority research in individual disciplines, and its recommendations for overall priorities for NASA-sponsored research across disciplinary boundaries. The final section outlines significant concerns in the program and policy arena and offers related recommendations.

PRIORITIES FOR RESEARCH

Taking into account budgetary realities and the need for clearly focused programs, the highest priority for NASA-supported research in space biology and medicine in the new century should be given to research meeting one of the following criteria:

1. *Research aimed at understanding and ameliorating problems that may limit astronauts' ability to survive and/or function during prolonged spaceflight.* Such studies include basic as well as applied research and ground-based investigations as well as flight experiments. NASA programs should focus on aspects of research in which NASA has unique capabilities or that are underemphasized by other agencies.

2. *Fundamental biological processes in which gravity is known to play a direct role.* As above, programmatic focus should emphasize NASA's capabilities and take into account the funding patterns of other agencies.

A lower priority should be assigned to areas of basic and applied research that are relevant to fields of high priority to NASA but are extensively funded by other agencies, and in which NASA has no obvious unique capability or special niche.

HIGH-PRIORITY
DISCIPLINE-SPECIFIC RESEARCH

Because the recommendations for research, and research priorities, in the discipline-specific chapters cover a wide range of fields relevant to space biology and medicine, the committee chose not to reproduce all of those recommendations in full in this executive summary. Instead the committee sought to capture the essence of what is recommended in Chapters 2 through 12, an approach that was best served by condensation, full quotation, or addition of supplemental detail as seemed useful to preserve the intent of the recommendations in their full form and context. The recommendations are numbered only in instances in which the committee considered that there was a clear priority order.

Cell Biology

Rapid advancement in the field of cell biology offers novel opportunities for studying the effects of spaceflight, including weightlessness, on cells and tissues. This possibility for progress stems both from developments in technology and advances in basic concepts of cell structure and function at the molecular level. Reasonable goals for the next period of NASA investigation are to clearly delineate the specific cellular phenomena that are affected by conditions of microgravity, to develop an understanding of the molecular mechanisms by which these changes are induced, and to begin to suggest strategies for countermeasures where indicated. Experience from previous in-flight and ground-based studies has highlighted certain pitfalls that must be avoided in the design and analysis of future experiments. Cellular systems should be emphasized that are known to be affected by gravitational force (e.g., bone, muscle, and vestibular systems in animals; gravitropic systems in plants) or by other aspects of the space environment (e.g., stress-induced phenomena). Consideration should be given to using molecular techniques for the analysis of gene expression and cell architecture and function, and to extending cell culture studies to the analysis of cellular physiology in intact tissues and whole organisms.

The committee makes the following specific recommendations for research in cell biology:

• General mechanisms of mechanoreception and pathways of signal transduction from mechanical stresses are areas of special opportunity and relevance for NASA life sciences. Studies of mechanisms of cellular mechanoreception should include identification of the cellular receptor, investigation of possible changes in membrane and cytoskeletal architecture, and analysis of pathways of response, including signal transduction and resolution in time and space of possible ion transients.

• Studies of cellular responses to environmental stresses encountered in spaceflight (e.g., anoxia, temperature, shock, vibration) should include investigation of the nature of cellular receptors, signal transduction pathways, changes in gene expression, and identification and structure and function analysis of stress proteins that mediate the response.

• The successful conduct of sophisticated cell biological experiments in space will require the development of highly automated and miniaturized instrumentation and advanced methodologies. NASA should work with the scientific community and industry to foster development of advanced instrumentation and methodologies for space-based studies at the cellular level.

Developmental Biology

The specific physiological systems in humans and animals for which gravity is likely to play a critical role in development and/or maintenance include the vestibular system, the multiple sensory systems that interact with the vestibular system, and the topographic space maps that exist throughout the brain. Major changes in perspective in recent years in the general field of developmental biology could greatly affect our ability to study and understand these systems. In particular, the use of saturation mutagenesis to identify genetic components of development, the recognition that molecular mechanisms are conserved across phylogeny, and the information provided by genome sequencing projects have transformed basic developmental studies since the publication in 1987 of the Goldberg report.[5] In the present report the committee stresses the importance of two types of studies, those looking at life cycles and those examining development of gravity-sensing systems such as the vestibular system.

Complete Life Cycles in Microgravity

• The committee recommends that key model organisms be grown through two complete life cycles in space to determine whether there are any critical events during development that are affected by space conditions. Because no critical effects have been seen in model invertebrates, the highest priority should be given to testing vertebrate models such as fish, birds, and small mammals such as mice or rats. If developmental effects are detected, control experiments must be performed on the ground and in space, including the use of a space-based 1-*g* centrifuge, to identity whether gravity or some other element of the space environment induces the developmental abnormalities.

Development of the Vestibular System

Analysis of the development of gravity-sensing systems, including the vestibular system and other systems that interact with it in vertebrates, should be carried out to determine the importance of gravity to their normal development and maintenance. The recommended investigations summarized below should be performed first in ground-based studies to identify appropriate experiments to be performed in space.

• Studies should be performed to define the critical periods for development of the vestibular system. Thus, the critical periods for cellular proliferation, migration, and differentiation and programmed cell death should be identified and the effects of microgravity on these processes assessed.

Neural Space Maps

Neurons composing the brainstem, hippocampal, striatal, and sensory and motor cortical space maps should be investigated as part of the following recommended studies:

• The role of otolithic stimulation on the development and maintenance of the different neural space maps should be investigated.
• Studies should be designed to address how neurons of the various sensory and motor systems interact with vestibular neurons in the normal assembly and function of the neural space maps. Factors should be identified that are supplied by and to the sensory neurons that produce the orderly assembly of these maps in precise coordinate registration.
• The influence of microgravity on the development and maintenance of the neural space maps should be studied.

Neuroplasticity

It is important to characterize neuroplasticity using multidisciplinary approaches that combine structural and molecular with functional investigations of identified cell populations. The process should be characterized at several different times following perturbation, in order to determine the sequence of intermediate events leading to the plastic change. Controls for the effects of nongravitational stresses of the types likely to be encountered in space (such as loud noise and vibration) must also be performed on the ground, so that the space-based experiments can be designed to isolate the effects of microgravity from the effects of other stresses. The committee makes the following recommendations for research on neuroplasticity, including one recommendation taken from Chapter 5, "Sensorimotor Integration."

• Studies are needed to determine whether the compensatory mechanisms that normally function in the vestibulomotor pathways are altered by exposure to microgravity. These experiments should be given the highest priority, because these compensatory mechanisms operate in astronauts entering and returning from space and may have a profound effect on their performance in space and their postflight recovery on Earth.
• Experiments are needed to critically test the role of gravity on the development and maintenance of the vestibular system's capability for neuroplasticity.

• Because the vestibulo-oculomotor system is capable of learning new motor patterns in response to sensory perturbations, it is important to determine if and how these mechanisms are affected by exposure to microgravity.

• Functional magnetic resonance imaging (fMRI) should be employed to investigate the following:

—Changes in sensory and motor cortical maps in human bed-rest studies mimicking different flight durations.

—The effects of microgravity on cortical maps in the human. Pre- and postflight fMRI studies should be conducted with astronauts.

Plants, Gravity, and Space

The study of plants in the space environment has been driven by three main needs: (1) learning how to grow plants successfully in space (space horticulture) either for research or for eventual use in long-term life support systems, (2) determining whether there are any plant developmental or metabolic processes that are critically dependent on gravity, and (3) learning how plants alter their patterns of growth and development to respond to changes in the direction of the gravity vector.

Space Horticulture

A major goal of the Advanced Life Support (ALS) program is to develop an effective, completely closed plant growth system capable of growing plants for a bioregenerative life support system. Toward this end, the committee makes the following recommendations:

• The ALS program should concentrate its ground-based research on developing a completely enclosed plant growth system. This effort will require close collaboration between engineers and plant environmental scientists.

• The ALS spaceflight program should focus on testing the potentially gravity-sensitive components of the closed plant growth system, such as the nutrient delivery system.

Role of Gravity in Plant Development

Whether gravity is required for any specific aspect of the development or metabolism of a plant can best be determined by growing a model plant in space through at least two successive generations (seed-to-seed experiment) and examining carefully the development of the resulting plants to ascertain whether any aspect of the development is altered by a lack of gravity. Specifically, the committee recommends the following:

• The seed-to-seed experiment should be the top priority in this area. The promising results obtained with *Brassica rapa* should be confirmed and extended, using *Arabidopsis thaliana* plants. This experiment must be conducted on the ISS, because the plants should be grown through at least two generations in space.

• To conduct a meaningful seed-to-seed experiment, NASA needs to develop the following:

—A superior plant growth unit, with adequate lighting, gas exchange, and water and/or nutrient delivery; and

—*Arabidopsis thaliana* plants that are insensitive to expected environmental stresses and that contain indicator genes for all the expected environmental stresses, such as high levels of CO_2, vibration, anaerobiosis, water stress, and temperature stresses.

• In the interim, before the ISS is functional, studies on specific stages of plant development in space should be limited to small plants with short life cycles (e.g., *Arabidopsis thaliana* or *Brassica rapa*). Whenever possible, a 1-*g* on-board centrifuge should be available.

Responses of Plants to Change in Direction of the Gravity Vector

Plants respond to the specific direction of the gravity vector in several ways. Among these are the direction of growth of stems and roots (gravitropism) and the swimming direction of some unicellular algae (gravitaxis). Among the committee's recommendations regarding this area of research, the following have the highest priority:

• A primary focus of NASA-sponsored research in plant biology should be on the mechanisms of gravitropism. In particular, modern cellular and molecular techniques should be used to determine the following:

—The identity of the cells that actually perceive gravity, and the role of the cytoskeleton in the process;

—The nature of the cellular asymmetry set up in a cell that perceives the direction of the gravity vector;

—The nature and mechanism of the translocation of the signals that pass from the site of perception to the site of reaction; and

—The nature of the response to the signal(s) that leads to alterations in the rate of cell enlargement.

• A secondary focus should be on the mechanisms of graviperception in single cells, including gravitropic responses of mosses and gravitaxic responses of algae.

Sensorimotor Integration

Sensorimotor integration is an essential element in the control of posture and locomotion, as well as in coordinated body activities such as manipulation of objects and use of tools. The transition from normal gravity to microgravity disrupts postural control and orientation mechanisms. Spatial illusions, and often motion sickness, occur until adaptation to the new force background is achieved. On reentry, severe disturbances of postural, locomotory, and movement control are experienced with reexposure to the normal terrestrial environment. Thresholds for angular and linear accelerations, vestibulo-ocular reflexes, postural mechanisms, vestibulo-spinal reflexes, and gaze control all have been studied extensively in humans, but the development of animal models has lagged. Some of these areas require additional study, and a number of new experimental questions arise, given current knowledge and the need to consider human performance during extended-duration space missions.

Spatial Orientation

Future work should emphasize mechanisms related to the active control of body orientation and movement rather than passive thresholds for the detection of angular or linear acceleration. Briefly summarized, the committee's research recommendations are as follows:

1. It is of critical interest to determine how microgravity and other unusual force environments, including rotating environments, affect the integrative coordination of eye, head, torso, arm, and leg movements.

2. It is important for the success of long-duration space missions to identify the sensory, motor, and cognitive factors that influence adaptation and retention of adaptation to different force environments, including rotating environments.

3. The influence of altered force levels, including microgravity, on spatial coding of position should be explored in parallel experiments with humans and animals.

Posture and Locomotion

The severe reentry disturbances of posture and locomotion experienced by astronauts and cosmonauts after even short-duration spaceflight pose potentially dangerous operational problems. These disturbances would be especially critical in long-duration missions that require accurate postural, locomotory, and manipulatory control during transitions in background force level. The committee recommends the following:

• The time course for adaptation of locomotion and posture to variations in background force level should be determined.

• Techniques should be developed to provide ancillary sensory inputs or aids to enhance postural and locomotory control during and after transitions between different force levels.

Vestibulo-Ocular Reflexes and Oculomotor Control

Considerable progress has been made in understanding how microgravity affects vestibulo-ocular reflexes, pursuit and saccadic eye movements, and control of gaze. The following studies, which can be carried out in parabolic flight, orbital flight, and rotating rooms, are recommended to achieve closure on understanding these critical functions.

• Systematic parametric studies of pursuit, saccadic, and optokinetic eye movements should be carried out as a function of background force level in humans from microgravity to 2 *g*.

• The coordination of eye-head-torso synergies in different force levels and their adaptation to changes in force level should be assessed, with the goal of developing a comprehensive three-dimensional model of the vestibulo-ocular reflex and cervical control of gaze.

Space Motion Sickness

Space motion sickness is an operational problem during the first 72 hours of flight, despite the use of medication, and is a hazard for initial transitions between force environments. The use of virtual environment devices in spaceflight to augment training in long-duration missions and for experimental purposes will likely exacerbate motion sickness. Research is recommended on the following:

• The relationship of motion sickness to altered sensorimotor control of the head and body in microgravity and greater than 1-*g* force backgrounds generated in parabolic flight and rotating rooms; and

• The relationship of the vestibular system to autonomic function, especially cardiovascular regulation.

Bone Physiology

One of the best-documented pathophysiological changes associated with microgravity and the spaceflight environment is bone loss, which can exceed 1 percent per month in weight-bearing bones even when an in-flight exercise regime is followed. Within the discipline of bone physiology, the phenomenon of bone loss in astronauts is clearly the issue of greatest concern to NASA. Both the extent and the reversibility of the bone loss are crucial questions for long-term crewed flights on the space station and for future space exploration and should be addressed by collecting data from each astronaut to build up the necessary database.

Studies on Humans

The committee recommends that questions about microgravity-induced bone loss in humans be studied as follows:

1. To obtain a detailed description of human bone loss in space, a record of skeletal changes occurring during microgravity and postflight should be generated for each astronaut and correlated with age and gender, muscle changes, hormonal changes during flight, diet, and genetic factors (e.g., suscep-tibility to osteoporosis) if and when these genetic factors become known.

2. Bone turnover studies should establish if bone loss is due to increased bone destruction (resorp-tion), decreased bone formation, or both.

3. To develop effective countermeasures, different modalities of mechanical stimulation, the use of exercise (e.g., impact loading), and pharmacological means to prevent bone loss should be evaluated.

Animal Models

If applicable to humans, a considerable amount of useful data on bone loss could be generated using animal models. The committee's priority recommendations are summarized as follows:

1. It should be determined if mechanisms of the bone changes produced by microgravity in animal models are similar to those in humans. Rodent models should include mice, given their smaller size and the availability of genetic variants and transgenic animals. Adult animals should be used. In-flight experiments should include animals exposed to centrifugal forces that reproduce 1-g conditions.

2. When an animal model is identified that mimics human bone changes in spaceflight, it should be used in ground-based models of microgravity, such as hindlimb-suspension unloading. If the ground-based model reproduces the changes observed under microgravity conditions, it should be used exten-sively to address questions of mechanisms.

Skeletal Muscle

A better understanding of the deleterious effects on skeletal muscle of spaceflight and reloading upon return to Earth is necessary to maintain performance and prevent injury. Even after missions of a few weeks, the locomotion of astronauts is very unstable immediately after they return to Earth, owing to a combination of orthostatic intolerance, altered otolith-spinal reflexes, reliance on weakened atro-phic muscles, and inappropriate motor patterns. The committee's high-priority research recommenda-tions are summarized below:

• Priority should be given to research that focuses on cellular and molecular mechanisms underly-ing muscle weakness, fatigue, incoordination, and delayed-onset muscle soreness.

• Ground-based models, including bed rest for humans and hindlimb unloading in normal and genetically altered rodents, should be used within and across disciplines to investigate the mechanisms underlying in-flight and postflight effects on muscle mass, protein composition, myogenesis, fiber type differentiation, and neuromuscular development.

• The mechanisms should be determined whereby muscle cells sense working length and the mechanical stress of gravity. Signal transduction pathways for growth factors, stretch-activated ion channels, regulators of protein synthesis, and interactions of extracellular matrix and membrane proteins with the cytoskeleton should be investigated.

Cardiovascular and Pulmonary Systems

The cardiovascular and pulmonary systems undergo major changes in microgravity, including reduced blood volume that is redistributed headward, increased heart volume, altered blood pressure and heart rate, and improved gas exchange in the lungs despite the surprising persistence of lung ventilation-perfusion inequalities. Many observational research questions have been answered. Future research should focus more on mechanisms. The committee developed a number of recommendations for specific research studies which are broadly summarized below.

Cardiovascular System

• Reevaluate current antiorthostatic countermeasures, and develop and validate new ones. Priority should be given to interventions that may provide simultaneous bone and/or muscle protection.

• Extend current knowledge regarding the magnitude, time course, and mechanisms of cardiovascular adjustments to include long-duration microgravity.

• Determine the mechanisms underlying inadequate total peripheral resistance observed during postflight orthostatic stress.

• Identify and validate appropriate methods for referencing intrathoracic vascular pressures to systemic pressures in microgravity.

Pulmonary System

• Characterize gravity-determined topographical differences of blood flow, ventilation, alveolar size, intrapleural pressures, and mechanical stresses in microgravity during rest and exercise.

• Determine the extent to which pulmonary vascular and microvascular pressures and lymphatic flow are altered by microgravity and whether these changes have any impact on either aging or disease processes.

• Examine patterns of aerosol deposition, and determine whether ventilatory and nonventilatory responses to particulate or antigen inhalation are altered by microgravity.

• Identify changes in pulmonary function that occur during extravehicular activity (EVA), and establish resuscitation procedures for crew members in the event of loss of cabin pressure or EVA suit pressure.

• Evaluate respiratory muscle structure and function in microgravity, at rest, and during maximal exercise.

Endocrinology

The endocrine, nervous, and immune systems regulate the human response to spaceflight and the readjustment processes that follow landing. The principal spaceflight responses to which there is a significant endocrine contribution are the fluid shifts, perturbation of circadian rhythms, loss of red blood cell mass, possible alterations in the immune system, losses of bone and muscle, and maintenance of energy balance. With the advent of the space station era, the focus shifts from early responses to spaceflight to the long-term adaptive responses. The three chronic responses that are areas of serious concern are bone loss, muscle atrophy, and possibly the question of maintaining energy balance at an acceptable level. Priority should be given to studies that are designed to do the following:

• Ensure adequate dietary input during spaceflight. Energy intake must meet needs, and physiological measurements must be made on subjects that are in approximate energy balance so that measurements are not confounded by an undernutrition response. The relationship between the amount of exercise and the protein and energy balance in flight should be investigated.
• Obtain a human hormone profile early and late in flight and, as a control, preflight measurements on the same individuals over an extended period of time.
• Determine the effects of spaceflight on human circadian rhythms. If significant degradation of performance is found and it can be attributed to the disturbed circadian rhythm, explore the use of countermeasures, including a combination of light and melatonin.

Immunology

As individuals stay longer in space, the potential effects of spaceflight on immune function become more significant. There is now convincing evidence that immunological parameters are affected by spaceflight, and important questions should be answered regarding both the biological and the medical significance of these effects and their mechanisms. Future immunological studies should concentrate on functional immunological changes that have been shown to be biologically and medically significant.

Animal Studies

Rodent studies can be used to help determine the biological and/or biomedical significance of spaceflight-induced changes in immune responses. Both short- and long-term studies should be carried out, with priority given to those briefly summarized below:

1. Resistance to infection should be examined in animals immediately after their return from spaceflight.
2. Acquired immune responses should be examined, including specific humoral and cellular immune responses.

Human Studies

Immunological measurements and testing of humans should be carried out to examine parameters with potential functional consequences. The recommended studies are briefly summarized below:

1. Acquired immune responses should be examined as described above for animals.
2. Innate immune responses should be examined, including natural killer cell and neutrophil function.
3. Epidemiological studies should be conducted, as the population of astronauts and cosmonauts increases, to assess the potential risk of infection and, in particular, of the development of tumors.

Radiation Hazards

Exposure of crew members to radiation in space poses potentially serious health effects that need to be controlled or mitigated before long-term missions beyond low Earth orbit can be initiated. The levels of radiation in interplanetary space are high enough and the missions long enough that adequate shielding is necessary to minimize carcinogenic, cataractogenic, and possible neurologic effects for crew members.

The knowledge needed to design adequate radiation shielding has both physical and biological components: (1) the distribution and energies of radiation particles present behind a given shielding material as a result of the shield being struck by a given type and level of incident radiation and (2) the effects of a given dose on relevant biological systems for different radiation types. Each component involves significant uncertainty that must be reduced to permit the effective design of shielding, given that the level of uncertainty governs the amount of shielding.[6]

The execution of the recommended strategies will require considerably more beam time at a heavy-ion accelerator than is currently available, and it is recommended that NASA explore various possibilities, including the construction of new facilities, to increase the research time available for experiments with high-atomic-number, high-energy (HZE) particles. Priority should be given to the following studies:

1. Determine the carcinogenic risks following irradiation by protons and HZE particles.

2. Determine how cell killing and induction of chromosomal aberrations vary as a function of the thickness and composition of shielding.

3. Determine whether there are studies that can be conducted to increase the confidence of extrapolation from rodents to humans of radiation-induced genetic alterations that in turn could enhance similar extrapolations for cancer.

4. Determine if exposure to heavy ions at the level that would occur during deep-space missions of long duration poses a risk to the integrity and function of the central nervous system.

5. Determine if better error analyses can be performed of all factors contributing to the estimation of risk by a particular method, and determine the types and magnitude of uncertainty associated with each method.

6. Determine how the selection and design of the space vehicle affect the radiation environment in which the crew has to exist.

Behavioral Issues

Long-duration missions in space are likely to produce significant changes in individual, group, and organizational behavior. Future missions in space will involve longer periods of exposure to features of the physical environment unique to space and features of the psychosocial environment characteristic of isolated and confined environments. Evidence from previous space missions and from analogue studies suggests that behavioral responses to these environmental stressors will be influenced by characteristics of the individuals, groups, and organizations involved in long-duration missions.

The following list broadly summarizes, in order of priority, the recommended research for behavior and performance during long-duration missions in space:

1. Develop noninvasive qualitative and quantitative techniques for the ongoing assessment of pre-flight, in-flight, and postflight behavior and performance.

2. Investigate the neurobiological and psychosocial mechanisms underlying the effects of physical and psychosocial environmental stressors on cognitive, affective, and psychophysiological measures of behavior and performance. Such research should be conducted both in space and in ground-based analogue environments.

—Research on environmental factors should include an assessment of affective and cognitive responses to microgravity-related changes in perceptual and physiological systems and behavioral responses to perceived physical dangers, restricted privacy and personal space, and physical and social monotony.

—Research on physiological factors should include studies of behavioral correlates of changes in circadian rhythms and sleep patterns; changes in and stability of individual physiological patterns in response to psychosocial and environmental stress and their applicability to measures of in-flight behavior and performance; and the relationship between self-reports and external performance-related and physiological symptoms of stress.

—Research on individual factors should include studies of specific coping strategies and behavioral and physiological indicators of coping-stage transitions during long-duration missions; associations between general and mission-specific personality characteristics and performance criteria of ability, stability, and compatibility; changes in problem-solving ability and other aspects of cognitive performance in flight; and changes in personality and behavior postflight.

—Research on interpersonal factors should include studies of the influence of crew psychosocial heterogeneity on crew tension, cohesion, and performance during a mission; factors affecting ground-crew interactions; and the influence of different styles of leadership and decision-making procedures on group performance.

—Research on organizational factors should include studies of the effect of differences in the cultures of the participating agencies on individual and group performance and behavior; the association between mission duration and changes in behavior and performance; and the organizational requirements for effective management of long-duration missions as they relate to task scheduling and workload and to the distribution of authority and decision making.

3. Evaluate existing countermeasures and develop new countermeasures that effectively contribute to optimal levels of crew performance, individual well-being, and mission success. These countermeasures include the following:

—Screening and selection procedures that are based on a "select-in" assessment of individual personality characteristics and interpersonally oriented psychological assessments of crew compatibility;

—Training programs that are team oriented and that enable crews to successfully address the social, cultural, and psychological issues likely to occur in flight;

—Organizational countermeasures for filling unstructured time and reducing boredom and monotony;

—Clinical countermeasures, such as the use of psychoactive medications in microgravity environments and the use of voice analysis for monitoring the interpersonal performance of crews; and

—Design of spacecraft interiors and amenities to maximize control over the physical environment and reduce the impacts of physical monotony on behavior and performance.

CROSSCUTTING RESEARCH PRIORITIES

This section summarizes the committee's recommendations for the highest-priority research across the entire spectrum of space life sciences. In the near term, until the research facilities of the International Space Station come online or an additional Spacelab mission is provided, NASA-supported research will necessarily be directed primarily toward ground-based investigations designed to answer fundamental questions and frame critical hypotheses that can later be tested in space. Indeed, as this report emphasizes, understanding the basic mechanisms underlying biological and behavioral responses to spaceflight is essential to designing effective countermeasures and protecting astronaut health and safety both in space and upon return to Earth. For these reasons, the following recommendations for high-priority areas of crosscutting research place emphasis on ground-based studies.

Physiological and Psychological Effects of Spaceflight

Priority should be given to research aimed at ameliorating problems that may limit astronauts' health, safety, or performance during and after long-duration spaceflight. The committee emphasizes that specific priorities may shift to a significant degree depending on the types of missions to be carried out in the future, particularly as related to long-term human exploration of space. For this reason, the recommended areas of research are not given an order of priority.

Loss of Weight-bearing Bone and Muscle

Bone loss and muscle deterioration are among the best-documented deleterious effects caused by spaceflight in humans and animals. Exercise has been only partially successful in preventing muscle weakness and bone loss. Development of effective countermeasures requires advances in several areas of research:

• Research should emphasize studies that provide mechanistic insights into the development of effective countermeasures for preventing bone and muscle deterioration during and after spaceflight.
• Ground-based model systems, such as hindlimb unloading in rodents, should be used to investigate the mechanisms of changes that reproduce in-flight and postflight effects.
• A database on the course of microgravity-related bone loss and its reversibility in humans should be established in preflight, in-flight, and postflight recording of bone mineral density.
• Hormonal profiles should be obtained on humans before, during, and after spaceflight.
• The relationship between exercise activity levels and protein energy balance in flight should be investigated.

Vestibular Function, the Vestibular Ocular Reflex, and Sensorimotor Integration

During the transitions in gravitational force that occur going into and returning from spaceflight, the vestibular system undergoes changes in activity that can result in debilitating symptoms in astronauts.

• The highest priority should be given to studies designed to determine the basis for the adaptive compensatory mechanisms in the vestibular and sensorimotor systems that operate both on the ground and in space.
• In-flight recordings of signal processing following otolith afferent stimulation should be made to determine how exposure to microgravity affects central and peripheral vestibular function and development.

• Motor learning should be investigated in spaceflight and the results compared with findings obtained in ground-based studies of this process.

Orthostatic Intolerance Upon Return to Earth Gravity

Orthostatic hypotension, present since the very earliest human spaceflights, still affects a high percentage of astronauts returning from spaceflights even of relatively short duration and is an even greater problem for shuttle pilots, who must perform complex reentry maneuvers in an upright, seated position. The problem remains despite the use of extensive antiorthostatic countermeasures by both U.S. and Russian space programs. Studies should focus on determining physiological mechanisms and developing effective countermeasures.

• Current knowledge of the magnitude, time course, and mechanisms of cardiovascular adjustments should be extended to include long-duration exposure to microgravity.
• The specific mechanisms underlying inadequate total peripheral resistance observed during postflight orthostatic stress should be determined.
• Current antiorthostatic countermeasures should be reevaluated to refine those that offer protection and eliminate those that do not. Priority should be given to interventions that may provide simultaneous bone and/or muscle protection.
• Appropriate methods for referencing intrathoracic vascular pressures to systemic pressures in microgravity should be identified and validated, given the observed changes in cardiac and pulmonary volume and compliance.

Radiation Hazards

The biological effects of exposure to radiation in space pose potentially serious health effects for crew members in long-term missions beyond low Earth orbit. High priority is given to the following recommended studies:

• Determine the carcinogenic risks following irradiation by protons and high-atomic-number, high-energy (HZE) particles.
• Determine if exposure to heavy ions at the level that would occur during deep-space missions of long duration poses a risk to the integrity and function of the central nervous system.
• Determine how the selection and design of the space vehicle affect the radiation environment in which the crew has to exist.
• Determine whether combined effects of radiation and stress on the immune system in spaceflight could produce additive or synergistic effects on host defenses.

Physiological Effects of Stress

The immune system interacts closely with the neuroendocrine system. Results indicate a close association between the neuroendocrine status of the host and host defense systems.

• The role that the host response to stressors during spaceflight plays in alterations in host defenses should be determined.

Psychological and Social Issues

The health, well-being, and performance of astronauts on extended missions may be negatively affected by many stressful aspects of the space environment. Mechanisms of response to physiological and psychosocial stressors encountered in spaceflight must be better understood in order to ensure crew safety, health, and productivity.

• Highest priority should be given to interdisciplinary research on the neurobiological (circadian, endocrine) and psychosocial (individual, group, organizational) mechanisms underlying the effects of physical and psychosocial environmental stressors. Cognitive, affective, and psychophysiological measures of behavior and performance should be examined in ground-based analogue settings as well as in flight.

• High priority should be given to evaluation of existing countermeasures (screening and selection, training, monitoring, support) and development of effective new countermeasures.

Fundamental Gravitational Biology

Mechanisms of Graviperception and Gravitropism in Plants

Plants respond to changes in the direction of the gravitational vector by altering the direction of the growth of roots and stems. The gravitropic response requires (1) perception of the gravitational vector by gravisensing cells; (2) intracellular transduction of this information; (3) translocation of the resulting signal to the sites of reaction, i.e., sites of differential growth; and (4) reaction to the signal by the responding cells, i.e., initiation of differential growth.

• Studies of graviperception should concentrate on three problems:
  —The identity of the cells that actually perceive gravity;
  —The intracellular mechanisms by which the direction of the gravity vector is perceived; and
  —The threshold value for graviperception—this will require a spaceflight experiment.

• Studies of gravitropic transduction should focus on the nature of the cellular asymmetry that is set up in a cell that perceives the direction of the gravity vector.

• Studies on the translocation step should concentrate on the nature and mechanism of the translocation of the signals that pass from the site of perception to the site of reaction.

• Studies on the reaction step should focus on the mechanism(s) by which gravitropic signals cause unequal rates of cell elongation, and on the possible effects of gravity on the sensitivity of these cells to the signals.

Mechanisms of Graviperception in Animals

It is known that in several systems sensory stimulation plays a role in the development of the neural connections necessary for normal processing of sensory information. The potential role of gravity in the normal development of the gravity-sensing vestibular system of animals is therefore an important area for ground- and space-based research.

• Ground-based studies should identify the critical periods in vestibular neuron development before initiation of experiments on the effects of microgravity on vestibular development.

• Pre- and postflight functional magnetic resonance imaging (fMRI) studies should be conducted with astronauts to determine the effects of microgravity on neural space maps.

Effects of Spaceflight on Reproduction and Development

To determine whether there are developmental processes that are critically dependent on gravity, organisms should be grown through at least two full generations in space.

• Key model animals should be grown through two life cycles; the highest priority should be given to vertebrate models. If significant developmental effects are detected, control experiments must be performed to determine whether gravity or some other element of the space environment induces these developmental abnormalities.

• An analogous experiment should be carried out with the model plant *Arabidopsis thaliana* to confirm results obtained on Mir with a preliminary experiment using *Brassica rapa*.

PROGRAMMATIC AND POLICY ISSUES

Although NASA has responded effectively to many of the programmatic and policy issues raised in the 1987 and 1991 reports,[7,8] significant concerns in the program and policy arena remain unresolved. These concerns focus on issues relating to strategic planning and conduct of space-based research; utilization of the International Space Station (ISS) for life sciences research; mechanisms for promoting integrated and interdisciplinary research; collection of and access to human flight data, specifically; publication of and access to space life sciences research in general; and professional education.

Space-based Research

Development of Advanced Instrumentation and Methodologies

Future life sciences flight experiments on the ISS will depend on the availability of advanced instrumentation to carry out the measurements and analyses required by the research questions and approaches described in this report. In addition, facile data and information transfer between space- and ground-based investigators are crucial.

• NASA should work with the broad life sciences community to identify and catalyze the development of advanced instrumentation and methodologies that will be required for sophisticated space-based research in the coming decade.

• NASA should take advantage of advanced instrumentation developed in other countries.

• The capability for direct, real-time communication between space-based experimenters and principal investigators at their home laboratories should be a high-priority objective for the ISS.

Utilization of the International Space Station for Life Sciences Research

Issues relating to the design and use of the ISS are a major concern of the committee. These issues include (1) changes in the design of the ISS, (2) the diversion of funds intended for scientific facilities and equipment into construction budgets, (3) the adequacy of power and transmission of data to and from Earth, (4) the availability of crew time for research, and (5) an extended hiatus in flight opportunities for life sciences research owing to delays in ISS construction. These issues have alarmed the life sciences communities.

- To better ensure that the ISS will adequately meet the needs of space life sciences researchers, NASA should continue to bring the external user community as well as NASA scientists into the planning and design phases of facility construction.
- NASA should make every effort to mount at least one Spacelab life sciences flight in the period between Neurolab and the completion of ISS facilities.
- NASA should determine whether continuation of shuttle missions for short-term flight experiments after the opening of ISS would be economically and scientifically sound.

Science Policy Issues

Peer Review

The Division of Life Sciences initiated a universal system of peer review in 1994 for all NASA-supported investigators. The new process has the committee's strong support.

- Responsibility for the establishment of peer review panels and for funding decisions should remain a function of the Headquarters Division of Life Sciences.
- NASA should regularly evaluate the composition of scientific review panels to ensure that the feasibility of proposed flight experiments receives appropriate expert evaluation.

Integration of Research Activities

- Principal investigators of projected flight experiments should be brought together with NASA managers and design engineers at the beginning of the planning process to function as an integrated team responsible for all phases of the planning, design, and testing. This integration should continue throughout the life of the project.
- NASA should regularly review and evaluate the NASA Specialized Centers of Research and Training (NSCORT) program to determine whether this mechanism provides the best way to foster interdisciplinary research and increase the scientific value of the life sciences research program.
- NASA should regularly review and evaluate the performance of the National Space Biomedical Research Institute and the impact of its funding on the overall life sciences research budget and program.

Human Flight Data: Collection and Access

The disciplinary chapters of this report repeatedly stress the need for improved, systematic collection of data on astronauts preflight, in space, and postflight.

- NASA should initiate an ISS-based program to collect detailed physiological and psychological data on astronauts before, during, and after flight.
- NASA should make every effort to promote mechanisms for making complete data obtained from studies on astronauts accessible to qualified investigators in a timely manner. Consideration should be given to possible modifications of current policies and practices relating to the confidentiality of human subjects that would ethically ensure astronaut cooperation in a more effective manner.

Publication and Outreach

An essential outcome of scientific research is publication—dissemination of results to the scientific community at large. The record of peer-reviewed publication, especially of spaceflight experiments, by funded investigators in NASA's life sciences programs needs to be improved, as does the usefulness of the Spaceline Archive to the scientific community.

- NASA should provide funding for data analysis and publication of flight experiments for a sufficient period to ensure analysis of the data and publication of the results.
- NASA should insist on timely dissemination of the results of space life sciences research in peer-reviewed publications. For investigators with previous NASA support, the publication record should be an important criterion for subsequent funding.
- NASA should take as a high priority the completion of data entry into the Spaceline Archive and should ensure that access to the archive is simple and transparent.

Professional Education

NASA should make every effort to ensure the professional training of graduate students and postdoctoral fellows in space and gravitational biology and medicine.

- NASA should take as high priority the support of a small, highly competitive program of postdoctoral fellowships for training in laboratories of NASA-supported investigators in academic and research institutions external to NASA centers.

REFERENCES

1. Space Science Board, National Research Council. 1987. A Strategy for Space Biology and Medical Science for the 1980s and 1990s. National Academy Press, Washington, D.C.
2. Space Science Board, 1987, A Strategy for Space Biology and Medical Science for the 1980s and 1990s, p. xi.
3. Space Science Board, 1987, A Strategy for Space Biology and Medical Science for the 1980s and 1990s, p. 4.
4. Space Studies Board, National Research Council. 1991. Assessment of Programs in Space Biology and Medicine 1991. National Academy Press, Washington, D.C.
5. Space Science Board, National Research Council. 1987. A Strategy for Space Biology and Medical Science for the 1980s and 1990s. National Academy Press, Washington, D.C.
6. Wilson, J.W., Cucinotta, F.A., Shinn, J.L., Kim, M.H., and Badavi, F.F. 1997. Shielding strategies for human space exploration: Introduction. Chapter 1 in Shielding Strategies for Human Space Exploration: A Workshop (John W. Wilson, Jack Miller, and Andrei Konradi, eds.). National Aeronautics and Space Administration.
7. Space Science Board, National Research Council. 1987. A Strategy for Space Biology and Medical Science for the 1980s and 1990s. National Academy Press, Washington, D.C.
8. Space Studies Board, National Research Council. 1991. Assessment of Programs in Space Biology and Medicine 1991. National Academy Press, Washington, D.C.

PART I

Overview

1

Introduction

The core of the National Aeronautics and Space Administration's (NASA's) life sciences research programs lies in understanding the effects of the space environment on human physiology and on gravitational biology in plants and animals. The space environment exposes occupants not only to microgravity, but also to many other potentially perturbing factors, ranging from radiation to the vicissitudes of a confined, enclosed environment in which noise, vibration, temperature, atmospheric quality, and other aspects of the surroundings are generally less than ideal. This report, like its predecessors,[1,2] emphasizes the importance of investigating fundamental mechanisms of the effects elicited by microgravity and other aspects of the space environment, in contrast with the largely descriptive studies of the earlier era of space biology and medicine.

To achieve an understanding of the mechanisms underlying relevant biological and biomedical phenomena, it will be necessary to approach the problems at all levels of biological organization—the molecule, the cell, the organ system, and the whole organism. For example, loss of bone mass from weight-bearing bones is one of the issues of greatest concern in maintaining astronaut health and safety during prolonged spaceflight and upon return to Earth. Understanding the mechanisms responsible for the observed bone loss and the development of effective physical and/or pharmacological countermeasures will require the following at a minimum: characterization of the effects of microgravity on the cells responsible for bone growth and bone resorption; analysis of the molecular and cellular mechanisms whereby cells in weight-bearing bone perceive and respond to the force of gravity; identification and analysis of possible effects of microgravity-induced changes in muscle activity and blood flow on bone metabolism; and determination and understanding of the changes in levels of the many hormones induced by stress and the environment that contribute to the regulation of bone metabolism, both positively and negatively. This example, which spans a range of experimental approaches from molecular biology to organismic physiology, illustrates the integrated, multidisciplinary approach necessary to meet the goals set for NASA life sciences research in the next decade.

HISTORY

It has been 10 years since publication of the report *A Strategy for Space Biology and Medical Science for the 1980s and 1990s* (known as the Goldberg report),[3] in which for the first time, the Space Studies Board (then the Space Science Board), through its Committee on Space Biology and Medicine (CSBM), outlined a broad-based scientific strategy for research in space biology and medical science. The strategy and the major scientific goals remain generally valid today. However, during the 1990s there has been an explosion of new scientific understanding catalyzed by advances in molecular and cell biology and genetics, and substantially more information derived from flight experiments. New opportunities for long-term experiments in space biology and medicine are anticipated on completion of the International Space Station, and there is renewed interest in human exploration of the solar system, which will require long-term survival in space. All of these developments indicate the desirability of reevaluating the opportunities and priorities for NASA-supported life sciences research into the new century. Recent advances in areas of biology and medicine relevant to space offer the promise of better identifying and understanding factors that may be important for astronauts who have to live and work effectively in space for several years at a time and readapt successfully to Earth's environment upon return.

The Goldberg strategy had two main purposes: "(1) to identify and describe those areas of fundamental scientific investigation in space biology and medicine that are both exciting and important to pursue; and (2) to develop the foundation of knowledge and understanding that will make long-term manned space habitation and/or exploration feasible."[4] To achieve these purposes, the Goldberg report identified four major goals of space life sciences:

"1. To describe and understand human adaptation to the space environment and readaptation upon return to Earth.

"2. To use the knowledge so obtained to devise procedures that will improve the health, safety, comfort, and performance of the astronauts.

"3. To understand the role that gravity plays in the biological processes of both plants and animals.

"4. To determine if any biological phenomenon that arises in an individual organism or small group of organisms is better studied in space than on Earth."[5]

The Goldberg report was noteworthy in its emphasis on basic research in gravitational biology and its call for vigorous ground-based programs aimed at addressing fundamental biological mechanisms that underlie observed effects of the space environment on human physiology and other biological processes. Five areas of clinical and basic research related to the effects of microgravity and the space environment were highlighted:

1. Sensorimotor integration, focusing on vestibular function and space motion sickness;

2. Cardiovascular adaptation, particularly the fluid shift induced by microgravity and the orthostatic intolerance experienced upon return to normal gravity;

3. Muscle remodeling and the loss of mass in weight-bearing muscles;

4. Bone and mineral metabolism as related to the demineralization of weight-bearing bone; and

5. Human behavior and the effects of stress.

In addition, two areas of fundamental research in gravitational biology were emphasized: plant gravitropism and developmental biology, especially the question of whether the space environment could support the normal reproduction and development of plants and animals, including mammals, for one or more complete generations.

A follow-up report from CSBM in 1991 (the Smith report)[6] assessed progress in reaching specific scientific goals identified in the Goldberg report and called attention to several additional areas of research, including radiation biology and aspects of immunology, endocrinology, and stress that merited new or renewed consideration. Interestingly, this report was the first to consider cell biology explicitly as a discipline relevant to space biology and medicine. Because of the continuing hiatus in the shuttle program during the intervening period, most NASA studies were ground-based, although international flights yielded some new flight data. The 1991 Smith report cited progress in the development and analysis of ground-based model systems (both animal and human) used to study the effects of weightlessness on the functioning of the cardiovascular and musculoskeletal systems, and it noted significant progress in understanding of the physiology of the sensorimotor system and vestibular function. However, little progress was apparent in elucidating the fundamental cellular mechanisms underlying the observed physiological phenomena. Similarly, progress in plant gravitropism was thought to suffer from limitations in flight opportunities and continued slow progress in identifying basic cellular and intercellular mechanisms responsible for gravity-dependent responses.

Since publication of the Smith report in 1991, resumption of the shuttle program has provided a substantial number of flight opportunities, notably the dedicated Spacelab flights SLS-1, SLS-2, the German D-2 mission, and the recent Neurolab mission. In addition, the development of cooperative programs between NASA and the Russian space agency has made newly available some data pertaining to the effects of long-duration (6- to 12-month) flights.

The 1991 Smith report offered as a major conclusion, and as the first priority in any relevant research strategy, the need to focus on ground-based research aimed at understanding basic mechanisms underlying microgravity-induced changes. The current report strongly reemphasizes that conclusion. Although the overall goal of safeguarding the health, safety, and performance of astronauts (most especially during long-term flight) is an eminently practical one, understanding the true nature of risks and developing maximally effective countermeasures will require analysis of the mechanisms whereby cells, tissues, and whole organisms respond to changes in the magnitude of the gravitational vector. Effective use of the entire armamentarium of contemporary molecular and cell biology, as well as contemporary physiological techniques, will be necessary to achieve this goal. Spaceflight experiments will continue to be required to test and validate mechanistic hypotheses and potential countermeasures developed in ground-based investigations. Although ground-based models of hypogravity have provided important insights into specific aspects of gravitational biology, no ground-based model gives an adequate representation of the microgravity environment. However, experiments carried out in space need a solid foundation in ground-based research that uses a limited number of carefully chosen model systems, resulting in formulation of key hypotheses and identification of the key experiments that require the space environment.

GRAVITY AND LOW GRAVITY

Gravity, Microgravity, and Weightlessness

Orbital spaceflight and parabolic flight generate conditions of weightlessness. These are generally termed "microgravity conditions" and for convenience are so designated in this report. However, they are not actually conditions involving low levels of gravity. In orbital spaceflight, it is the force of Earth's gravity that keeps the vehicle in an orbital path. The gravitational attraction between Earth and the vehicle (and its occupants) provides the centripetal force (mv^2/r) that maintains the vehicle's orbital path about Earth, where r is the distance from the center of Earth to the orbit, v is the velocity of the vehicle, and m is the vehicle's mass. The variation of Earth's gravity g at altitude h above Earth's surface is governed by $g = g_0 \times r_e^2 / (r_e + h)^2$, where g_0, Earth gravity at sea level, is 9.8 m/s^2 and r_e, the radius of Earth, is 6.38×10^6 m. At an orbital altitude of 250 miles (4.4×10^5 m), g would be about 8.6 m/s^2, or 87.5 percent of Earth's gravity at sea level.

Orbital flight represents a condition of free fall in which the balance of forces acting between the craft and its occupants is effectively zero, except when the occupants apply forces to the vehicle so that they can move or operate within it. To interpret experiments carried out in such "weightless" conditions, the importance of contact forces acting on objects under terrestrial conditions must be realized. Einstein's principle of equivalence states that no measuring device can distinguish between inertial and gravitational forces. The same is true of the sensors of the human or animal body. It is the contact surface on which a person or animal stands that provides reaction forces to gravitational or inertial accelerations, allowing the person or animal to attain posture and locomotory control. It is this reaction force acting on an object that a scale measures as weight.

Direct and Indirect Effects of Microgravity

In considering the possible effects of microgravity conditions on biological systems, it is necessary to distinguish between *direct* effects of microgravity, in which the system perceives changes in the gravitational force per se, and *indirect* effects, in which the system responds not to the gravitational force itself, but rather to changes in the local environment that are induced by microgravity conditions. The major effect of low-gravity environments is a reduction in gravitational body forces, thus decreasing buoyancy-driven flows, rates of sedimentation, and hydrostatic pressure. Under such conditions other gravity-independent forces, such as surface tension, assume greater importance. Transport processes, such as heat transfer and solute mixing, are reduced in the gas phase and at interfaces with solids.

Alterations in fluid dynamics in the low-gravity environment have significant implications for the behavior of biological systems, both isolated cells as well as intact organisms. Thus, for example, in cell culture experiments, the diffusion of nutrients, oxygen, growth factors, and other regulatory molecules to the plasma membrane, as well as the diffusion of waste products and CO_2 away from the cell, will be reduced in the near absence of convection unless countered by stirring or forced flow of medium. Such effects might account, at least in part, for the reduced growth rates or decreased rates of glucose utilization sometimes encountered in cultured cells in flight. It is also possible that increased accumulation of CO_2 adjacent to the plasma membrane could result in deleterious changes in local pH. Examples of indirect effects of microgravity on intact plants are also known. Effective delivery of water to roots is compromised by the greater tendency of water droplets to cohere and ball up because of the now dominant effects of surface tension, so that roots may become either water deficient or waterlogged. Similarly, reduced rates of gas exchange at the leaf surface decrease the availability of CO_2 for photosynthesis and thus perturb the plant's carbohydrate and energy production. In addition, decreased diffusion of transpired water away from the leaf may result in excessive local humidity and waterlogging. The physiological importance of the reduction in air convection in microgravity conditions has been dramatically demonstrated by Musgrave's recent experiments on flower development and sexual reproduction in *Arabidopsis thaliana* (see also the section "Role of Gravity in Plant Processes" in Chapter 4).[7]

If the purpose of the 1987 Goldberg report was to guide the infant field of space biology through its next phase of development, the current report may be said to have the purpose of guiding an adolescent field in its further development toward maturity. The body of this report aims to summarize recent advances in fields relevant to space biology and medicine; recommends directions for future research, emphasizing integrative, multidisciplinary approaches to both ground-based and space-based investigations; and considers program and policy issues that may affect the nature and quality of NASA research programs in the life sciences.

REFERENCES

1. Space Science Board, National Research Council. 1987. A Strategy for Space Biology and Medical Science for the 1980s and 1990s. National Academy Press, Washington, D.C.
2. Space Studies Board, National Research Council. 1991. Assessment of Programs in Space Biology and Medicine 1991. National Academy Press, Washington, D.C.
3. Space Science Board, 1987, A Strategy for Space Biology and Medical Science for the 1980s and 1990s.
4. Space Science Board, 1987, A Strategy for Space Biology and Medical Science for the 1980s and 1990s, p. xi.
5. Space Science Board, 1987, A Strategy for Space Biology and Medical Science for the 1980s and 1990s, p. 4.
6. Space Studies Board, National Research Council. 1991. Assessment of Programs in Space Biology and Medicine 1991. National Academy Press, Washington, D.C.
7. Musgrave, M.E., Kuang, A., and Matthews, S.W. 1997. Plant reproduction during spaceflight: Importance of the gaseous environment. Planta 203: S177-S184.

PART II

Physiology, Gravity, and Space

2

Cell Biology

INTRODUCTION

The discipline of cell biology examines biological processes at the level of the basic unit of biology, the cell. While increasingly incorporating a molecular viewpoint on the one hand and extending to the tissue level on the other, investigations in cell biology focus principally on events intrinsic to individual cells and on cellular responses to environmental factors. Cell biology therefore provides the underpinning for other disciplines relevant to space biology, including developmental biology; muscle, bone, and mineral metabolism; cardiopulmonary and other homeostatic systems; immunology; and sensorimotor integration (see individual chapters of this report). Each of these areas of inquiry at the tissue and organism levels ultimately depends on the normal function of individual cells and their integration into physiological networks.

The Goldberg report[1] devoted no specific chapter to the role of cell biology in space research, but instead incorporated cell biology within chapters on various physiological systems. That this report begins with a discussion of cell biology is a testament to the dramatic growth during the past decade in our understanding of fundamental biological processes at the cellular level and to the increasing applicability of cell biology to physiology at all levels. This growth in understanding has derived from a number of major technical developments and powerful new approaches to the study of the eukaryotic cell, among them the following:

• *Molecular genetics and molecular biology, and their application to systems ranging from single cells to intact multicellular organisms.*[2,3] The introduction of new and improved techniques, including subtractive hybridization, differential display, and the polymerase chain reaction (PCR), have revolutionized the isolation and identification of both new genes and new members of known gene families. Similarly, approaches using both naturally occurring mutations and experimentally modified genomes (generated by [over]-expressing genetic constructs, including antisense or dominant-negative elements, or modification of the genome by homologous recombination [i.e., "knock-outs"]) have shed substantial light on many basic processes. Previously recognized cellular genes (e.g., growth factor receptors, adhesion molecules, signal transduction molecules) are being cloned and their regulation and gene products examined in increasing detail. In addition, the Human Genome Project and completion of the genome sequences of several model organisms are providing powerful tools. The increased recognition of the conservation of genetic homologies across evolution, and a heightened appreciation of the importance of such homologies in unraveling complex biological processes, emphasizes the significance of these projects. The ever-increasing sophistication of gene sequence databases, including those that consider not only primary sequence but also the predicted three-dimensional structure of the product proteins, provides new insights into protein function. Finally, there are ongoing advances in the design of reporter genes and sophisticated, sensitive detection apparatus for assessing transcriptional activity. Associated with this, detection hardware is being developed that can, in principle, be linked by telecommunications technology to remote sites for real-time analysis.

• *Imaging technology.*[4-7] Confocal microscopy and real-time video microscopic methods (e.g., green fluorescent protein-tagged molecules), coupled with increasingly sophisticated computer-based image enhancement and processing techniques, are active areas of development, providing novel views of cellular processes, often visualized with the help of antibodies tagged with increasingly sensitive fluorochromes.

• *Cell culture methodology.* Growth factor biology, including the interactions of cells with extracellular matrix components, cell cycle regulation, and genetic mechanisms of programmed cell death, offer new insights into requirements for cell survival and differentiation. These advances allow researchers to study primary cultures of cells isolated directly from animal tissue, thereby alleviating the need to study transformed clonal cell lines. Progenitor cells of many cell types, including bone and muscle, can be grown in culture under conditions leading to their regulated differentiation.[8] Dissociated cell cultures are increasingly being augmented by studies of microsphere, micromass, and bioreactor culture technology,[9] as well as explant, slice, and reaggregating cultures, providing intermediate steps between dissociated cell populations and intact tissue. Recognition of marker molecules to identify sites of initial appearance, migration, and differentiation of specific progenitor cells allows analyses of lineage-specific events of cells both in culture and in vivo, including microenvironmental regulation of origins, migration, and differentiation. Questions regarding the extent to which necessary culture conditions can be maintained in the space environment, and thus the usefulness of culture techniques in the analysis of spaceflight biology, are considered below.

• *Protein chemistry.*[10-12] Micro- and submicromethods for the detection and analysis of proteins have also advanced. These include immunochemical and immunohistological techniques, genetic modification of proteins to incorporate immunological tags or fluorescent markers, chemical crosslinking methods to detect and identify protein-protein interactions, and microsequencing to allow the synthesis of degenerate oligonucleotide probes for identifying and cloning the corresponding gene from genetic libraries.

• *Macromolecular structure determination.*[13,14] Advances in instrumentation, including electron and x-ray diffraction, magnetic resonance, and mass spectrometry, coupled with dramatic developments in computer-based computational analysis and molecular modeling, have revolutionized structural biology. These results offer a new level of understanding about the way proteins function to produce specific physiological responses.

As in other disciplines, this rapid and dramatic progress in what *can* be studied has meant that old questions can be asked in new ways, and asked at a level of experimental refinement that was previously unimaginable. In response to this expanded experimental capacity, a number of themes have emerged as major foci of cell biological research:

• *How do cells replicate and maintain their genomes, including the regulation of their proliferative capacity and survival?* These are questions that impinge not only on normal growth and development, including aging, but also on cancer biology. Identification of specific tumor suppressor pathways (e.g., p53), their interaction with genetically defined programs for regulating cell survival and death (e.g., ICE and bcl-2), and the elucidation of the regulation of telomere length and its influence on gene expression[15,16] are a few examples.

• *How do individual cells carry out genetically defined programs of differentiation and development into specialized tissues and multicellular organisms?* Answers to these questions will help define the critical genetic and epigenetic events at the cellular level that lead to choices of specific lineage pathways in development.[17,18]

• *How do cells generate and maintain their complicated internal cytoarchitecture, including the cytoskeleton and the host of specialized membrane-bound organelles and membrane domains, thereby regulating both growth and form?* This specific cellular substructure strictly defines both basic and differentiated cell function. Local perturbation at cell membrane focal adhesion complexes may cause profound effects to be telegraphed throughout the cell. This is done through the integration of individual components into a complex network coupling the cell surface to the cytoskeleton and to the nuclear matrix.[19]

• *How do cells synthesize and maintain organelle substructure?* Differentiated function generally depends on specialized cell structure, as exemplified by such diverse cell types as myofibrils, neurons, polarized epithelia, and gravisensing cells in plants and the mammalian vestibular system. Our understanding of how cells address and target cellular material (including both proteins and RNAs) to produce and maintain the subcellular structure is also rapidly advancing, and specific molecular addresses and their receptors are being identified. Biophysical and biochemical principles underlying the assembly of multimolecular complex structures, regulation of assembly and disassembly, mechanisms for delivery of molecules to the specific cellular sites, and the role of the intracellular cytoskeleton are among the subjects being actively investigated.[20,21]

• *How do organisms respond at the cellular level to changes in their extracellular environments?* This question integrates central, diverse concerns of cellular research. Cells interact with their environments through two major signal pathways: one based on soluble growth factors and their receptors (including those for hormones, growth factors, chemokines, and other small molecule effectors), and the second involving molecules responsible for direct interactions between cells (e.g., cadherins, selectins, CAMs)[22] and extracellular matrix (integrins, collagens, glycosaminoglycans, laminin, fibronectin, and so on), including focal adhesion complexes. These two systems interact at the cell surface. The ligand-receptor interactions activate multiple interactive signaling complexes and pathways of intracellular signal transduction, leading to signal amplification.[23] Finally, these metabolic cascades bring about changes in gene expression that govern cell physiology and behavior. Specific points of inquiry include mechanisms of reception or detection and response to extracellular chemical signals, environmental stress, and mechanical forces (e.g., touch and stretch-sensitive ion channels, fluid flow-shear forces, muscle and bone loading and unloading), including gravity, by mechanoreceptors.

PREVIOUS CELL BIOLOGICAL RESEARCH IN SPACE

Studies of effects of the space environment on mammalian cells (in particular, cells growing in culture) began in the earliest days of spaceflight and have continued to the present. A number of compendia and summaries of these experiments have been published. For example, Dickson[24] has presented a useful summary of all flight experiments using isolated cells (microbial and plant, as well as mammalian) through 1990, and Sahm and co-workers[25] and Moore and Cogoli[26,27] have summarized many studies conducted through 1994. In addition, data from IML-1, SL-J, and IML-2[28] are now becoming available.

A broad range of effects of spaceflight on the cellular physiology of human lymphocytes, embryonic lung cell lines, and other cell types have been observed. These include changes in proliferation, genetic expression, signal transduction, morphology, motility, and cell-cell interaction and energy metabolism. A detailed evaluation of this extensive body of data is beyond the capacity of this report; however, several general problems are evident. Overall, the results have often proved inconsistent or contradictory. Many of the experiments have not been adequately controlled or replicated. Engineering restraints have often compromised experimental rigor. Unexpected technical difficulties have arisen. In many cases, the experimental findings have not been published in peer-reviewed journals.

Fully evaluating the true physiological significance of many of the reported effects is often difficult. Moreover, our understanding of mechanisms responsible for these changes is still limited. In particular, which of these effects should be attributed to direct effects of gravity on cells, and in contrast, which are the result of indirect effects resulting from alterations of the cellular environment? In retrospect, it is likely that many of the described cellular responses of single cells to spaceflight (especially those growing in culture) can be attributed to indirect effects of microgravity and/or other aspects of the flight environment.[29,30] The lack of sedimentation and thermal convection in microgravity is of particular concern because it results in the formation of stationary boundary layers around cells, seriously reducing nutrient uptake, gas exchange, and the removal of toxic products. The importance of providing adequate nutrient and gas exchange has been convincingly demonstrated by Musgrave and co-workers[31] for intact *Arabidopsis* plants in flight; similar considerations would surely hold for cells in culture. Decreased use of glucose noted in cells undergoing spaceflight compared with ground controls can also most likely be accounted for by limitations in the accessibility of nutrients and/or oxygen in the absence of convective flow, which in turn result in decreased cellular growth rates. An alternative hypothesis[32] attributed decreased glucose utilization to a lack of a requirement for the maintenance of position of subcellular structures, or positional homeostasis. In addition, the induction of the cellular stress response by other perturbing factors in the environment (including strong hypergravity and vibration levels during launch and reentry) is undoubtedly important. Finally, the potential for enhanced radiation damage must be considered[33] (see Chapter 11), although appropriate shielding and short-duration studies can minimize such effects. These and other problems are likely to be confounding variables that are difficult to evaluate by ground controls and that preclude unambiguous identification of microgravity per se as the agent responsible for a given effect.

Nevertheless, the question of whether single cells might sense gravity *directly* and if so, how, has received theoretical consideration. Some researchers have proposed that single normal-sized cells (e.g., ~10-μm diameter) simply do not weigh enough for gravitational forces to compete with much larger molecular forces.[34,35] The question then is, Can forces acting on single cells be amplified to a physiologically significant signal? Examples have been proposed: Interaction of plasma membrane integrins with the extracellular matrix or substratum may transduce a reorganization of the intracellular cytoskeleton, with profound effects on cell behavior and gene expression;[36] inherent amplification and adaptation (detection of relative changes, potentially over a wide range of input) of cellular biochemical networks are being analyzed in ever greater detail.[37] Other hypotheses based on the nonlinearity of cellular molecular processes have suggested that single cells might amplify weak gravitational forces by virtue of nonlinear state transitions.[38] The development of novel theoretical approaches to gravisensing by single cells could, in principle, lead to specific, testable hypotheses.

In summary, experience from numerous previous studies, both in flight and ground based, has highlighted certain pitfalls that can and must be avoided in the design and analysis of future experiments. Thus, the design of experiments in cell biology should evolve from the extensive and growing context of basic cell biological research. There is a strong current trend toward the investigation of cell biology at the level of molecular mechanism. Similarly, space cell biology should emphasize the identification of molecular mechanisms by which cells and tissues respond to spaceflight conditions. A few well-chosen model systems should be used to facilitate comparisons among experiments and create a reliable baseline of data. Experiments should begin with extensive ground-based analyses of normal (i.e., nonspaceflight-stressed) cell biology. Growing consideration should be given to studies on cells in situ, using steadily improving techniques for the analysis in intact tissues and organisms of cellular physiology from initial gene expression to cell architecture (e.g., in situ hybridization, single-cell PCR, confocal microscopy, patch-clamping, etc.)

Finally, there have been difficulties in critically evaluating some studies in spaceflight that used cell culture techniques. To minimize these problems, experiments with cells in culture should be carefully evaluated before they are conducted, with thought given to their theoretical and practical justification, the availability of fully tested hardware, the capacity to carry out appropriate controls, adequate sample sizes, and the potential for repetition. The problem of the lack of sedimentation and fluid and gas convection in weightlessness must be considered. Differences in the fragility inherent among various cell strains and types also should be noted in choosing and comparing model systems. Generally, single cell culture models should be analyzed in ground-based studies.

OPPORTUNITIES FOR NASA-SUPPORTED RESEARCH IN CELL BIOLOGY

As biomedical research as a whole moves toward the goal of understanding the molecular mechanisms underlying physiology, NASA-sponsored research over the next decade should focus on cellular and molecular mechanisms responsible for specific physiological phenomena in which microgravity or other stressful aspects of the space environment are significant. In other words, the space environment should be a potentially significant variable. For example, mechanisms by which cells respond to mechanical forces such as shear and gravity, and those related to environmental stress, are areas of special interest and opportunity for continued NASA emphasis. (See also the sections in this report on bone, muscle, cardiovascular physiology, immunology, and neuroendocrine and sensorimotor integration for additional focused discussions.)

Mechanisms of Cellular Response to Mechanical Force

Weight-bearing bone and skeletal muscle require the gravity-dependent mechanical force of compression on bone and contraction of muscle to maintain homeostasis of bone and muscle mass (see Chapters 6 and 7). Nevertheless, the molecular and cellular mechanisms whereby these tissues respond to modulate bone or muscle synthesis and resorption, turnover, and remodeling in response to gravitational forces are not well understood. These questions are central to space biology and medicine at the cellular level.

Similar considerations hold for the investigation of other systems that respond directly to gravitational stimuli. These include the cellular mechanisms within the vestibular system involved in gravity reception and transduction of gravitational signals into perception of spatial orientation; gravitational response in plants (see Chapters 4 and 5); and activation or deactivation of touch receptors, stretch receptors, and other receptors sensitive to membrane distortion that are transduced into programmed changes in cell behavior and function.

General mechanisms of mechanoreception and pathways of signal transduction from mechanical stresses are therefore recommended as areas of special opportunity and relevance for NASA life sciences. Many different cell types respond to various molecular and physical stimuli via a limited number of transduction pathways. Therefore, even though the final physiological effects may differ with stimulus and cell type, the probability is high that the tissue-specific response pathways to gravitational force will share common mechanistic features with other kinds of force-sensing pathways.

Recommendation

Studies of mechanisms of cellular mechanoreception should include identification of the cellular receptor, investigation of possible changes in membrane and cytoskeletal architecture, and analysis of pathways of response, including signal transduction and resolution in time and space of possible ion transients.

Cellular Response to Environmental Stress

Space is a stressful environment. Organisms have developed highly sophisticated mechanisms for coping with stress, not only at the organismic and physiological levels but also at the level of single cells. The mechanisms by which cells perceive, respond, and adapt to various kinds of environmental stress are a major research emphasis in cell biology and are relevant to understanding the mechanisms of response to the space environment.[39,40]

Recommendation

Studies of cellular responses to environmental stresses encountered in spaceflight (e.g., anoxia, temperature shock, vibration) should include investigation of the nature of cellular receptors, signal transduction pathways, changes in gene expression, and identification and structure and function analysis of stress proteins that mediate the response.

Development of Advanced Instrumentation and Methodologies

Conducting sophisticated cell biology experiments in space successfully will require the development of highly automated and miniaturized instrumentation and advanced methodologies, many of which will be equally useful for ground-based research. The recommendations of a previous report that "dedicated microprocessors should be used for process control, data storage, or both, and rapid communication in real time with ground-based teams should be a goal" remain valid.[41] This goal can be realized in the current research climate and is being pursued with significant success.[42,43]

Recommendation

NASA should work with the scientific community and industry to foster development of advanced instrumentation and methodologies for space-based studies at the cellular level.

REFERENCES

1. Space Science Board, National Research Council. 1987. A Strategy for Space Biology and Medical Science for the 1980s and 1990s. National Academy Press, Washington, D.C.

2. Sambrook, J., Fritsch, E.F., and Maniatis, T. 1989. Molecular Cloning. Cold Spring Harbor Laboratory Press, Cold Spring Harbor, N.Y.

3. Ausebel, F.M., Brent, R., Kingston, R.E., Moore, D.D., Seiodman, J.G., Smith, J.A., and Struhl, K. 1991. Current Protocols in Molecular Biology. John Wiley & Sons, New York.

4. Ivins, J.K., Clark, S.M., and Fraser, S.E. 1993. Biological microscopy: The emergence of digital microscopy. Curr. Opin. Biotechnol. 4: 69-74.

5. Maxfield, F.R. 1994. Introduction: Optical microscopy in physiological investigations. FASEB J. 8: 571-572.

6. Shotton, D.M. 1995. Electronic light microscopy: Present capabilities and future prospects. Histochem. Cell Biol. 104: 97-137.

7. Asai, D.J., ed. 1993. Antibodies in cell biology. Methods Cell Biol., Vol. 37. Academic Press, San Diego.

8. Weiss, S., Reynolds, B.A., Vescovi, A.L., Morshead, C., Craig, C., and Gvan der Kooy, D. 1996. Is there a neural stem cell in the mammalian forebrain? Trends Neurosci. 19: 387-393.

9. Duray, P.H., Hatfill, S.J., and Pellis, N.R. 1997. Tissue culture in microgravity. Sci. Med. (May/June): 46-55.

10. Coligan, J.E., Dunn, B.M., Ploegh, H.L., Speicher, D.W., and Wingfield, P.T. 1995. Current Protocols in Protein Science, Vol. 1 and 2. John Wiley & Sons, New York.

11. Karge, B.L., and Hancock, W.S. 1996. High resolution separation and analysis of biological macromolecules. Methods Enzymol., Vol. 270. Academic Press, San Diego.

12. Karge, B.L., and Hancock, W.S. 1996. High resolution separation and analysis of biological macromolecules. Methods Enzymol., Vol. 271. Academic Press, San Diego.

13. Riddihough, G., ed. 1997. Nat., Struct. Biol. 4 (Suppl.): 841-865.

14. Carter, C.W., Jr., and Sweet, R.M. 1997. Macromolecular crystallography. Methods Enzymol., Vol 276. Academic Press, San Diego.

15. Bodnar, A.G., Ouellette, M., Frolkis, M., Holt, S.E., Chiu, C.-P., Morin, G.B., Harley, C.B., Shay, J.W., Lichtsteiner, S., and Wright, W.E. 1998. Extension of life-span by introduction of telomerase into normal human cells. Science 279: 349-352.

16. Myerson, M., Counter, C.M., Eaton, E.N., Ellisen, L.W., Steiner, P., Caddle, S.D., Ziaugra, L., Beijersbergen, R.L., Davidoff, M.J., Liu, Q., Bacchetti, S., Haber, D.A., and Weinber, R.A. 1997. hEST2, the putative human telomerase catalytic subunit gene, is up-regulated in tumor cells and during immortalization. Cell 90: 785-795.

17. Gilbert, S.F. 1994. Developmental Biology. Sinauer, Sonderland, Mass.

18. Hunter, T. 1995. Protein kinases and phosphatases: The yin and yang of protein phosphorylation and signaling. Cell 80: 225-236.

19. Ingber, D.E. 1997. Tensegrity: The architectural basis of cellular mechanotransduction. Ann. Rev. Physiol. 59: 575-599.

20. Rothman, J.E., and Wieland, F.T. 1996. Protein sorting by transport vesicles. Science 272: 227-234.

21. Simons, K., and Ikonen, E. 1997. Functional rafts in cell membranes. Nature 387: 569-572.

22. Fannon, A.M., and Colman, D.R. 1996. A model for central synaptic junctional complex formation based on the differential adhesive specificities of the cadherins. Neuron 17: 423-434.

23. Bray, D. 1997. Protein molecules as computational elements in living cells. Nature 376: 307-312.

24. Dickson, K.J. 1991. Summary of biological spaceflight experiments with cells. ASGSB Bull. 4: 151-260.

25. Sahm, P.R., Keller, M.H., and Schiewe, B., eds. 1995. Scientific Results of the German Spacelab Mission D-2. Wissenschaftliche Projectführung D-2, DLR, Köln, Germany.

26. Moore, D., and Cogoli, A. 1996. Gravitational and space biology at the cellular level. Pp. 1-106 in Biological and Medical Research in Space (D. Moore, P. Bie, and H. Oser, eds.). Springer, New York.

27. Cogoli, A., and Cogoli-Greuter, M. 1997. Activation of lymphocytes and other mammalian cells in microgravity. Adv. Space Biol. Med. 6: 33-79.

28. Cogoli, A., ed. 1996. Biology under microgravity conditions in Spacelab IML-2. J. Biotechnol. 47: 65-403 (27 papers).

29. Albrecht-Buehler, G. 1991. Possible mechanisms of indirect gravity sensing by cells. ASGSB Bull. 4: 25-34.

30. Menningmann, H.D. 1995. Cell functions and biological processing in a micro-gravity environment. Pp. 539-543 in Scientific Results of the German Spacelab Mission D-2 (P.R. Sahm, M.H. Keller, and B. Schiewe, eds.). Wissenschaftliche Projectführung D-2, DLR, Köln, Germany.

31. Kuang, A., Musgrave, M.E., and Matthews, S.W. 1996. Modification of reproductive development in Arabidopsis thaliana under spaceflight conditions. Planta 198: 588-594.

32. Nace, G.W. 1983. Gravity and positional homeostasis of the cell. Adv. Space Res. 3: 159-168.

33. National Aeronautics and Space Administration. 1997. Modeling Human Risk: Cell and Molecular Biology in Context. Report No. LBNL-40278. Space Radiation Program, NASA, Washington, D.C.

34. Pollard, E.C. 1965. Theoretical studies on living systems in the absence of mechanical stress. J. Theor. Biol. 8: 113-123.

35. Albrecht-Buehler, G. 1991. Possible mechanisms of indirect gravity sensing by cells. ASGSB Bull. 4: 25-34.

36. Ingber, D.E. 1997. Tensegrity: The architectural basis of cellular mechanotransduction. Ann. Rev. Physiol. 59: 575-599.

37. Bray, D. 1995. Protein molecules as computational elements in living cells. Nature 376: 307-312.

38. Baxter, D.A., and Byrne, J.H. 1997. Complex oscillations in simple neural systems. Biol. Bull. 192: 167-169.

39. Hartl, F.U. 1996. Molecular chaperons in cellular protein folding. Nature 381: 571-580.

40. Deshaies, R.J. 1995. Make it or break it: The role of ubiquitin-dependent proteolysis in cellular regulation. Trends Cell Biol. 5: 428-434.

41. Space Studies Board, National Research Council. 1991. Assessment of Programs in Space Biology and Medicine 1991. National Academy Press, Washington D.C.

42. Plautz, J.D., Day, R.N., Dailey, G.M., Welsch, S.B., Hall, J.C., Halpain, S., and Kay, S.A. 1996. Green fluorescent protein and its derivative as versatile markers for gene expression in living Drosophila melanogaster, plant and mammalian cells. Gene 173: 83-87.

43. Brandes, C., Plautz, J.D., Stranewsky, R., Jamison, C.F., Straume, M., Wood, K.V., Kay, S.A., and Hall, J.C. 1996. Novel features of Drosophila period transcription revealed by real-time luciferase reporting. Neuron 16: 687-692.

3

Developmental Biology

INTRODUCTION

Traditionally, developmental biologists are concerned with the processes and mechanisms responsible for the development of the zygote into a primordial set of cell types, as well as the later developmental events that produce the mature organism, including organogenesis, histogenesis, and cellular differentiation. In effect, every process from conception to aging and death could be considered a component of development. In the Goldberg report published in 1987,[1] two primary concerns were stressed regarding future investigations conducted in space on developmental processes: Can organisms undergo normal development in microgravity? and, Are there developmental phenomena that can be studied better in microgravity than on Earth? Since 1987 research has partly answered the first question, but some important issues must still be addressed. With regard to the second point, the distinct possibility remains that the space environment may be useful for understanding certain biological phenomena occurring in specific systems identified below.

This chapter first reviews the major changes in perspective during the last few years in the general field of developmental biology and then discusses the importance of pursuing complete life cycles in space. The specific systems for which gravity is likely to play a critical role in development and/or maintenance include the vestibular system (that part of the ear and nervous system controlling posture and balance) and also the multiple sensory systems that interact with the vestibular system. In addition, gravity should influence the topographic neural space maps that exist throughout the brain. Space maps represent areas of the brain containing orderly structural and/or functional representations of either sensory or motor systems found within the brainstem, hippocampus, sensory and motor cortical areas, and corpus striatum. Neuroplasticity is discussed both generally and specifically as it pertains to gravity-sensitive systems. Neuroplasticity refers to long-term changes in neuron structure and function in response to changes in their activity. Finally, aspects of plant development that have special pertinence to microgravity are covered in Chapter 4.

PROGRESS IN DEVELOPMENTAL BIOLOGY

Three major advances have transformed basic cellular and developmental studies during the past 10 to 15 years. These advances have, in turn, affected our ability to seek a molecular understanding of nearly every aspect of the response to microgravity by organisms in space.

Developmental Genetics

Saturation mutagenesis has been applied to identify key genetic components involved in specific developmental events using *Drosophila melanogaster, Caenorhabditis elegans,* and *Arabidopsis thaliana.* For the first time, these experiments are providing a molecular basis for understanding the mechanism underlying the step-by-step progression that occurs during development. Although these initial genetic screens revealed broad outlines of developmental mechanisms, new genetic screens are being tested that are sensitized to detect interacting genetic components.[2] Techniques to produce clones of cells in which the phenotype of mutations can be analyzed in a subset of cells[3-5] have led to the identification of many constitutively important functions. These studies have shown that multiple genes are required to act collectively in a variety of developmental processes, so that mutations in any of these may be lethal. For example, the molecular cascade responding to growth factor signaling is used in many other receptor-mediated responses. In addition, by analyzing the effect of mutations in a clone of cells, researchers avoid the complexities that arise from the participation of these genes in essential functions in a wide variety of cells. Outstanding examples of the power of this genetic approach include recent advances in identifying the complex set of interacting systems involved in the development of the vulva in nematodes[6] and the eye in fruitflies,[7] including the receptor tyrosine kinase-*ras-raf*-pathway. These studies have transformed the understanding of developmental progression and may provide a basis for similar analyses of vertebrate systems.

Similar saturation genetic approaches have been applied with considerable success to zebrafish, *Danio rerio.*[8,9] However, in this organism, identifying genes at the molecular level is still difficult. In mammals, the laboratory mouse, *Mus musculus,* has taken a central role in the genetic analysis of gene function. Two major advances have propelled this organism to the forefront of developmental studies. First, the development of embryonic stem (ES) cells, in which homologous recombination can be accomplished efficiently, has made targeted mutagenesis possible.[10] Mutated ES cells can be transplanted into genetically marked blastocysts, where they populate the germ line of the host embryo. By appropriate crosses, embryos can be produced that completely lack certain gene functions. Second, the introduction of ethyl-N-nitrosurea as a mutagen has allowed mutations to be produced in specific genes, either as alleles of known genes or as genes contained within deficiencies.[11] The availability of multiple alleles of known mutations complements the approach of targeted mutagenesis with ES cells. Although this is still in the future, it should soon be possible to use these technologies to investigate the genetic and molecular basis for inner ear function in mice, and then in humans. While the analysis of developmental events in vertebrates is more difficult to achieve and more expensive because of the animals' larger size and longer generation times, the potential for understanding specific developmental events in these systems is great, and may provide critical models for detailed analyses of human genetic defects.

Molecular Conservation

The second major advance has been the recognition that molecular mechanisms are conserved across phylogeny. The initial discovery of the homeobox sequence in *Drosophila* and its conservation in vertebrates in 1984 alerted the biological community that basic mechanisms for establishing polarity in the anteroposterior axis were conserved in multicellular organisms. This discovery was quickly followed by the recognition that control of the cell cycle in vertebrates used the same set of genes identified in yeast. New examples of the conservation of genetic function between simple model organisms, vertebrates, and humans have since been identified almost on a regular basis.[12] For example, the organization of the arthropod embryo with a ventral nervous system and the vertebrate embryo with a dorsal neural tube have been shown to depend on the same molecular mechanisms for establishing polarity.[13] The difference in organization is explained by differences in the formation of the mouth (stomodeum) rather than in differences in the dorsal-ventral patterning of the embryo. Similarly, the development of compound eyes in insects, squid, and certain vertebrates—animals that were once thought to depend on fundamentally different developmental processes and to have evolved independently of each other—have recently been shown to involve the same conserved gene, *eyeless* in *Drosophila* and *small eye* in the mouse.[14,15] The genetic circuitry used for appendage formation in flies is used in a remarkably similar manner to control development of vertebrate limbs.[16] Molecular experiments have even shown conservation of some developmentally associated mechanisms in organisms as distantly related as plants and animals. Even though plants and animals are thought to have evolved their developmental processes independently, it has recently been found, for example, that the proteins that maintain cellular memory of positional information are conserved, even though initially established by different types of genes in plants and animals.[17]

Although these examples seem diverse, they represent a general theme now apparent in the field of developmental biology, which has transformed the experimental approach applied to understanding cell differentiation and development. Also, this conservation of molecular mechanisms offers the opportunity of using systems that can be manipulated experimentally in organisms whose genetics would be intractable. This includes systems such as the limb bud of the chick embryo and Spemann's organizer in amphibian embryos. Thus, newly discovered genetic interactions now can be integrated quickly into a developmental model that can be tested by direct experimentation (e.g., Quiring et al., 1994).[18]

Genome Sequencing Project

The third advance in new information has been provided by the genome sequencing projects that are transforming the identification of the components of cellular and developmental processes. Today, the field of developmental biology is focused on a small number of model organisms that allow for detailed analysis of development at the cellular, genetic, and molecular levels. In combination with the considerable new information available as a consequence of the genome project, comparative molecular studies of yeast, plants, worms, flies, fish, and mammals will become the dominant theme for future studies. Because of the power of this type of analysis, most future studies of developmental phenomena will be conducted on organisms for which there is complete knowledge of the genome. In the field of development, where sea urchins, frogs, and chicks (three classical experimental organisms) are not subjects for the genome project, these models will lack the level of completeness that can be achieved from organisms such as yeast, flies, and worms, whose full genetic and molecular composition have been analyzed. However, the existence of genetic homologies makes investigations of the classical organisms still practical, as recently exemplified by the study of chick limb development.[19] New technologies, such as differential displays and differential cDNA libraries of cells and tissues at various stages of development, also provide potential avenues for integrating the classical organisms into the genetic analysis of development.

MAJOR ISSUES IN SPACE DEVELOPMENTAL BIOLOGY

As stated in the 1987 Goldberg report,[20] the two objectives for developmental biology studies remain unchanged: to identify areas of fundamental research in space biology that are important to pursue in space, and to develop a knowledge base for long-term manned space habitation and/or exploration. Although significant progress has been made on a number of research projects proposed in that report, it is clearly impossible to test the wide range of possible effects of microgravity on biological development. In this report the importance of two types of future studies is stressed. First, complete life cycles in space should be used as an approach to determining whether there are developmental events affected by reduced gravity. In addition, continued analysis of the development of the gravity-sensing systems, including the vestibular system and other systems that interact with it in vertebrates, should be carried out to determine the importance of gravity in their normal development and maintenance.

Complete Life Cycles in Microgravity

Critically testing hypotheses regarding the effect of microgravity on specific developmental processes continues to be extremely difficult because of the engineering demands and the difficulty of repeating experiments in space. The latter reflects the high cost of performing experiments and the limited access to in-flight experimental animals. For example, a number of land-based experiments conducted on frog embryos during the first cell cycle after fertilization had suggested that the reorganization of yolk during rotation of the newly fertilized egg is responsible for establishing polarity of the early embryo. In experiments in which the rotation was prevented, axis formation was impaired. This led to the hypothesis that a gravity-induced movement of yolk was important for establishing the initial polarity. This was a natural experiment to test in microgravity. Costly in-flight experiments were performed in which the frog eggs were fertilized in space, and the embryos were allowed to develop to the tadpole stage before returning to Earth. Surprisingly, most of these embryos were normal, and some even developed to the adult stage and were fertile.[21] Thus, although there was a clear indication that this developmental step depended on gravity, experiments in microgravity had ruled this out. Unfortunately, because of the difficulty of observing experimental animals in space, it is not yet known whether the reorientation of yolk occurred in microgravity through an endogenous contractile system, or whether the polarity was established independently of yolk movement.

This example illustrates both the utility and the limitations of conducting experiments in space relating to specific developmental issues. Because of the extraordinary expense and technical limitation of in-flight experiments, requiring extensive engineering backup and preflight experiments, it is clear that only a few critical tests of hypotheses concerning the role of gravity in normal developmental events can be performed.

An alternative approach is to maintain a number of key organisms in space for two complete generations. The basis for this is the assumption that if organisms can grow successfully and reproduce in space, there are no specific stages during the life cycle that rely critically on a gravity-dependent process. A second generation is required to ensure that the processes of gamete production (particularly the development of the oocyte) can be accomplished in microgravity. Of course, in applying this approach small effects of gravity may be overlooked, such as the finding that amphibian embryos grown in microgravity possess a thicker blastula roof than normal.[22] However, this difference must be compensated for later on, since these embryos did develop into normal tadpoles. The maintenance of life and the capability to reproduce do not necessarily ensure that higher brain functions have developed normally and are functional in organisms raised in space (see specific systems below). This would be most apparent in vertebrate animals. A major focus for space experiments should therefore be to maintain animals through successive generations, followed by detailed analyses to determine whether deficiencies detected are produced routinely.

During the past 10 years fruitflies[23] and nematodes[24] have been grown successfully in microgravity through more than one generation with no significant developmental abnormalities. Although further repetition of the experiments in flies might be necessary because some abnormalities were observed, these results basically indicate that no major developmental process in these simple organisms critically depends on gravity. There are no comparable results with vertebrate animals, and such experiments should have high priority. The zebrafish, the current favorite of developmental geneticists for performing studies in lower vertebrates, may be difficult to rear in space because of its finicky environmental requirements. But the medaka fish, which is the favorite of the developmental biology community in Japan, may be more suitable for this experiment. In terms of avian development, some efforts have been made to grow chick or quail embryos in space, but these have been unsuccessful when embryos were brought into space during the earliest stages.[25] It seems that the high content of yolk in these eggs requires the embryos to reach a stage of organization at which the yolky mass is segregated within membranes before normal development can be achieved in space.

Recommendation

Key model organisms should be grown through two complete life cycles in space to determine whether there are any critical events during development that are affected by space conditions. Because no critical effects have been seen in model invertebrates, the highest priority should be given to testing vertebrate models such as fish, birds, and small mammals such as mice or rats. If developmental effects are detected, control experiments must be performed on the ground and in space, including the use of a space-based 1-g centrifuge, to identity whether gravity or some other element of the space environment induces the developmental abnormalities.

Development of the Vestibular System

Neurobiologists working on space research are concerned about whether that part of the vestibular system that is sensitive to gravity can develop in microgravity. The vestibular sensory receptors that are sensitive to gravity are called, collectively, the otolith organs. This includes the utricle and saccule. A principal role of the vestibular system is to relay signals from the otoliths regarding linear acceleration, and from the semicircular canals regarding rotation or angular acceleration, to the brain in order to control the motor output of the extrinsic eye muscles and those muscles in the neck (collic) and body (vestibulospinal) concerned with posture and balance.[26] In every other sensory system known, especially those that make up the neural space maps in the brainstem,[27] sensory stimulation has been implicated in the initial specification of the connections[28] and physiological properties[29] of the constituent neurons. Another example is the development of the visual system, where activity in the retinal pathway influences the specification of the connections determining how visual information is processed in the cerebral cortex.[30,31] Only in the otolithic gravitational pathway has it been impossible to study the role of sensory deprivation, because there is no way to deprive the system of gravitational stimulation on Earth. Therefore, experiments should be planned to be carried out in space that will test the hypothesis that gravity itself plays a role in the development and maintenance of the components of the peripheral and central vestibular system.

Structures whose development may depend at least in part on exposure to gravitational stimuli include the vestibular sensory receptors of gravity themselves—the hair cells situated in the utricle and saccule, vestibular ganglion cells that form synapses with vestibular hair cells and vestibular neurons in the brainstem vestibular nuclei, vestibular nuclei neurons themselves, and motor neurons receiving input from axons of vestibular nuclei neurons composing the vestibular reflex pathways. These reflex pathways include the vestibulo-ocular, vestibulospinal and vestibulocollic fibers whose pathways, connections, and development have been investigated most extensively within the last 10 years.[32-34] Moreover, other sensory systems are known to interact with the vestibular pathway. For example, the gravity-driven otoliths function in producing basic postural adjustments by interacting with signals from the semicircular canals.[35,36] The vestibular system also receives inputs from the proprioceptive system, involved in the control of muscle length and tension,[37] and from the visual system, involved in the control of eye movements.[38] Little is known about the exact nature of these interactions and virtually nothing concerning the development of these connections. As a consequence, there is a need to understand more about these basic topics through ground-based studies before evaluating the potential for space-based experiments. In addition, there is some evidence that the vestibular system plays a role in regulating the autonomic nervous system.[39] In particular, the vestibular system may influence cardiovascular output[40] and pulmonary function.[41] The pathways subserving vestibuloautonomic connections are complex and have not yet been fully characterized either structurally or functionally. Moreover, how these connections develop is still not understood.

Recommendation

The following experiments should be performed first in ground-based studies to identify appropriate experiments to be performed in space. Because in ovo rather than in utero experiments afford the possibility of manipulating the embryo, using an avian system is most appropriate.

Studies should be performed to identify the critical periods in vestibular neuron development, including proliferation, migration, differentiation, and programmed cell death. This information is essential to design and interpret subsequent flight experiments on the effects of microgravity on vestibular development.

Neural Space Maps

Neural Space Maps in the Brainstem

Vertebrate brains form and maintain multiple neural maps of the spatial environment, which provide distinctive, topographical representations of different sensory and motor systems. For example, visual space is mapped onto the retina in a two-dimensional coordinate plan.[42] This plan is then remapped in the central nervous system in several places, including the superior colliculus (optic tectum).[43] Likewise, the surface of the body is mapped in a somatotopic plan onto the superior colliculus.[44] In addition, there is a map relating the localization of sounds in space[45,46] and one that corresponds to oculomotor activity.[47] During development, all of these maps must be established. In the adult, the neural maps must be maintained in register so that appropriate perceptual and motor adjustments can occur.

There are several reasons to suspect that gravitational stimuli may have special importance in the development and maintenance of space maps and neural tracts that act in position-sensing in mammalian brains. Apparently this system of neural maps must have appropriate information regarding the location of the head in the gravitational field, and so it follows that the vestibular system must play a key role in the organization of these maps. Thus, from a theoretical point of view, it can be predicted that important interactions must occur between sensory and motor systems on the one hand, and the vestibular system on the other. To understand how the status of the space maps relates to gravitational perturbations, it will be necessary to acquire detailed information from ground-based investigations on their structure and function in different species, including those with specialized adaptations (e.g., owls). This should permit researchers to frame testable hypotheses on the role of gravity in the assembly and maintenance of neural space maps.

Neural Space Maps in the Hippocampus

An analogous multisensory space map has been demonstrated in the mammalian hippocampus, which has the important function of providing short-term memory for an animal's location in a specific spatial venue.[48] This neural map is particularly focused on body position and makes use of proprioceptive as well as visual cues. It is used by the animal to resume its location at a previous site. For theoretical reasons, it can be predicted that head position must be involved in this function. It is important to determine how the vestibular system, especially the otolithic component, is related to this neural space map.

Neural Space Maps in the Sensory and Motor Cortices and the Corpus Striatum

Other topographical maps of the body are found in the neocortex[49,50] and neostriatum,[51] which could be expected to interact with the vestibular system. In particular, the caudate nucleus of the corpus striatum has been implicated in vestibular-related locomotion.[52]

Recommendations

• *The role of otolithic stimulation on the development and maintenance of the different neural space maps—including those within the brainstem, hippocampus, sensory and motor cortices, and corpus striatum—should be investigated.*
• *Studies should be designed to address how neurons of the various sensory and motor systems interact with vestibular neurons in the normal assembly and function of the neural space maps.*
• *Studies should be performed to determine the influence of decreased stimulus (as experienced in microgravity) on the development and maintenance of the neural space maps.*

Neuroplasticity

Neuroplasticity refers to the long-lived alterations in structure and function that neurons may undergo following changes in their activity. It occurs in both developing and in mature neurons. However, the degree of alteration depends in part on the type of lesion or change,[53] the age of the animal at the time of the lesion or change (younger animals are usually more sensitive to loss),[54] and the time during the development of the system that the change occurs (i.e., whether it occurs during critical developmental events).[55,56] Such alterations range from gross to microscopic and molecular, involving cell death,[57] cell atrophy,[58] loss of dendrites,[59] synaptic reorganization[60] and long-term potentiation (LTP) or depression (LTD) of synaptic transmission or its efficacy (synaptic modifiability).[61] Various types of changes in stimuli may lead to neuroplasticity of the target neurons. These include eyelid closure in the visual system,[62] cochlear deprivation[63] or overstimulation in the auditory system,[64] labyrinthectomy in the vestibular system,[65] whisker removal in the somatosensory system,[66] or simple nerve transection of any afferent input, in addition to changes in the frequency of stimulation of the afferent fibers.[67] Neuroplasticity is especially important to characterize, because it may result in changes in function not only of a single target neuron but also of the entire pathway or system in which the target neuron participates.[68-70] Thus, the effects on the organism can be profound. In some cases, neuroplasticity can result in enhanced performance. For example, LTP is thought to represent a mechanism subserving learning and/or memory.[71,72]

An important type of plasticity is exhibited by the central vestibular system in response to decreased inputs from the labyrinths, as with exposure to microgravity. Typically, this process is called vestibular compensation.[73] Following diminished inputs from the labyrinths, the organism exhibits multiple symptoms such as disequilibrium, disorientation, and locomotor deficits, but after about a week the symptoms diminish or disappear to some extent in most adult mammals. The mechanisms underlying vestibular compensation are thought to be mediated by modifications in some synapses, most likely those within the vestibular brainstem nuclei,[74,75] although other brainstem neurons and parts of the cerebellum have been implicated.[76]

It is important to distinguish between overstimulation and understimulation (sensory deprivation) as perturbational factors. Research on sensory systems (e.g., the auditory system) has shown that over-stimulation can produce wholesale destruction of the sensory receptor cells and a toxic condition in neurons (excitotoxicity)[77] resulting from overaccumulation of excitatory neurotransmitters such as glutamate.[78] This can lead to catastrophic changes in calcium fluxes and even cell death.[79] Long-term changes may be manifested in phenomena such as tinnitus,[80] which probably reflects a persistent state of hyperactivity or supersensitivity in the central nervous system. Neurological disturbances following overstimulation are seen in adults as well as juveniles. In contrast, the neurological effects of under-stimulation or sensory deprivation are generally quite different. In adults, the changes in the nervous system have been relatively minor[81] compared with the effects on developing systems.[82] In the adult (e.g., following auditory deprivation) there is some atrophy of neuronal cell size but seldom cell death and degeneration.[83] In infants or embryos, however, such deprivation can result in cell death in the central nervous system and seriously disrupt the establishment of neural connections. [84,85]

In consideration of these observations made in many other sensory systems, it is important to study how microgravity, which probably represents decreased stimulation of the vestibular system, affects the vestibular neural pathway. Furthermore, it appears that the concept of a dose-response curve does not adequately reflect how the nervous system responds biologically to changes in gravity. Consequently, it is unlikely that the mechanisms activated in hypergravity will resemble those operating in microgravity. In conclusion, the concept of a dose response does not adequately reflect the biology of the nervous system in its response to changes in gravity.

Recommendations

• *From ground-based studies, researchers know that there are compensatory mechanisms that function normally in the vestibulomotor pathways, and these mechanisms occur in space.[86,87] What is the basis for the compensation on Earth and in space, and are the mechanisms the same? These experiments should be given the highest priority, because these compensatory mechanisms operate in astronauts entering and returning from space and may have a profound effect on their performance in space and their postflight recovery on Earth.*

• *Experiments are needed to critically test the role of gravity on the development and maintenance of the vestibular system's capability for neuroplasticity. The process needs to be characterized at several different times following perturbations to determine the sequence of intermediate events leading to the plastic change.*[88] *Controls for the effects of nongravitational stresses likely to be encountered in space (such as loud noise and vibration) must also be performed on the ground, so that space experiments can be designed to isolate the effects of microgravity from the effects of other stresses.*

   • *From ground-based experiments, the vestibulo-oculomotor system is known to be capable of learning new motor patterns in response to sensory perturbations.*[89] *Future investigation should focus on determining if and how these mechanisms are affected by exposure to microgravity.*

REFERENCES

1. Space Science Board, National Research Council. 1987. A Strategy for Space Biology and Medical Science for the 1980s and 1990s. National Academy Press, Washington, D.C.

2. Wilson, C., Pearson, R.K., Bellen, H.J., O'Kane, C.J., Grossniklaus, U., and Gehring, W.J. 1989. P-element-mediated enhancer detection: An efficient method for isolating and characterizing developmentally regulated genes in *Drosophila*. Genes Dev. 3: 1301-1313.

3. Golic, K., and Lindquist, S. 1989. The FLP recombinase of yeast catalyzes site-specific recombination in the *Drosophila* genome. Cell 59: 499-509.

4. Xu, T., and Rubin, G.M. 1993. Analysis of genetic mosaics in developing and adult *Drosophila* tissues. Development 117: 1223-1237.

5. Feil, R., Brocard, J., Mascrez, B., LeMeur, M., Metzger, D., and Chambon, P. 1996. Ligand-activated site-specific recombination in mice. Proc. Natl. Acad. Sci. U.S.A. 93: 10887-10890.

6. Kenyon, C. 1994. A perfect vulva every time: Gradients and signaling cascades in *C. elegans*. Cell 82: 171-174.

7. Dickson, B., and Hafen, E. 1993. Genetic dissection of eye development in *Drosophila*. Pp. 1327-1362 in The Development of *Drosophila melanogaster* (M. Bates and A. Martinez Arias, eds.). Cold Spring Harbor Laboratory Press, Cold Spring Harbor, N.Y.

8. Haffter, P., Granato, M., Brand, M., Mullins, M.C., Hammerschmidt, M., Kane, D.A., Odenthal, J., Van Eeden, F.J.M., Jiang, J.-J., Heisenberg, C.-P., Kelsh, R.N., Furutani-Seiki, M., Vogelsang, E., Beuchle, D., Schach, U., Fabian, C., and Nusslein-Volhard, C. 1996. The identification of genes with unique and essential functions in the development of the zebra fish, *Danio rerio*. Development 123: 1-36.

9. Driever, W., Solnica-Krezel, L., Schier, A.F., Neuhauss, S.C.F., Malicki, J., Stemple, D.L., Stainier, D.Y.R., Zwartkruis, F., Abdelilah, S., Rangini, Z., Belak, J., and Boggs, C. 1996. A genetic screen for mutations affecting embryogenesis in zebra fish. Development 123: 37-46.

10. Mansour, S.L., Thomas, K.R., and Capecchi, M.R. 1998. Disruption of the proto-oncogene int-2 in mouse embryo-derived stem cells: A general strategy for targeting mutations to non-selectable genes. Nature 336: 348-352.

11. Cordes, S., and Barsh, G.S. 1994. The mouse segmentation gene kr encodes a novel basic domain-leucine zipper transcription factor. Cell 79: 1025-1034.

12. Gerhart, J., and Kirschner, M. 1997. Cells, Embryos, and Evolution. Blackwell Science, Malden, Mass.

13. Holley, S.A., and Ferguson, E.L. 1997. Fish are like flies are like frogs: Conservation of dorsal-ventral patterning mechanisms. Bioessays 19: 281-284.

14. Tomarev, S.I., Callaerts, P., Kos, L., Zinovieva, R., Halder, G., Gehring, W., and Piatigorsky, J. 1997. Squid Pax-6 and eye development. Proc. Natl. Acad. Sci. U.S.A. 94: 2421-2426.

15. Quiring, R., Walldorf, U., Kloter, U., and Gehring, W.J. 1994. Homology of the eyeless gene of Drosophila to the small eye gene in mice and Aniridia in humans. Science 265: 785-789.

16. Laufer, E., Dahn, R., Orozco, O.E., Yeo, C.-Y., Pisenti, J., Henrique, D., Abbott, U.K., Fallon, J.F., and Tabin, C. 1997. Expression of radical fringe in limb-bud ectoderm regulates apical ectodermal ridge formation. Nature 386: 366-373.

17. Goodrich, J., Puangsomlee, P., Martin, M., Long, D., Meyerowitz, E.M., and Coupland, G. 1997. A Polycomb-group gene regulates homeotic gene expression in *Arabidopsis*. Nature 386: 44-51.

18. Quiring, R., Walldorf, U., Kloter, U., and Gehring, W.J. 1994. Homology of the eyeless gene of *Drosophila* to the small eye gene in mice and Aniridia in humans. Science 265: 785-789.

19. Laufer, E., Dahn, R. Orozco, O.E., Yeo, C.-Y., Pisenti, J., Henrique, D., Abbott, U.K., Fallon, J.F., and Tabin, C. 1997. Expression of radical fringe in limb-bud ectoderm regulates apical ectodermal ridge formation. Nature 386: 366-373.

20. Space Science Board, National Research Council. 1987. A Strategy for Space Biology and Medical Science for the 1980s and 1990s. National Academy Press, Washington, D.C.

21. Souza, K.A., Black, S.D., and Wassersug, R.J. 1995. Amphibian development in the virtual absence of gravity. Proc. Natl. Acad. Sci. U.S.A. 92: 1975-1978.

22. Souza, K.A., Black, S.D., and Wassersug, R.J. 1995. Amphibian development in the virtual absence of gravity. Proc. Natl. Acad. Sci. U.S.A. 92: 1975-1978.

23. Johnson Marco, R., Benguria, A., Sanchez, J., and de Juan, E. 1996. Effects of the space environment on *Drosophila melanogaster* development. Implications of the IML-2 experiment. J. Biotechnol. 47: 179-189.

24. Johnson, T. E., and G. A. Nelson. 1991. *Caenorhabditis elegans*: A model system for space biology studies. Exp. Gerontol. 26: 299-309.

25. Kenyon, R.V., Kerschmann, R., Sgarioto, R., Jun, S., and Vellinger, J. 1995. Normal vestibular function in chicks after partial exposure to microgravity during development. J. Vestib. Res. 5: 289-298.

26. Kelly, P.K. 1991. The sense of balance. Principles of Neural Science (E.R. Kandel, J.H. Schwartz, and T.M. Jessell, eds.). Elsevier, New York.

27. Olsen, J.F., Knudsen, E.I., and Esterly, S.D. 1989. Neural maps of interaural time and intensity differences in the optic tectum of the barn owl. J. Neurosci. 9: 2591-2605.

28. Carr, C.E., and Boudreau, R.E. 1996. Development of the time coding pathways in the auditory brainstem of the barn owl. J. Comp. Neurol. 373: 467-483.

29. Knudsen, E.I., and Konishi, M. 1979. Sound localization by the barn owl (*Tyto alba*). J. Comp. Physiol. 133: 1-11.

30. Hubel, D.H., and Wiesel, T.N. 1970. The period of susceptibility to the physiological effects of unilateral eye closure in kittens. J. Physiol. (Lond.) 206: 419-436.

31. Shatz, C.J. 1990. Impulse activity and the patterning of connections during CNS development. Neuron 5: 745-756.

32. Glover, J.C., and Petursdottir, G. 1988. Pathway specificity of reticulospinal and vestibulospinal projections in the 11-day chicken embryo. J. Comp. Neurol. 270: 25-38.

33. Cox, R.G., and Peusner, K.D. 1990. Horseradish peroxidase labeling of the efferent and afferent pathways of the avian tangential vestibular nucleus. J. Comp. Neurol. 296: 324-341.

34. Graf, W., Spencer, R., Baker, H., and Baker, R. 1997. Excitatory and inhibitory vestibular pathways to the extraocular motor nuclei in goldfish. J. Neurophysiol. 77: 2765-2779.

35. Telford, L., Seidman, S.H., and Paige, G.D. 1996. Canal-otolith interactions during vertical and horizontal eye movements in the squirrel monkey. Exp. Brain Res. 109: 407-418.

36. Angelaki, D.E., and Hess, B.J. 1996. Three-dimensional organization of otolith-ocular reflexes in rhesus monkeys, II: Inertial detection of angular velocity. J. Neurophysiol. 75: 2425-2440.

37. Keshner, F.A., and Peterson, B.W. 1995. Mechanisms controlling human head stabilization, I: Head-neck dynamics during random rotations in the horizontal plane. J. Neurophysiol. 73: 2293-2301.

38. Scudder, C.A., and Fuchs, A.F. 1992. Physiological and behavioral identification of vestibular nucleus neurons mediating horizontal vestibulo-ocular reflex in trained rhesus monkey. J. Neurophysiol. 68: 244-264.

39. Steinbacher, B.C., Jr., and Yates, B.J. 1996. Brainstem interneurons necessary for vestibular influences on sympathetic outflow. Brain Res. 720: 204-210.

40. Yates, B.J., and Miller, A.D. 1994. Properties of sympathetic reflexes elicited by natural vestibular stimulation: Implications for cardiovascular control. J. Neurophysiol. 71: 2087-2092.

41. Miller, A.D., Yamaguchi, T., Siniaia, M.S., and Yates, B.J. 1995. Ventral respiratory group bulbospinal inspiratory neurons participate in vestibular-respiratory reflexes. J. Neurophysiol. 73: 1303-1307.

42. Sperry, R.W. 1944. Optic nerve regeneration with return of vision in anurans. J. Neurophysiol. 7: 57-69.

43. Gaze, R.M., and Jacobson, M. 1963. A study of the retino-tectal projection during regeneration of the optic nerve in the frog. Proc. R. Soc. London, Ser. B. 157: 420-448.

44. Wallace, M.T., and Stein, B.E. 1996. Sensory organization of the superior colliculus in cat and monkey. Prog. Brain Res. 112: 301-311.

45. Olsen, J.F., Knudsen, E.I., and Esterly, S.D. 1989. Neural maps of interaural time and intensity differences in the optic tectum of the barn owl. J. Neurosci. 9: 2591-2605.

46. Knudsen, E.I., and Konishi, M. 1979. Sound localization by the barn owl (*Tyto alba*). J. Comp. Physiol. 133: 1-11.

47. Groh, J.M., and Sparks, D.L. 1996. Saccades to somatosensory targets, II: Motor convergence in primate superior colliculus. J. Neurophysiol. 75: 428-438.

48. Morris, R.G.M., Garrud, P., Rawlings, J., and O'Keefe, J. 1982. Place navigation impaired in rats with hippocampal lesions. Nature 297: 681-683.

49. Felleman, D.J., Wall, J.T., Cusick, C.G., and Kaas, J.H. 1983. The representation of the body surface in S-1 of cats. J. Neurosci. 3: 1648-1649.

50. Cusick, C.G., Wall, J.T., Felleman, D.J., and Kaas, J.H. 1989. Somatotopic organization of the lateral sulcus of owl monkeys: Area 3b, S-II, and a ventral somatosensory area. J. Comp. Neurol. 282: 169-190.

51. Selemon, L.D., and Goldman-Rakic, P.S. 1985. Longitudinal topography and interdigitation of corticostriatal projections in the rhesus monkey. J. Neurosci. 5: 776-794.

52. Levy, R., Friedman, H.R., Davachi, L., and Goldman-Rakic, P.S. 1997. Differential activation of the caudate nucleus in primates performing spatial and nonspatial working memory tasks. J. Neurosci. 17: 3870-3882.

53. Li, H., Godfrey, D.A., and Rubin, A.M. 1995. Comparison of surgeries for removal of primary vestibular inputs: A combined anatomical and behavioral study in rats. Laryngoscope 105: 417-424.

54. Levi-Montalcini, R. 1949. The development of the acoustico-vestibular centers in the chick embryo in the absence of afferent root fibers and of descending fiber tracts. J. Comp. Neurol. 91: 209-241.

55. Hubel, D.H., and Wiesel, T.N. 1970. The period of susceptibility to the physiological effects of unilateral eye closure in kittens. J. Physiol. (Lond.) 206: 419-436.

56. Knudsen, E.I., Knudsen, P.F., and Esterly, S.D. 1984. A critical period for the recovery of sound localization accuracy following monaural occlusion in the barn owl. J. Neurosci. 4: 1012-1020.

57. Peusner, K.D., and Morest, D.K. 1977. Neurogenesis in the nucleus vestibularis tangentialis of the chick embryo in the absence of the primary afferent fibers. Neuroscience 2: 253-270.

58. Powell, T.P.S., and Erulkar, S.D. 1962. Transneuronal cell degeneration in the auditory relay nuclei. J. Anat. (Lond.) 91: 249-268.

59. Sanes, D.H. 1992. The influence of inhibitory afferents on the development of postsynaptic dendritic arbors. J. Comp. Neurol. 321: 637-644.

60. Jean-Baptiste, M., and Morest, D.K. 1975. Transneuronal changes of synaptic endings and nuclear chromatin in the trapezoid body following cochlear ablations in cats. J. Comp. Neurol. 162: 111-134.

61. Bernard, C.L., Hirsch, J.C., Khazipov, R., and Ben-Ari, Y. 1997. Redox modulation of synaptic responses and plasticity in rat CA1 hippocampal neurons. Exp. Brain Res. 113: 343-352.

62. Shatz, C.J., and Stryker, M.P. 1978. Ocular dominance in layer IV of the cat's visual cortex and the effects of monocular deprivation. J. Physiol. (Lond.) 281: 267-283.

63. Garden, G.A., Redeker-DeWulf, V., and Rubel, E.W. 1995. Afferent influences on brainstem auditory nuclei of the chicken: Regulation of transcriptional activity following cochlea removal. J. Comp. Neurol. 359: 412-423.

64. Morest, D.K., and Bohne, B.A. 1983. Noise-induced degeneration in the brain and representation of inner and outer hair cells. Hear. Res. 9: 145-151.

65. Smith, P.F., and Darlington, C.L. 1992. Comparison of the effects of NMDA antagonists on medial vestibular nucleus neurons in brainstem slices from labyrinth-intact and chronically labyrinthectomized guinea pigs. Brain Res. 590: 345-349.

66. Killackey, H.P., Belford, G., Ryugo, R., and Ryugo, K.K. 1976. Anomalous organization of thalamocortical projections consequent to vibrissae removal in the newborn rat and mouse. Brain Res. 104: 309-315.

67. Bliss, T.V.P., and Collingridge, G.L. 1993. A synaptic model of memory: Long term potentiation in the hippocampus. Nature 361: 31-39.

68. Catalano, S.M., Robertson, R.T., and Killackey, H.P. 1995. Rapid alteration of thalamocortical axon morphology follows peripheral damage in the neonatal rat. Proc. Natl. Acad. Sci. U.S.A. 92: 2549-2552.

69. Killackey, H.P., Chiaia, N.L., Bennett-Clarke, C.A., Eck, M., and Rhoades, R.W. 1994. Peripheral influences on the size and organization of somatotopic representations in the fetal rat cortex. J. Neurosci. 14: 1496-1506.

70. Erzurumlu, R.S., and Jhaveri, S. 1990. Thalamic axons confer a blueprint of the sensory periphery onto the developing rat somatosensory cortex. Brain Res. Dev. Brain Res. 56: 229-234.

71. Bernard, C.L., Hirsch, J.C., Khazipov, R., and Ben-Ari, Y. 1997. Redox modulation of synaptic responses and plasticity in rat CA1 hippocampal neurons. Exp. Brain Res. 113: 343-352.

72. Bliss, T.V.P., and Collingridge, G.L. 1993. A synaptic model of memory: Long term potentiation in the hippocampus. Nature 361: 31-39.

73. Smith. P.F., and Curthoys, I.S. 1989. Mechanisms of recovery following unilateral labyrinthectomy: A review. Brain Res. Rev. 14: 155-180.

74. deWaele, C., Abitbol, M., Chat, M., Menini, C., Mallet, J., and Vidal, P.P. 1994. Distribution of glutamatergic receptor and GAD mRNA-containing neurons in the vestibular nuclei of normal and hemilabyrinthectomized rats. Eur. J. Neurosci. 6: 565-576.

75. Li, H., Godfrey, T.G., Godfrey, D.A., and Rubin, A.M. 1996. Quantitative changes of amino acid distributions in the rat vestibular nuclear complex after unilateral vestibular ganglionectomy. J. Neurochem. 66: 1550-1564.

76. Goto, M.M., Romero, G.G., and Balaban, C.D. 1997. Transient changes in flocculonodular lobe protein kinase C expression during vestibular compensation. J. Neurosci. 171: 4367-4381.

77. Kim, J., Morest, D.K., and Bohne, B.A. 1997. Degeneration of axons in the brainstem of the chinchilla after auditory overstimulation. Hear. Res. 103: 169-191.

78. Kato, B.M., Lachica, E.A., and Rubel, E.W. 1996. Glutamate modulates intracellular Ca^{+2} stores in brain stem auditory neurons. J. Neurophysiol. 76: 646-650.

79. Zirpel, L., and Rubel, W.W. 1996. Eighth nerve activity regulates intracellular calcium concentration of avian cochlear nucleus neurons via a metabotropic glutamate receptor. J. Neurophysiol. 76: 4127-4139.

80. Attias, J., Pratt, H., Reshef, I., Bresloff, I., Horowitz, G., Polyakov, A., and Shemesh, Z. 1996. Detailed analysis of auditory brainstem responses in patients with noise-induced tinnitus. Audiology 35: 259-270.

81. Powell, T.P.S., and Erulkar, S.D. 1962. Transneuronal cell degeneration in the auditory relay nuclei. J. Anat. (Lond.) 91: 249-268.

82. Born, D.E., Durham, D., and Rubel, E.W. 1991. Afferent influences on brainstem auditory nuclei of the chick: Nucleus magnocellularis neuronal activity following cochlea removal. Brain Res. 557: 37-47.

83. Powell, T.P.S., and Erulkar, S.D. 1962. Transneuronal cell degeneration in the auditory relay nuclei. J. Anat. (Lond.) 91: 249-268.

84. Born, D.E., Durham, D., and Rubel, E.W. 1991. Afferent influences on brainstem auditory nuclei of the chick: Nucleus magnocellularis neuronal activity following cochlea removal. Brain Res. 557: 37-47.

85. Peusner, K.D., and Morest, D.K. 1977. A morphological study of neurogenesis in the nucleus vestibularis tangentialis in the late chick embryo. Neuroscience 2: 209-227.

86. Oman, C.M., and Balkwill, M.D. 1993. Horizontal angular VOR, nystagmus dumping, and sensation duration in Spacelab SLS-1 crewmembers. J. Vestib. Res. 3: 315-330.

87. Viéville, T., Clément, G., Lestienne, F., and Berthoz, A. 1986. Adaptive modifications of the optokinetic vestibulo-ocular reflexes in microgravity. Pp. 111-120 in Adaptive Processes in Visual and Oculomotor Systems (E.L. Keller and D.S. Zee, eds.). Pergamon Press, New York.

88. Peusner, K.D., and Morest, D.K. 1977. A morphological study of neurogenesis in the nucleus vestibularis tangentialis in the late chick embryo. Neuroscience 2: 209-227.

89. Lisberger, S.G., and Pavelko, T.A. 1986. Vestibular signals carried by pathways subserving plasticity of the vestibulo-ocular reflex in monkeys. J. Neurosci. 6: 346-354.

4

Plants, Gravity, and Space

INTRODUCTION

Studies of plants and their responses to gravity and the space environment have concentrated on three main themes. The first is space horticulture, the study of how to grow plants successfully in space, either for experimental purposes or for human consumption. This involves assessments of the conditions needed for optimal crop yield, the best plants to grow in space, and the problems inherent in growing plants in a gravity-free and totally enclosed system.

The second theme is the necessity for gravity, or whether there is any facet of a plant's growth, development, and metabolism that is impaired if there is no gravity. In other words, are there plant processes where the mere presence of gravity is essential, regardless of the actual direction of the gravity vector?

Finally, there is the response to the direction of the gravity vector, or how plants respond to specific directions of gravity by altering their pattern of growth and development. For example, plant stems and roots will alter their direction of growth to maintain a set angle with the gravity vector (gravitropism).

Research on each of these three themes has been supported by NASA in the past decade, but each needs additional experimentation and study.

SPACE HORTICULTURE

Reasons for Studies on Space Horticulture

There are three reasons for studying space horticulture. First, if the microgravity environment of spacecraft is to be used to carry out scientific experiments on basic mechanisms of plant responses to gravity, the plants must be grown under optimal conditions. Sick plants are not suitable scientific subjects.

Second, if a lunar (or Mars) colonization base is ever to be set up, it will have to be able to grow its own food. It is doubtful that anyone has actually succeeded in growing crops in a totally enclosed system such as would be needed on the moon. The problems that have been encountered in closing the plant growth system can best be worked out on Earth or in a spacecraft and cannot be delayed until it is time to start building a lunar colony. Ground-based studies may be more relevant than spaceflight studies here, because plants on the moon will be subjected to substantial gravity (0.16 g), even if it is less than on Earth.

The third often-stated reason is to provide food for astronauts during prolonged spaceflights. An as yet unsettled question is the minimum length of the mission that would justify the weight of onboard crop-growing facilities, since for shorter flights it will be more weight efficient to carry food rather than grow it in space. Some current estimates have suggested that the spaceflight mission must exceed 2.6 years before space horticulture would be of value,[1,2] and this estimate assumes that there will be sufficient power to provide optimal illumination of the plants. It is therefore unlikely that space horticulture will actually be used to provide food aboard a spacecraft in the near future, but the development of the needed technology is worth pursuing now, to prepare for later space endeavors.

Accomplishments

Until 1997, studies in space horticulture in NASA were primarily located in the Closed Ecological Life Support System (CELSS) program. In 1997, the CELSS program was merged with the nonbiological life support studies into the Advanced Life Support (ALS) program. The CELSS program expended considerable effort to obtain basic information needed for space horticulture. One focus of activity has been to determine the maximum usable biomass per square meter that can be obtained with selected crop plants. When selected varieties of wheat were grown in dense stands on Earth under high light intensities (150 moles $m^{-2} d^{-1}$), yields as high as 60 g $m^{-2} d^{-1}$ were obtained.[3] These are among the highest plant yields ever recorded on Earth. High yields under similar circumstances have been obtained with potatoes as well.[4] These studies, supported by the CELSS program, have indicated that maximum efficiency may best be achieved by using stands of crop plants far denser than normally used in agriculture, as long as there is sufficient light available. Experiments have reduced the estimate of the amount of growing space needed to support an astronaut in space from 25 m^2 to as low as 10 m^2, assuming that comparable yields can be obtained in space.[5] Algae have often been suggested as a potential food source in space and they can be grown in a spacecraft, but to date the difficulty is producing palatable food from them.

A major problem has been how to close the plant growth unit completely to the outside environment. A second major problem has been developing the lighting systems to provide maximum photosynthetically active radiation with minimum power. Studies are ongoing to improve and adapt light-emitting diode (LED) and microwave lights for use in plant spacecraft facilities. By closed conditions it is meant that the plants must be grown in a system in which there is no exchange of gases with the outside and no input of water or nutrients after the start of the cultivation, and one in which all wastes are recycled or destroyed. The large environmental chamber (known as the Breadboard project) at Kennedy Space Center has come as close as any to reaching that goal,[6] but there are still many problems to be overcome. The experience with the Biosphere 2 project in Arizona has shown just how difficult it is to close such a system successfully. Plant experiments conducted in space so far have not used a closed system; there is constant exchange of gases with the cabin atmosphere, and often the addition of water and nutrients from outside sources. But the system will have to be completely closed if it is to be used for farming on the moon. The technology to recycle water, nutrients, and gases must be developed first.

For a crop plant to be useful as a seed source in space, it must be able to go through a complete life cycle. A number of attempts at this have been made, generally with rather limited success. For example, wheat plants were recently grown to maturity on the Mir space station, but while many heads were produced, none contained seeds.[7] The reasons for the failures are believed to be primarily hardware problems, such as low light intensities, inability to control atmospheric contaminants, and problems with nutrient and water delivery. On the other hand, plants of the rapid-cycling brassica (*Brassica rapa*) have been successfully raised through two generations on the Mir, although the growth and number of buds produced were significantly lower in space in the second-generation seeds (M.E. Musgrave, personal communication, 1998). Normal-appearing potato microtubers have been grown on potato stem cuttings in space, but no attempt has yet been made to grow a real potato crop in space.

Future Directions

Future research on space horticulture, supported by NASA, should be focused along two lines. The first and highest-priority objective is to develop a system on Earth for growing plants under completely closed conditions. This would require developing the most effective methods for removing biologically active atmospheric pollutants (such as ethylene) that might be emitted by the plant, and learning how best to maintain the O_2 and CO_2 levels in the plant chamber. The technology for recycling of water from both the atmosphere and the root environment as well as the nutrients from the soil solution and the unused parts of the harvested plants needs to be perfected. This technology development should be concentrated on a system that can be used for basic scientific experiments aboard the International Space Station (ISS); this will mean that the size of the growth chamber will of necessity be small. It will be necessary, then, to focus attention initially on small plants such as *Arabidopsis* or on plants that can thrive under low light intensities, because power will be a limiting factor on the ISS. The engineering problems associated with developing a totally closed plant growth system are not trivial, but the need is pressing.

Certain engineering problems will require tests in the microgravity environment. In particular, the delivery system for water and nutrients to the roots must be tested in microgravity; it is not possible to validate such systems in the presence of 1 *g*. Such tests are already under way and should be continued and expanded. However, flight experiments should be limited to those needed to develop the closed plant growth system for use in space. When a plant growth system suitable for growing small test plants in space has been perfected, it will then be time to attempt to grow crops of seed- or tuber-producing food plants in space.

A second, lower-priority focus of activity should be to study the responses of plants to specific stresses that plants in spaceflight might undergo. In particular, the response of plants to elevated CO_2 needs to be studied. Plants on Earth grow in the presence of 0.035 percent CO_2, but in the ISS the CO_2 levels will be as high as 0.7 percent for up to 180 days.[8] In general, plants can tolerate CO_2 levels as high as 0.1 percent, but their ability to tolerate the levels of CO_2 expected onboard ISS is unknown. One solution is to keep the atmosphere in the plant chambers completely separate from the cabin atmosphere; another is to develop strains of plants capable of handling such high CO_2 levels. The CELSS program should consider supporting studies on the adaptation of plants to super-high CO_2 levels.

A second stress of importance is vibration. Plants transported to and from orbit on the shuttle are subjected to rather considerable vibration for short periods. Both the immediate effects of this vibration and the long-term responses of the plants need to be understood so that they will not be mistaken for responses to the lack of gravity. NASA is already supporting research in this area (e.g., Xu et al.[9]), but more experimentation is needed. These studies can best be conducted on the ground.

Recommendations

• *The ALS program should concentrate its ground-based research on developing a completely enclosed plant growth system. This effort will require close collaboration between engineers and plant environmental scientists.*

• *The ALS spaceflight program should focus on testing the potentially gravity-sensitive components of the closed plant growth system, such as the nutrient delivery system.*

• *A lower-priority concern is a ground-based study of the problems and mechanisms associated with stresses that plants in space might encounter, including super-high CO_2 levels and vibrations.*

ROLE OF GRAVITY IN PLANT PROCESSES

Scientific Problems

Plants have evolved under conditions of constant 1 *g*, so there is a possibility that they have taken advantage of this influence to modulate some of their processes. For example, differences in density of cellular organelles might lead to a stratification of the cytoplasm that would influence the polarity of the cells. In the absence of gravity, then, the development of a polarity in plant cells might be adversely affected. This hypothesis has led to the repeated suggestion (e.g., in the Goldberg report[10]) that plants should be grown for protracted periods in microgravity to determine whether any such gravity-requiring step can be detected.

The basic experiment in this case is to grow plants in space through several generations (the "seed-to-seed experiment"). If there is any major process that is adversely affected by the lack of gravity, it should manifest itself by an obvious change in the pattern of development. However, the alteration may be minor in significance and difficult to spot unless directly targeted. Critical developmental steps may need to be examined in detail to discover this. For example, the location of the position of root hairs depends on the polarity of root epidermal cells;[11] if the polarity of these cells is altered in microgravity, the location of the root hairs may be altered. This seed-to-seed experiment needs to be run for at least two generations. It is possible that in the absence of 1 *g*, maternal effects on the seeds would only be manifested in the second generation. If substantial effects are not found in two generations with a small number of species, it should be possible to conclude that gravity itself is not of major importance in plant growth and development.

One problem in the interpretation of seed-to-seed experiments is that there may be effects on plant growth in microgravity that are only indirectly the result of the lack of gravity. For example, in the absence of gravity, air convection around leaves may be significantly reduced, with subsequent reduction in uptake of CO_2 needed in photosynthesis and in the transpiration of water vapor from leaves. As a result, the water use efficiency of plants (CO_2 assimilated/H_2O transpired) in microgravity may be considerably different from that in 1 *g*. Similarly, the lack of gravity may adversely affect root aeration, because in the absence of gravity water tends to pool in large, unbroken masses, causing an excess of water in some places with a concomitant lack of oxygen.

A second problem is that the plant growth unit in which this seed-to-seed experiment is conducted must be as "artifact-free" as possible. This means that the light intensity must be high enough so that the plants are actually carrying out net photosynthesis. This value can vary from 1 to 20 μmoles m^{-2} sec^{-2}, depending on the plant used. The atmosphere in the chamber must be controlled and monitored to eliminate complications from ethylene or from unnaturally high levels of CO_2.

It has been frequently assumed that the effect of weightlessness on plants can be mimicked by placing them on a horizontal clinostat, rotating at a slow speed.[12]   If this were so, it would not be necessary to make use of the space environment to test the effects of a lack of gravity on plant processes. For example, the seed-to-seed experiment could be conducted on a clinostat rather than in space. However, the clinostat does not create a weightless environment; instead it produces a constant movement of the gravity vector so that the net effect cancels out the directional response to the gravity vector. A major question, then, has been the extent to which a clinostat actually simulates 0 g, and if so, at what rotation it should be used and whether it should be a horizontal or a 3-D clinostat.

Accomplishments

Attempts to grow any plant from seed to seed in space have met with limited success and much frustration. The earlier Russian attempts were unsuccessful, largely because of technical problems. For example, it appears that the roots were not given sufficient aeration, and the plants suffered severe waterlogging.[13]   In a recent Mir experiment, the wheat plants' inability to form seeds may have been due to enhanced levels of atmospheric ethylene.[14]   A series of three flight experiments by Musgrave, however, has demonstrated that the problems caused by secondary effects of microgravity can be at least partly overcome.[15]   She examined the ability of *Arabidopsis* plants to undergo one part of the life cycle: flower development and sexual reproduction. The technical problems that beset the first experiment, in which sexual reproduction in the plants failed, were partly corrected in a second experiment, which had partial success. A third, improved experiment resulted in apparently normal sexual reproduction in the plants. These experiments showed that the apparent inability of *Arabidopsis* to undergo seed formation in space was an artifact of the poor experimental conditions (for example, low air convection in the growth chamber, remediable by forced air circulation) and not a direct consequence of microgravity.

Rapid-cycling brassica (*Brassica rapa* L) plants have recently been grown through two complete generations on Mir (M.E. Musgrave, personal communication, 1998). Plants grown in space from the first, space-raised generation were generally smaller than comparable plants grown from seeds brought up from Earth. Some of these plants were frozen and brought back to Earth for analysis. A thorough analysis of these space-grown plants and a repeat of this experiment with both *Brassica rapa* and *Arabidopsis thaliana* plants will be needed to assess possible effects of a lack of weightlessness on plant development.

Some developmental abnormalities have been recorded with plants grown in space. For example, a number of root cells, fixed either in space or after return to Earth, showed chromosomal abnormalities.[16]   Whether these were a response to the lack of gravity or to some other stress the plants experienced while on board the spacecraft, such as elevated levels of ethylene or CO_2, cannot yet be determined. Another abnormality is the apparent failure of decapped maize roots to reform caps while in space.[17]   The balance between starch and sugar in plant leaves has also been reported to be altered in plants grown in microgravity compared with the ground controls,[18] but this may also be the result of secondary effects of low gravity.

One approach to the study of the role of gravity has been to examine plant response on a horizontal clinostat. Although the development of plants on a clinostat is generally similar to that of plants grown in space, some small differences have been detected (e.g., Brown[19]). A careful analysis of clinostats in Germany and Japan indicates that the best simulation of weightlessness is obtained using a 3-D clinostat rather than a horizontal one, and with the speed of rotation increased above 1 rpm.[20]   Clinostat experiments have proven useful for obtaining preliminary information about the role of gravity in plant processes, but they cannot substitute for an actual spaceflight experiment on how plants develop in the total absence of gravity.

Future Directions

The most pressing need in space horticulture is for the successful completion of a plant seed-to-seed experiment. Because a complete generation for either *Arabidopsis thaliana* or *Brassica rapa* requires a minimum of 60 days, this experiment is not feasible aboard the space shuttle but can only be performed on a space station. However, a definitive experiment will only be possible when the plant growth apparatus has been sufficiently perfected so that extraneous stresses do not interfere with the detection of any gravity-requiring processes. At the present time at least four competing plant growth units are being designed and supported by NASA. This seems an unnecessary duplication of effort. NASA should consider concentrating its efforts on obtaining one superior plant growth apparatus.

The results of the *Brassica rapa* seed-to-seed experiment need to be confirmed and extended. This experiment should first be carried out with *Arabidopsis*, which has an equally short generation time and the advantages of a small genome size, a wealth of available genetic mutants, and an already well-characterized pattern of development. If no major effects on growth or development are seen in these experiments, it will be apparent that, for at least two species, gravity is not a major requirement for development. In that case, it is doubtful that any additional effort should be expended in exploratory experiments, looking for effects that are not fundamental to all plant species.

If major effects on growth and development are found, it will be necessary to exclude secondary effects of a microgravity environment (such as gravity effects on convection, and thus on gas exchange and heat conduction) before concluding that gravity is important to plant growth. It may be difficult to develop a plant growth unit for use in the seed-to-seed experiment that does not expose the plants to any stress other than that of lack of gravity. This is not a fatal flaw as long as researchers know exactly what stresses the space-grown plants have encountered and how much. Physical measurements of stresses are difficult and require extensive instrumentation.

There are three possible solutions. One is to use an onboard 1-*g* centrifuge as a control. If small plants such as *Arabidopsis* are used, the centrifuge can be one with a relatively small radius rather than the large-diameter centrifuge planned for the ISS. Several small centrifuges have been produced or are currently under development. However, differences in air convection will still exist between plants grown in space which are on a centrifuge versus those not on a centrifuge.

A second approach is to use mutant plants that are insensitive to specific environmental stresses—for example, using ethylene-insensitive mutants to eliminate complications from exogenous ethylene. Another possible solution is to use indicator *Arabidopsis* plants that have been transformed with indicator genes coupled to specific stress-induced promoters. These specific promoters have already been or are being developed for stresses such as temperature extremes, ethylene, water stress, vibration, and anaerobiosis.[21] It should be possible to detect and perhaps measure the amount and types of stresses the space-grown *Arabidopsis* undergoes by use of these plants.

The root development of *Arabidopsis* follows a definite pattern so that cell fates are predictable.[22] It may be possible to determine whether the microgravity conditions cause chromosomal abnormalities sufficient to adversely affect development. If there is any significant cell death, it would be recorded in the pattern of root cell walls and could be visualized by examining the roots.

Recommendations

• *The seed-to-seed experiment, using* Arabidopsis thaliana *and* Brassica rapa *plants, should be a top priority for the ISS. This experiment must be conducted on the ISS, because the plants should be grown through at least two generations in space.*

• *To conduct a meaningful seed-to-seed experiment, NASA needs to develop the following:*

—A superior plant growth unit, with adequate lighting, gas exchange, and water and/or nutrient delivery; and

—Arabidopsis thaliana plants that are insensitive to expected environmental stresses (such as ethylene) and that contain indicator genes for all the expected environmental stresses, such as high levels of CO_2, vibration, anaerobiosis, water stress, and temperature stresses.

• *In the interim, before the ISS is functional, studies on specific stages of plant development in space should be limited to small plants with short life cycles (e.g.,* Arabidopsis thaliana *or* Brassica rapa*). Whenever possible, a* 1-g *on-board centrifuge should be available.*

RESPONSES OF PLANTS TO A CHANGE IN THE DIRECTION OF THE GRAVITY VECTOR

Known Responses

Plants have three major responses to changes in the gravity vector. First, plant stems and roots can alter the direction of their growth in relation to the direction of gravity (gravitropism). An organ will assume a specific direction at some defined angle to the gravity vector, and if displaced from that direction, will curve until the original direction is resumed. Gravitropism is by far the most extensively studied plant response to the gravity vector. In the second response (gravitaxis), the swimming direction of some unicellular algae is directed by the gravity vector. Finally, plant development can be influenced by the direction of the gravity vector. For example, the weight of a tree limb alters the pattern of formation of new wood to provide additional support for the limb. This extra strengthening, called reaction wood, is in response to the tissue stresses, which are directly related to the direction of the stresses within the tissues, as well as to the direction of gravity itself.

Gravitropism

The Component Parts

Gravitropism consists of a series of sequential events leading to an alteration in the growth pattern of the affected organ. In all organs studied so far, the gravity-induced curvature of the organ is a result of a differential in the rate of cell enlargement on the organ's two sides, rather than a change in the number of cells on the two sides.[23] Over the years gravitropism has attracted the attention of a wide range of biologists, including Charles Darwin.[24] One reason for this is that a change in the direction of the gravity vector is one of the few methods that causes changes in plant development in a fast, reversible, but noninvasive way. The graviresponses of plants are a powerful system with which to determine the mechanisms controlling plant development, information that will be of fundamental value in agricultural sciences.

In general, gravitropism can be broken down into four steps: perception, intracellular transduction, translocation, and reaction. The perception step involves the cell's perception of the direction of the gravity vector—the displacement of some component of a cell, up or down. This component can be the whole mass of the protoplast as it presses against and interacts with the cell wall, or it may be a sedimenting heavy organelle (statolith), such as an amyloplast. In the cellular transduction step, some biochemical or structural asymmetry is set up in the perceiving cell so that the cell is polarized with respect to the direction of gravity. The result of cellular transduction is the movement of some growth-altering factor laterally. This sets up an asymmetry of this factor across the organ, coupled with longitudinal movement of the factor from the site of perception to the site of reaction. In the reaction phase, the unequal concentration of growth factors creates an unequal rate of cell elongation on the two sides of the organ.

Gravitropism in Single Cells

The simplest gravitropic system, in theory, is one in which the whole process takes place within a single cell. There is no need for a translocation step, because the reaction occurs in the gravity-perceiving cell. Several such systems are known. For example, the curvature of the rhizoid of the green alga *Chara* occurs within the same cell that perceives the gravity,[25] as does the gravitropism of the protonema of the moss *Ceratodon*.[26]

In both systems, it appears that the direction of gravity is perceived by the sedimentation of statoliths, tethered by the cytoskeleton. The result is a change in the location of the tip growth by which such cells elongate. The steps in between are largely a mystery. Possible involvement of localized cytoplasmic calcium and cytoskeletal elements such as actin have been suggested.[27] To understand how this gravitropic response is controlled, the process of tip growth will need to be studied in greater detail to learn what controls the location of this growth. Root hairs and pollen tubes also grow by tip growth, so learning about the control of tip growth in these gravitropic systems could have major benefits to other areas of plant science. To do this, the enzymes, cell wall components, and vesicle trafficking involved need to be identified. It would help greatly if mutants altering the gravitropic process in tip-growing cells were available and if the genetics of these organisms were better understood. Nevertheless, these tip-growing gravitropic systems hold great promise for some comprehensive study of the cell biology involved and may prove to be the systems that lead to an understanding of gravitropism in more complex, multicellular systems.

Gravitropism in Multicellular Plants

Gravitropic curvature in response to a change in the gravity vector can occur in most stems and roots as well as some flowers and fruits, but rarely in leaves. The resultant direction of growth can be either vertical (orthogravitropic) or at some specific and set angle to the gravity vector (plageotropic). Most research into the gravitropism of higher plants has concentrated on the response of roots, although earlier work focused more on coleoptiles. The present state of knowledge and the areas where research is needed are summarized below.

Perception. In the past decade researchers have made a concerted effort to determine whether the perception of the gravity vector's direction is mediated by sedimenting statoliths interacting with some cellular component, by the tension on the cytoskeleton exerted by tethered statoliths, or by the mass of the whole protoplast acting on plasma membrane attachments to the cell wall.[28] In the giant cells of the alga *Chara*, the direction of gravity is apparently perceived by the whole cell. This information is then used to cause different rates of protoplasmic streaming on the two sides of the cell.[29] In higher plants, the evidence strongly supports the idea that amyloplasts play a major role in gravity perception. For example, removal of the root cap (the only part of the growing root that has amyloplasts) renders the root insensitive to the direction of gravity.[30] But it is unclear whether the falling amyloplast must interact with some cellular component such as the endoplasmic reticulum, or whether the amyloplasts create tensions in the cytoskeleton which are transmitted to the plasma membrane. Attempts to resolve this question by disrupting and dissolving the cytoskeleton from gravity-perceptive cells have produced equivocal results, because of the difficulty in determining whether or not the whole cytoskeleton has been removed.[31]

Starch-containing amyloplasts are apparently not required for graviperception in *Arabidopsis* roots. Starchless mutants of *Arabidopsis* retain some ability to undergo root gravitropism,[32] as do roots in which amyloplast-containing root cap cells have been eliminated by laser ablation.[33] This suggests that roots may have a second, residual method for detecting the direction of the gravity vector. The importance of the dense starch grains in graviperception can be partly assessed by comparing the *g*-threshold for starch-containing versus starchless *Arabidopsis* roots. Surprisingly, the *g*-threshold for a tropic response in normal starch-containing roots is still uncertain.[34,35] This *g*-threshold should be determined in space for both starch-containing and starchless *Arabidopsis* roots, using a variable-*g* centrifuge.

The ability of starchless *Arabidopsis* roots to have a tropic response raises the important question of the extent to which nonstatolith-containing plant cells, in general, can perceive the direction of the gravity vector. It may be that all cells perceive gravity, but the magnitude of their perception depends on the density of the cellular components. Cells with dense objects such as amyloplasts are highly sensitive to the gravity vector and are the normal gravity-perceptive cells. But other cells may be weakly gravity perceptive or may use the information about the direction of the gravity vector to do something other than change the direction of growth. If decapped maize roots, which grow well but fail to curve in response to 1 *g*, are subjected to higher *g* forces, would they then be able to perceive the direction of the gravity vector? Any cell that perceives gravity must become polarized with respect to the gravity vector. It is the transduction steps that hold the key to an understanding of gravity responses in plants.

Intracellular Transduction of Gravity Vector Information. The perception of gravity must produce an asymmetry in the perceiving cell, either in its structure or in its biochemistry. Despite the importance of this transduction step, almost nothing is known for certain about it. A number of suggestions have been made: an asymmetry in cellular calcium,[36] an activation of plasma membrane ion channels localized at one position in the cell,[37] a change in the structure and distribution of cytoskeletal elements, or a change in the gating of plasmodesmata between cells.[38] Any or all of these might be possible, but there is no direct evidence to support any of them. Whatever the nature of the asymmetry, it must be capable of being induced at 4°C and must be "remembered" for at least an hour at 4°C.[39] None of the current theories can explain these observations.

A major emphasis of NASA-sponsored research should be on the nature of this transduction step. Techniques are available to examine intracellular localization of calcium or of cytoskeletal elements. Studies should focus on comparing cells containing amyloplasts, since these cells at least have the potential for detecting the direction of gravity, with cells with starchless amyloplasts or with cells lacking amyloplasts. There is no reason why these experiments should not be conducted on single cells or even protoplasts. Any relevant asymmetry should change when the direction of the gravity vector shifts and should be either absent or greatly reduced in cells that lack amyloplasts.

The Translocation Steps. The transduction event induces a change in the perceiving cells so that some signal is translocated differentially to the site of reaction on the two sides of an organ. The Cholodny-Went hypothesis[40] proposed that the signal is auxin, and that it is first transported transversely across a horizontal stem or root, and then translocated longitudinally to the site of reaction. This is clearly a gross oversimplification. The fact that inhibitors of polar auxin transport block gravitropism of both stems and roots indicates that auxin plays some role[41] and that it must be polarly transported during gravitropism in both organs. In stems, at least, auxin is not actually translocated from one side of the stem to the other; but there is apparently a differential transport of auxin from the vascular tissue to the neighboring outer cell layers.[42] In roots, the signal is translocated from the root cap through the meristem to the distal elongation zone (DEZ).[43] However, it is not certain that the translocated signal is auxin or that there is just one signal. Researchers have proposed that there may be rapid signals, which might be electrical or hydraulic, as well as slow signals that are hormonal.[44] Rapid gravity-induced changes in membrane potential have been detected in both root cap[45] and DEZ cells,[46] but their significance is open to speculation. Asymmetric concentrations of auxin have been detected in the DEZ area of maize roots,[47] but whether the auxin was translocated from the root cap or from the basal portions of the root is not known.

This area of signal translocation is ripe for additional studies. The cellular pathway for auxin transport needs to be ascertained in both stems and roots. The signal may move from cell to cell through the extracellular wall solution, or intracellularly through the plasmodesmata. The interaction between the gravity vector and auxin transport has yet to be determined. It is difficult to see how an asymmetric gradient of a chemical signal can persist if the signal is free to pass through the plasmodesmata, as auxin would be expected to do. For that reason, the effect of gravity on the conductance of plasmodesmata merits study. In roots, a more extensive study of gravity-actuated electrical signals is needed, coupled with precise measurements of the timing of the arrival of electrical signals versus the start of gravitropic curvature.

The Reaction Phase. The paradigm of the reaction phase is that an asymmetric gradient of auxin, with more on the lower side of either stem or root, causes the differential in the rate of cell elongation. In stems, the greater auxin level on the lower side leads to enhanced growth; in roots, it leads to an inhibition of growth.[48] There is strong evidence for such auxin asymmetry in stems and coleoptiles[49] and some evidence for such an auxin gradient in roots.[50] And though the cell layers that control elongation of stems are clearly the outer layers, the location of the comparable controlling layers in roots is uncertain, although it may be the pericycle.[51]

In addition to the asymmetry of auxin, there may also be a gravity-induced differential sensitivity of the cells in the reaction zone to auxin. The signal moving from the site of perception might not be auxin but some other factor that alters the sensitivity of the reacting cells to it. There is some evidence of enhanced sensitivity to auxin on the lower side of horizontal soybean hypocotyls.[52] The fact that the auxin-induced small auxin up-regulated (SAUR) genes are more strongly induced on the lower side of horizontal tobacco stems[53] is compatible with there being either more auxin or more sensitivity to auxin on the lower side. Further investigations into the effect of the gravity vector on the sensitivity of reacting cells to auxin are needed.

The mechanism by which auxin controls the rate of cell enlargement has been extensively studied. There is mounting evidence that auxin stimulates elongation of stem and coleoptile cells in part by promoting the export of protons from the cells, with the lowered wall pH activating wall loosening proteins such as the expansins.[54] But why does auxin slow root cell elongation? There are still large gaps in knowledge here, and until these are filled in, it will not be possible to fully comprehend how gravitropic curvature occurs in roots.

Needs for Future Research on Gravitropism. During the past decade the extensive research into gravitropism has greatly enhanced knowledge about this process. But future advances could be hastened by following several principles:

1. Research on gravitropism should be concentrated on only a few systems. Because systems differ, the use of large numbers of different experimental systems has created confusion and diluted effort. By concentrating efforts on a few systems, results obtained by one research group would have a synergistic effect on the research of others using the same system.

Gravitropism in single cells might best be concentrated on the *Chara* rhizoid and the *Ceratodon* protonema systems. For roots of higher plants, there are advantages to working with either *Zea mays* roots, which are well studied and large, or on *Arabidopsis* roots, with a range of known mutants, and a whole genome that will be known within a few years. Research on gravitropism in stems might well concentrate on *Arabidopsis*, tomato, and peas, in which gravitropic mutants exist. Coleoptiles of *Avena* or *Zea mays* have the advantage that most of their cells contain amyloplasts and are good subjects for studies on perception, especially the transduction phase.

2. Mutants should be used wherever possible. The power of genetics to dissect the steps in gravitropism is immense, and the information that could be obtained by sequencing genes directly involved in this process could lead to an understanding of how the pieces fit together. Using complementation analysis, it should be possible to locate agravitropic mutants that are blocked specifically in each of the four main steps in gravitropism. Mutants have only just started to be used effectively in such studies. Another group of useful mutants are those that alter the angle at which the organs grow in response to the direction of gravity or that change this direction in response to the presence or absence of light.[55]

3. The research should concentrate on the *basic* questions and not be waylaid by less important ones. For any research supported by NASA, the PI should demonstrate that the results would enhance an understanding of the overall process. Perhaps the most important question is the nature of the asymmetry that occurs in any cell that perceives the direction of the gravity vector.

4. Studies of gravitropism must make use of the most advanced techniques available, including extensive use of genetics and transgenic plants. Using advanced microscopic techniques, such as confocal microscopy to study cytoplasmic calcium distribution, could provide information not obtainable by other methods.

Recommendations

• *A primary focus of NASA-sponsored research in plant biology should be on the mechanisms of gravitropism. In particular, modern cell and molecular techniques should be used to determine the following:*

 —*The identity of the cells that actually perceive gravity, and the role of the cytoskeleton in the process;*

 —*The nature of the cellular asymmetry set up in a cell that perceives the direction of the gravity vector;*

 —*The nature and mechanism of the translocation of the signals that pass from the site of perception to the site of reaction; and*

 —*The nature of the response to the signal(s) that leads to alterations in the rate of cell enlargement.*

• *Maximum use should be made of mutants and gene sequencing to identify specific proteins involved in gravitropism.*

• *A secondary focus should be on the mechanisms of graviperception and graviresponse in single cells, especially the algae and mosses.*

Gravitaxis

Some flagellated single-celled protists can orient their swimming in response to the direction of gravity. For example, young *Euglena* cells, in the absence of light signals, swim upward, whereas older cells swim downward.[56] The perception of gravity appears to be by the whole *Euglena* protoplast, because the direction of swimming can be reversed if the cells are placed in a medium whose density is greater than that of the cells. The microgravity environment of the shuttle was used effectively to show that the threshold for perception of gravity by *Euglena* cells is about 0.16 *g*.[57]

No research on gravitaxis is currently being supported by NASA. This area that holds great promise for understanding the response of cells to gravity and might well be one that NASA should consider supporting.

Recommendation

As a lower priority, NASA should consider supporting research on algal gravitaxis.

Effects of Gravity-induced Tissue Stresses on Plant Development

The above-ground organs of multicellular plants can undergo considerable stress in their tissues due to suspended weight. Cells on the top of such suspended organs are under tension, whereas those on the bottom are compressed. Horizontal tree limbs undergo changes in the pattern of deposition of new secondary xylem in response to the weight of the limb. The reaction wood that is produced supports the limb and counteracts the effects of gravity. In conifers, the reaction wood forms on the lower side of the limb and forces it upward; in hardwoods, it forms on the upper side and contracts to pull the limb upward. The result is that the branch maintains a set angle with respect to gravity, regardless of the branch's weight. The formation of reaction wood has been extensively studied by forestry scientists, because the physical properties of reaction wood are different from those of regular wood. The mechanisms involved in sensing the stresses are still largely unknown.[58] However, it is not recommended that NASA support work in this area.

REFERENCES

1. Schwartzkopf, S.H. 1992. Design of a controlled ecological life support system. BioScience 42: 526-535.
2. Aeronautics and Space Engineering Board, National Research Council. 1998. Advanced Technology for Human Support in Space. National Academy Press, Washington, D.C.
3. Bugbee, B.G., and Salisbury, F.B. 1988. Exploring the limits of crop productivity: I. Photosynthetic efficiency of wheat in high irradiance environments. Plant Physiol. 88: 869-878.
4. Wheeler, R.M., and Tibbitts, T. 1987. Utilization of potatoes for life support systems in space: IV. Effect of CO_2. Am. Potato J. 66: 25-34.
5. Hoff, J.E., Howe, J.M., and Mitchell, C.A. 1982. Nutritional and cultural aspects of plant species selection for a controlled ecological life support system. NASA-CR-166324. National Aeronautics and Space Administration, Washington, D.C.
6. Corey, K.A., and Wheeler, R.M. 1992. Gas exchange in NASA's biomass production chamber: A preprototype closed human life support system. BioScience 42: 503-509.
7. Strickland, D.R., Campbell, W.F., Salisbury, F.B., and Bingham, G.E. 1997. Morphological assessment of reproductive structures of wheat grown on Mir. Gravit. Space Biol. Bull. 11: 14.
8. Board on Environmental Studies and Toxicology, National Research Council. 1996. Spacecraft Maximum Allowable Concentrations for Selected Airborne Contaminants, Vol. 2. National Academy Press, Washington, D.C.

9. Xu, W., Durugganan, M.M., Polisensky, D.H., Anatosiewicz, D.M., Fry, S.C., and Braam, J. 1995. Arabidopsis TCH4, regulated by hormones and the environment, encodes a xyloglucan endotransglycosylase. Plant Cell 7: 1555-1567.

10. Space Science Board, National Research Council. 1987. A Strategy for Space Biology and Medical Science for the 1980s and 1990s. National Academy Press, Washington, D.C.

11. Dolan, L. 1996. Pattern in the root epidermis: An interplay of diffusible signals and cellular geometry. Ann. Bot. 77: 547-553.

12. Brown, A.H., Johnsson, A., Chapman, D.K., and Heathcote, D. 1996. Gravitropic responses of the *Avena* coleoptile in space and on clinostats: IV. The clinostat as a substitute for space experiments. Plant Physiol. 98: 210-214.

13. Halstead, T.W., and Dutcher, F.R. 1987. Plants in space. Annu. Rev. Plant Physiol. 38: 317-345.

14. Ivanova, T., Kostov, P., Sapunova, S., Dandolov, I., Sytchev, V., Podolski, I., Levinskskh, M., Meleshko, G., Bingham, G., and Salisbury, F. 1998. From fresh vegetables to the harvest of wheat plants grown in the "Svet" space greenhouse onboard the Mir orbital station. J. Gravit. Physiol. 4: 71-72.

15. Musgrave, M.E., Kuang, A. and Matthews, S.W. 1997. Plant reproduction during spaceflight: Importance of the gaseous environment. Planta 203: S177-S184.

16. Krikorian, A.D., Levine, H.G., Kann, R.P., and O'Conner, S.A. 1992. Pp. 491-555 in Advances in Space Biology and Medicine (S.L. Bonting, ed.). JAI Press, Greenwich, Conn.

17. Moore, R., McClelen, C.E., and Wang, C.-L. 1987. The influence of microgravity on root-cap regeneration and the structure of columella cells in Zea mays L. Am. J. Bot. 74: 218-223.

18. Brown, C.S., Obenland, D.M., and Musgrave, M.E. 1993. Spaceflight effects on growth, carbohydrate concentration and chlorophyll content in Arabidopsis. ASGSB Bull. 7: 83.

19. Brown, A.H. 1996. Gravity related features of plant growth behavior studied with rotating machines. J. Gravit. Physiol. 3: 69-74.

20. Hoson, T., Kamisaka, S., Masuda, Y., Yamashita, M., and Buchen, B. 1997. Evaluation of the three-dimensional clinostat as a simulator of weightlessness. Planta 203: S187-S197.

21. Baker, S.S., Wilhelm, K.S., and Thomashow, M.F. 1994. The 5' region of *Arabidopsis thaliana* Cor15A has *cis* acting elements that confer cold regulated, drought regulated and ABA regulated gene expression. Plant Mol. Biol. 24: 701-713.

22. Dolan, L., Janmaat, K., Villemsen, V., Linstead, P., Poethig, S., Roberts, K., and Scheres, B. 1993. Cellular organization of the *Arabidopsis thaliana* root. Development 119: 71-84.

23. Wilkins, M.B. 1984. Gravitropism. Pp. 163-185 in Advanced Plant Physiology (M.B. Wilkins, ed.). Pitman Publishing, London.

24. Darwin, C. 1880. The Power of Movement in Plants. John Murray, London.

25. Hejnowicz, Z., and Sievers, A. 1981. Regulation of the position of statoliths in Chara rhizoids. Protoplasma 108: 117-137.

26. Schwuchow, J., Sack, F.D., and Hartmann, E. 1990. Microtubule distribution in gravitropic protonemata of the moss *Ceratodon*. Protoplasma 159: 60-69.

27. Sievers, A., Kramer-Fischer, M., Braun, M., and Buchen, B. 1991. The polar organization of the growing *Chara* rhizoid and the transport of statoliths are actin-dependent. Bot. Acta 104: 103-109.

28. Barlow, P.W. 1995. Gravity perception in plants: A multiplicity of systems derived by evolution? Plant Cell Environ. 18: 951-962.

29. Staves, M.P., Wayne, R., and Leopold, A.C. 1995. Detection of gravity-induced polarity of cytoplasmic streaming in *Chara*. Protoplasma 188: 38-48.

30. Wilkins, M.B. 1984. Gravitropism. Pp. 163-185 in Advanced Plant Physiology (M.B. Wilkins, ed.). Pitman Publishing, London.

31. Baluska, F., and Hasenstein, K.H. 1997. Root cytoskeleton: Its role in perception of and response to gravity. Planta 203: S69-S78.

32. Kiss, J.Z., and Sack, R.D. 1989. Reduced gravitropic sensitivity in roots of a starch-deficient mutant of *Nicotiana sylvestris*. Planta 180: 123-130.

33. Blancoflor, E.B., Fasano, J.M., and Gilroy, S. 1997. Using laser ablation to probe the functional role of cap cells in Arabidopsis root gravitropism. Gravit. Space Biol. Bull. 11: 47.

34. Perbal, G., Driss-Ecole, D., Tewinkel, M., and Volkmann, D. 1997. Statocyte polarity and gravisensitivity in seedling roots grown in microgravity. Planta 203: S57-S62.

35. Laurinavicius, R., Svegzdiene, D., Sievers, A., Buchen, B., and Tairbekov, M. 1997. Statics and kinetics of statolith positioning in cress root statocytes (Bion-11 mission). Gravit. Space Biol. Bull. 11: 24.

36. Slocum, R.D., and Roux, S.J. 1983. Cellular and subcellular localization of calcium in gravistimulated oat coleoptiles and its possible significance in the establishment of tropic curvature. Planta 157: 481-492.

37. Pickard, B.G., and Ding, J.P. 1993. The mechanosensory calcium-selective ion channel: A key component of a plasmalemmal control centre? Aust. J. Plant Physiol. 20: 439-459.

38. Cleland, R.E. 1997. General discussion on graviresponses. Planta 203: S170-S173.

39. Fukaki, H., Fujisawa, H., and Tasaka, M. 1996. How do plant shoots bend up? The initial step to elucidate the molecular mechanisms of shoot gravitropism using *Arabidopsis thaliana*. J. Plant Res. 109: 123-137.

40. Wilkins, M.B. 1984. Gravitropism. Pp. 163-185 in Advanced Plant Physiology (M.B. Wilkins, ed.). Pitman Publishing, London.

41. Katekar, G.F., and Giessler, A.E. 1980. Auxin transport inhibitors: IV. Evidence for a common mode of action for a proposed class of auxin transport inhibitors: The phytotropins. Plant Physiol. 66: 1190-1195.

42. Bandurski, R.S., Schulze, A., Jensen, P., Desrosiers, M., Epel, B., and Kowalczyk, S. 1992. The mechanism by which an asymmetric distribution of plant growth hormone is attained. Adv. Space Res. 12: 203-210.

43. Moore, R., Evans, M.L., and Fondren, W.M. 1990. Inducing gravitropic curvature of primary roots of *Zea mays* cv Agotropic. Plant Physiol. 92: 310-315.

44. Cleland, R.E. 1997. General discussion on graviresponses. Planta 203: S170-S173.

45. Behrens, H.M., Gradmann, D., and Sievers, A. 1985. Membrane-potential responses following gravistimulation in roots of *Lepidium sativum* L. Planta 163: 463-472.

46. Ishikawa, H., and Evans, M.L. 1990. Gravity-induced changes in intracellular potentials in elongating cortical cells of mung bean roots. Plant Cell Physiol. 31: 457-462.

47. Young, L.M., Evans, M.L., and Hertel, R. 1990. Correlations between gravitropic curvature and auxin movement across gravistimulated roots of *Zea mays*. Plant Physiol. 92: 792-796.

48. Evans, M.L. 1985. The action of auxin on plant cell elongation. CRC Crit. Rev. Plant Sci. 2: 317-365.

49. Harrison, M.A., and Pickard, B.G. 1989. Auxin asymmetry during gravitropism by tomato hypocotyls. Plant Physiol. 89: 652-657.

50. Young, L.M., Evans, M.L., and Hertel, R. 1990. Correlations between gravitropic curvature and auxin movement across gravistimulated roots of *Zea mays*. Plant Physiol. 92: 792-796.

51. Bjorkman, T., and Cleland, R.E. 1988. The role of the epidermis and cortex in gravitropic curvature of maize roots. Planta 176: 513-518.

52. Rorabaugh, R.A., and Salisbury, F.B. 1989. Gravitropism in higher plant shoots: VI. Changing sensitivity to auxin in gravistimulated soybean hypocotyls. Plant Physiol. 91: 1329-1338.

53. Li, Y., Hagen, G., and Guilfoyle, T.J. 1991. An auxin-responsive promoter is differentially induced by auxin gradients during tropisms. Plant Cell 3: 1167-1175.

54. Cleland, R.E. 1995. Cell elongation. Pp. 214-227 in Plant Hormones: Physiology, Biochemistry and Molecular Biology (P.J. Davies, ed.). Kluwer Academic Press, Dordrecht, The Netherlands.

55. Lomax, T.L. 1997. Molecular genetic analysis of plant gravitropism. Gravit. and Space Biol. Bull. 10: 75-82.

56. Häder, D.-P. 1987. Polarotaxis, gravitaxis and vertical phototaxis in the green flagellate, Euglena gracilis. Arch. Microbiol. 147: 179-183.

57. Häder, D.-P., Rosum, A., Schäfer, J. and Hemmersbach, R. 1995. Gravitaxis in the flagellate *Euglena gracilis* is controlled by an active gravireceptor. J. Plant Physiol. 146: 474-480.

58. Timell, T.E. 1986. Compression Wood in Gymnosperms. Springer-Verlag, Berlin.

5

Sensorimotor Integration

INTRODUCTION

On Earth, the force of gravity is an omnipresent influence on the physiology and behavior of organisms. Sensorimotor integration plays a key role in posture and movement control, locomotion, object manipulation, and tool use in an environment containing gravity torques that vary with body orientation and configuration. In the Goldberg report,[1] the section on sensorimotor integration focused on five areas, including mechanisms of spatial orientation, postural control, vestibulo-ocular reflexes, vestibular processing, and space motion sickness. During the last 10 years, extensive experimental research has been performed in all of these areas. In particular, considerable progress has been made in understanding how the vestibulo-ocular reflex, gaze control, and thresholds for angular and linear accelerations are affected by exposure to microgravity in humans. This chapter reviews the major advances in these fields, followed by recommendations for future investigations on these and related areas of research.

SPATIAL ORIENTATION

Accurate determination of the body's position relative to the external environment is critical in controlling body movements and posture, as well as for interacting with objects in the environment. On Earth, gravity plays a fundamental role in spatial orientation, accelerating the body downward. Multimodal sensory stimuli[2] and motor feedback[3] are important factors in the appreciation and regulation of body orientation. Extensive research has already been performed to evaluate the relative contributions of visual, vestibular, and tactile stimuli to the perception of self-orientation and motion before, during, and after spaceflight. In terrestrial experiments, moving visual stimuli are commonly used to induce illusions of self-motion and tilt in stationary subjects. These experiments permit an evaluation of the relative contributions visual and vestibular cues make in the perception of body orientation. The influence of simple rotating visual patterns (dots and stripes) in the frontal plane on the perceived vertical axis and perceived self-motion has been studied systematically in spaceflight and parabolic flight.[4,5] Although individual differences are considerable, the basic finding is that weightlessness enhances the magnitude of visually induced apparent roll of the body. Some individuals experience full 360° roll, which happens rarely on Earth. In orbital flight experiments, the enhanced effectiveness of visual stimuli for inducing apparent self-motion is thought to be related to the altered orientation cues from the otolith organs. On Earth, the otoliths respond to gravitoinertial acceleration by changing the pattern of their output signals, which provide information about head orientation relative to gravity. In microgravity, the otoliths are effectively unloaded and cannot provide information about static head orientation. The influence of tactile cues to the soles of the feet during rotary visual stimulation has been tested in spaceflight.[6] In this situation, tactile stimulation is generated by a bungee cord apparatus that pulls the astronaut against the spacecraft deck. It was found that bungee loading could attenuate the sense of self-motion and tilt.

In microgravity, astronauts often spontaneously experience spatial orientation illusions, especially during their initial exposure to weightlessness.[7] In the second manned USSR flight, the cosmonaut Titov was the first to report an illusion of "feeling upside down" while in orbit. Other astronauts have since reported a variety of illusions involving virtually all possible combinations of self-orientation and vehicle orientation. Some of these effects are dependent on body orientation in relation to architecturally specified horizontals and verticals, such as the deck or walls of the spacecraft. If oriented with his or her head near the architectural "down," an astronaut may feel inverted in an upright craft. For some astronauts, the direction in which they perceive "down" corresponds to the direction their feet are pointing. For these individuals, changing the position of their feet shifts the apparent "down" direction. There are significant individual differences in the extent to which these orientation illusions are experienced, as well as fluctuations over time in their precise makeup.[8,9] Researchers have attempted to relate the occurrence of orientation illusions to otolith function, especially the function of the saccule.[10-12] At present, it is thought that a combination of visual and cognitive factors, tactile input, vestibular asymmetries, and the positioning of the feet all contribute to varying degrees in producing orientation illusions for different individuals, making it difficult to predict when orientation illusions will be experienced.[13]

The otolith organs' sensitivity to linear acceleration in weightlessness has been evaluated for the three primary axes using linear sleds that can hold an astronaut in different sled orientations and provide controlled acceleration patterns.[14-16] Overall, the results are variable, with in-flight decreases in the threshold for detection of linear acceleration in some individuals and increases in others, relative to preflight values. Postflight assessments show similar variability. In spaceflight, the detection time for reporting linear acceleration (i.e., the period of time between being accelerated and reporting a sensation of motion) seems to be lowered for the x and y axes and elevated for the z-axis.[17] This could reflect decreased sensitivity of the sacculus, since the saccular receptor end organ is aligned with the z-axis of the head and is subject to constant shear when the head is in the upright position on Earth. On the other hand, somatosensory stimuli generated by contact forces between the astronaut's body and the sled apparatus could also play a role.

Studies on the position sense of the limbs have been carried out both in parabolic flights and spaceflight.[18,19] Position sense or limb proprioception is derived from afferent signals of the muscle spindles, and possibly also Golgi tendon organs and joint receptors, which are interpreted in relation to ongoing patterns of muscle activity. On Earth, muscle spindle sensitivity is influenced by head orientation, which is detected by the otoliths and (through spinal cord connections) modulates the antigravity muscles of the body. Proprioception has been assessed by investigating the tonic vibration reflexes, which can be measured to determine muscle spindle gain. These reflexes are elicited by mechanically vibrating a muscle or its tendon with a physiotherapy vibrator. In parabolic flight tests, tonic vibration reflexes show immediate decreases on transition to 0 *g*, and increases in 2 *g* relative to 1-*g* straight-and-level flight.[20,21] Kinematic studies of arm movement control, in which subjects attempt during parabolic flight to repeat arm movements practiced preflight, also suggest diminished muscle spindle gain in microgravity, at least during initial exposure. Rhythmic movements made in 0 *g* tend to be of smaller amplitude and have more frequent dynamic overshoots than in 1 *g*, both for horizontal and vertical forearm orientations (relative to the subject).[22] This pattern is consistent with decreased spindle gain, and consequently decreased damping of limb movements. Diminished muscle spindle sensitivity is consistent with the frequent reports that position sense of the limbs is degraded in weightless conditions.[23] During spaceflight, there may be a gradual change in the perceptual interpretation of proprioceptive signals. Initially, vibration of leg muscles of a test subject, who is attached to the deck with foot supports and restrained in position, leads to illusory tilt of the body. On the ground, this would occur under comparable conditions. However, later in flight such vibration leads to sensations that the deck is tilting or rising under the subject's feet, depending on the muscles stimulated.[24]

Lifting and manipulating objects is also an important aspect of limb movement control. Studies of the ability to discriminate between differences in the mass of objects having similar size and appearance by hefting them show degradations of performance in microgravity conditions.[25,26] If the hefting frequency is increased, performance improves considerably.[27] Rapid arm movements are known to be less dependent for their accurate execution on muscle spindle feedback and are less impaired than slow movements in microgravity. Consequently, the degradation in mass discrimination associated with slow movements likely reflects errors in resolution of limb trajectory, at least in part. Subjects experience postflight increases in the apparent heaviness of hefted objects and of the body and limbs. This change points to central nervous system reinterpretations of the apparent effort associated with supporting the limbs or holding up objects against gravity.

Summary and Recommendations

Considerable progress has been made in determining thresholds for perception of linear and angular acceleration under conditions of passive body motion in microgravity. The relative influences of visual and tactile stimulation on perceived orientation under passive conditions have been determined. Researchers have also made progress in understanding how microgravity affects limb position sense. In the future, crewed space travel, especially interplanetary exploration, will probably involve transitions between different force backgrounds, with the need for adequate sensorimotor performance and orientation control in each force level. To meet such demands will require an understanding of how human spatial behavior is affected during active body movement by transitions to different force levels and how adaptation is achieved. This will also be an important issue when Space Station Freedom becomes a permanent orbiting facility and astronauts travel back and forth between it and Earth. The deleterious effects of microgravity on bone and muscle physiology despite exercise countermeasures (described in other sections of this report) also raise the possibility that some form of artificial gravity will be necessary to ensure the success of long-term missions.

1. There is a critical need to evaluate the influence of microgravity and other non-1-g force levels on the integrative coordination of complex body activities, including reaching and locomotory movements involving combinations of eye, head, torso, arm, and leg activity. Studies should be performed before, during, and after spaceflight so that initial disruptions and the time course of adaptation and readaptation can be identified.

2. It is important for the success of long-duration space missions (e.g., crewed flights to Mars) to determine the sensory, motor, and cognitive factors influencing the ability to adapt and to retain adaptation to different force backgrounds. These experiments can be conducted on the International Space Station, in rotating environments, and in parabolic flight.

3. Neural coding of spatial navigation may be affected by a change from a 1-g to a microgravity force background and may relate to some of the orientation illusions experienced by astronauts. The influence of altered force levels on orientation and geographical localization should be explored in parallel experiments with humans and animals. Ground-based centrifuges, body-loading paradigms, parabolic flight conditions, virtual reality conditions, and eventually orbital flight should be utilized for testing.

4. The relative contribution of different sensory modalities and motor factors in influencing perceived orientation and body configuration in various force environments should be determined for conditions involving active postural control in astronauts.

POSTURE AND LOCOMOTION

The control of stance and locomotion is complex and involves vestibular, somatosensory, proprioceptive, visual, and motor mechanisms. Regulation of stance differs greatly under terrestrial and spaceflight conditions. Static balance on Earth requires maintaining the projection of the body's center of mass within the support area defined by the feet. When arm movements are made during static balance, anticipatory innervations of leg muscles compensate for the impending reaction torques and the changes in location and projection of the center of mass associated with the voluntary arm movements. Arm raising has been used in flight with the subject's feet secured to the deck to determine whether exposure to microgravity affects the pattern of anticipatory compensations usually associated with arm-raising maneuvers.[28,29] Initially, similar patterns are seen preflight and in flight in terms of the muscle groups and timing of activity involved. This means that compensation both for dynamic and static influences of rapid arm raising still occurs early in spaceflight, although the latter is functionally unnecessary. On later flight days, the activation of the extensor component (e.g., soleus and gastrocnemius) is reduced, indicating that the compensation for a shift of center of mass (which is physically unnecessary in weightless conditions) no longer occurs.

On Earth, rapidly bending the trunk forward and backward at the waist is accompanied by backward and forward displacements of the hips and knees to maintain balance. Similar trunk movements made in flight with the feet attached to the deck show the same compensatory movements of the hips and knees, with kinematics corresponding to those that occur under terrestrial conditions. These in-flight movements must reflect reorganized patterns of muscle activation, because the innervations necessary to achieve these axial synergies in microgravity are considerably different from those needed on Earth, given the absence of effective gravity torques in orbital flight.[30,31]

Initially, the basic stance patterns adopted in flight when the feet are attached to the deck differ little from those on Earth. However, later on in flight a more pitched forward posture appears that reflects decreased extensor tonus and increased flexor activity.[32] This change is consistent with the gradual shift mentioned above in the interpretation of proprioceptive signals associated with leg muscle vibration.[33] In microgravity, deep knee bends or raising and lowering the whole body with the feet anchored can evoke illusions of deck displacement.[34] The deck seems to move downward as the body moves toward it and upward as the body moves away from it. The reverse pattern occurs with exposure to high force levels. These illusions probably result from the altered relationship between motion of the body, the muscle forces necessary to produce that motion, and the associated muscle spindle feedback from the muscles.[35]   With repeated movements, adaptation takes place so that apparent stability of the "support surface" is regained during such deep knee bends. This adaptation may entail a remapping of muscle spindle and muscle innervation signals in the generation of limb position sense.

Two reflexes that test the status of otolith spinal modulation of the antigravity muscles have been evaluated in spaceflight: the H-reflex (Hoffman's reflex) and the otolith-spinal reflex. The H-reflex is an index of the excitability of alpha motoneurons and is elicited by electrical stimulation in the popliteal fossa of muscle spindle afferent fibers from the gastrocnemius muscle. Initially in flight, the H-reflex shows little change from 1-*g* baseline during bungee cord-induced "falls" in which the astronaut is briefly accelerated toward his or her feet, but later on the amplitude of the astronaut's reflex response diminishes.[36,37]   In addition, there are subjective changes that make later "falls" feel stronger and faster. The otolith-spinal response to bungee cord "falls" is lower than preflight and continues to diminish over subsequent flight days.[38,39]   The H-reflex has not been tested immediately after an astronaut lands back on Earth. However, the otolith-spinal reflex, measured by electromyography from the gastrocnemius-soleus muscles during the astronaut's sudden acceleration toward his or her feet, was inhibited as soon as weightlessness was achieved and declined further during the flight, but it was unchanged from preflight levels when measured shortly after the astronaut's return to Earth.[40]   The in-flight decreases in the otolith-spinal and H-reflexes are consistent with observations made in transient periods of weightlessness in parabolic flight.

Postflight measurements of posture and locomotion were first made during the Apollo and Skylab missions.[41]   In these early missions, the astronauts were parachuted to Earth in a reentry capsule and recovered at sea. The severity and operational seriousness of reentry disturbances of posture and locomotion have become apparent in the space shuttle missions. Some astronauts (especially those returning from longer missions) have been unable to walk unaided immediately after landing and could not make a rapid egress in case of emergency.

Postflight static posture exhibits a considerable increase in sway amplitude. On rail tests (i.e., standing heel-to-toe on rails of various widths to determine the least area necessary for balance), performance is decreased for more than 7 days postflight relative to preflight levels.[42-44]   Tests of postural responses to displacements of the support surface in translation and in rotation show considerable postflight decrements in terms of overshoots of the trunk and undershoots of the hips, and an increase in time needed to settle into a new stable posture.[45,46]   Tests in which platform motion and the visual surround can be manipulated independently or with sway referencing of vision (platform and visual motion spatially linked) show serious postflight decrements. This pattern of results is consistent with decreased vestibular and ankle proprioceptive contributions to balance and with increased reliance on vision.[47,48]   Recovery of balance control seems to follow a double exponential time course, with considerable improvement in balance within the first 12 hours or so after landing and a much more gradual return toward preflight baseline thereafter. Full recovery can take weeks.

Postflight studies of locomotion involving treadmill walking indicate that the timing of both toe-off and heel-strike components are much more variable immediately after landing.[49,50]   These are phases of the step cycle in which accurate neural control is important because of the great energy transfer involved. Astronauts returning from their first spaceflight also show greater eye and head instability during treadmill locomotion than more experienced astronauts. These gaze instabilities have important practical implications for performing a rapid and safe egress after reentry, as do the alterations in locomotory transport timing. Head movements, especially in pitch and roll, can evoke apparent visual motion and disorientation during and after reentry.[51,52]   Alterations in the motor control of the head and its relationship with oculomotor control occur in spaceflight because the head is weightless and adaptive changes in control are necessary in the absence of gravity torques. On reentry, the return of a *g*-load on the head is probably responsible for many aspects of postflight gaze instabilities.

Postflight locomotory disturbances are commonplace and include feelings of bobbing of the ground during stepping, an inability to walk in a straight line with eyes closed, loss of balance when turning corners, and the sense that movements require extra "effort."[53,54] The difficulty in making turns has a perceptual component known as the "giant hand" effect, which feels as if some external force is making the body deviate from its intended course. An analogous phenomenon can be experienced during disorientation in aircraft, when the controls do not feel as if they are responding appropriately because a "giant hand" is tilting the aircraft. Similar processes may be involved in the two types of illusions.[55]

Attempts have been made to relate postflight disturbances of head and locomotory control to central reinterpretations of otolith function.[56-59] The basic notion is that static head tilt in spaceflight is no longer associated with modulation of otolith output. Consequently, the central nervous system reinterprets all changes in otolith activity as indicating linear translation and oculomotor, postural, and subjective responses are remapped in accordance with the reinterpreted otolith output. After return from spaceflight, persistence of this remapping would mean that tilting movements of the head would initially be interpreted as linear translation—for example, a forward head tilt would be interpreted as backward translation of the head, because the same otolith displacement would result from an actual linear acceleration of the head backward. Such reinterpretation of otolith output, although conceptually appealing, is unlikely to be the sole factor responsible for the postflight illusions evoked by head movements, because the magnitude, character, and time course of the perceptual responses are not completely appropriate. Some of the confounding factors that need to be considered in interpreting reentry disturbances—especially for flights lasting more than several days—are the alterations in muscle fiber type and the decreases in muscle strength and muscle mass that occur during spaceflight. Such changes (along with alterations in sensorimotor control mechanisms, brain maps of motor control, and somatosensation) could all be contributing factors to postflight postural and locomotory deficits.

Summary and Recommendations

The severe reentry disturbances of posture and locomotion that astronauts and cosmonauts exhibit after even relatively brief spaceflights pose potentially dangerous operational problems. These disturbances would be especially critical in long-duration missions involving transitions between force levels on arrival, coupled with the need for accurate postural, locomotory, and prehension control.

• *The time course of postural and locomotory adaptation to variations in background force level and to variations in effective body weight should be determined. Attention should be paid to determining whether the rate of adaptation is affected by age.*

• *Techniques should be developed to provide ancillary sensory inputs or aids to heighten postural and locomotory control and to hasten adaptation during transitions between gravitational force environments.*

• *Animal models of reentry disturbances could be developed to elucidate the underlying physiological processes.*

VESTIBULO-OCULAR REFLEXES AND OCULOMOTOR CONTROL

The vestibulo-ocular reflexes (VORs) are important for controlling the direction of gaze (eye and head position relative to space) during active and passive movements of the head, as well as those of the head and body. They are mediated by the semicircular canals of the inner ear, which are activated by angular acceleration; by the otolith organs, which are sensitive to gravitoinertial acceleration; and (to a lesser extent) by neck proprioceptive inputs. Both semicircular canal and otolith-driven eye movements have been studied in spaceflight. Caloric irrigation of the external ear canal (with hot or cold water, or with air) can be used under terrestrial conditions to stimulate the horizontal semicircular canals and other sources, depending on the mode of body motion. Irrigation leads to a pattern of rhythmic eye movements known as nystagmus, consisting of slow and fast phases. Such nystagmus is thought to result from density differences between the endolymph and cupula of the canal created by the thermal stimulus, thus rendering the canal "sensitive" to gravity. Caloric irrigation was first used in the Space-lab-1 flight, and nystagmus was elicited.[60,61] This response was surprising, since the generally accepted thermoconvection theory of Barany predicts that density differences should be irrelevant in weightlessness. Subsequent observations, made with controls for noise levels and other possible artifacts, have confirmed the original finding and support the view that there may be a direct thermal effect on canal activity. [62,63]

Swivel or yaw head movements stimulate the semicircular canals and elicit compensatory eye movements in the direction opposite to the displacement of the head. The yaw VOR has been studied with voluntary head movements during attempted visual fixation at frequencies from 0.25 to 1.0 Hz.[64] Few changes are apparent in flight, other than a slight decrease in gain. The yaw VOR for passive rotation has been studied systematically in parabolic and orbital flight.[65,66] The basic finding is that gravitoinertial force level has no influence on the peak slow-phase velocity of the nystagmus elicited by sudden deceleration to rest. This means that the peripheral response of the semicircular canals is not influenced by alterations in linear background force—at least not by brief exposures. The time constant of nystagmus decay has been studied in weightlessness and at greater than 1-g force levels to evaluate "velocity storage."[67] Velocity storage refers to the observation that the nystagmus response to a velocity step outlasts the physical return of the cupula-endolymph system of the semicircular canal to resting levels. Velocity storage is thought to reflect the midbrain integration of a velocity signal originating from the semicircular canals.[68] The overall eye movement response is thought to reflect the contribution of a "direct pathway" from the canal as well as velocity storage activity from an "indirect pathway." When the otolith organs are ablated, velocity storage is abolished.[69] Interestingly, under microgravity conditions in spaceflight and the microgravity phases of parabolic flights, the time constant of nystagmus decay is significantly shorter than on Earth.[70,71] This suggests that the velocity storage mechanism is sensitive to linear acceleration. The time constant is also shorter in a 1.8-g force background, showing that departures in either direction from 1 g tend to suppress storage.[72] The three-dimensional organization of velocity storage is also influenced by gravity. The attenuation of postrotary nystagmus by tilting of the head (so-called "dumping" of velocity storage) appears to be absent during brief as well as prolonged exposure to microgravity.[73,74] Velocity storage can also be generated by visually driven eye movements. Motion of a large portion of the visual field elicits tracking movements of the eyes followed by rapid eye flick, known as optokinetic nystagmus. If the background illumination is eliminated, eye movements will persist in darkness for some seconds. This phenomenon of optokinetic after-nystagmus is another index of velocity storage.[75] Optokinetic after-nystagmus is also attenuated and spatially drifts in microgravity—further evidence for a gravity dependence of velocity storage.[76]

VOR responses to voluntary head oscillation in pitch vary and are hard to interpret because the data have so much intra- and interindividual variability.[77,78] Adequate on-axis studies to separate otolith and canal contributions have not yet been done in spaceflight.

On Earth, static tilts of the head evoke counterrolling of the eyes in the direction opposite the head tilt. This effect is mediated by the otolith organs. Dynamic tilts of the head stimulate the semicircular canals as well and elicit a torsional VOR.[79] Linear acceleration along the y-axis of the head (the orientation that runs through the ears) can also evoke torsion of the eyes. Astronauts and cosmonauts show diminished ocular counterrolling during static head tilts after return from spaceflight.[80,81] The reduction can persist for 10 days or more following flights lasting more than several weeks. Postflight assessments using a linear sled also show diminished torsional responses.[82] Rhesus monkeys show postflight reduction of ocular counterrolling as well.[83] Observations of the torsional VOR elicited by voluntary head movements during spaceflight suggest initial decreases in gain followed by increases to levels even greater than preflight values.[84]

In general, all of the postflight VOR assessments both for angular and linear stimulation are consistent with a decreased contribution of otolith function to the responses.

Optokinetic, voluntary pursuit, and saccadic eye movements have all been studied in spaceflight and parabolic flight.[85-90] Optokinetic responses tend to vary, with some individuals showing changes and others not. There are occasional reports of astronauts and cosmonauts exhibiting a spontaneous vertical or horizontal nystagmus during spaceflight.[91]

Studies of pursuit eye movements so far have found that horizontal pursuit is not affected by weightlessness either for head-fixed or head-free test conditions.[92] Striking changes can occur in vertical pursuit, however. Upward pursuit is accomplished in flight mainly by saccadic eye movements and downward pursuit by a combination of smooth pursuit and saccades. Saccadic eye movements tend to display increased latencies and decreased peak velocities relative to preflight control values.

Summary and Recommendations

Considerable progress has been made in understanding how microgravity affects vestibulo-ocular behavior and control of gaze in humans, and experiments using primates have also made gains. Future studies should be directed toward bringing experimental and theoretical closure to understanding vestibulo-ocular function under different background force levels.

- *Systematic parametric studies of pursuit, saccadic, and optokinetic eye movements should be carried out as a function of background force level in humans from microgravity to 2 g.*
- *Human and animal experiments should be directed toward developing an adequate three-dimensional model of the VOR and velocity storage for both angular and linear accelerations in both space and ground-based environments.*
- *The coordination of eye-head-torso synergies should be evaluated in humans under different force backgrounds, including microgravity.*

VESTIBULAR PROCESSING DURING MICROGRAVITY

Morphological data concerning potential changes in the vestibular end organs as a consequence of exposure to microgravity are scarce so far but point to significant changes that occur. In the rat, about a twofold increase in ribbon synapses of Type II hair cells and a 50 percent increase in Type I cell synapses of the otoliths have been found after 2 weeks of exposure to microgravity, compared with ground-based controls. The increased synaptic levels were still apparent in animals assayed 14 days postflight.[93] By contrast, animals exposed to increased g levels through centrifugation exhibit decreased synaptic density levels.[94] The functional significance of these synaptic changes has not been determined. Data on whether there are alterations in otoconia are less clear, with some studies suggesting that there are changes while others do not.

Few data are available concerning the physiological activity of vestibular afferents in microgravity. Early observations did not have adequate controls that would allow conclusions to be drawn. Recent work is preliminary, with few animals tested so far.[95] Two studies performed on Rhesus monkeys have reported a postflight increase in gain of the horizontal semicircular canal afferents after a 14-day mission, but no change in the gain of the horizontal VOR postflight or in velocity storage relative to preflight values. In contrast, experiments performed on two other monkeys from a different mission found a decrease in afferent gain postflight (M.J. Correia, personal communication, August 1996). Physiological data concerning activity at other levels of the vestibular pathways are lacking.

Summary and Recommendations

Very little is known about how exposure to weightlessness affects peripheral and central vestibular function and how development of the vestibular system (especially the otolith organs) would be affected. These are areas that warrant extensive study using animal models and state-of-the-art electrophysiological, morphological, and molecular biological approaches.

- *The effect of altered calcium regulation in microgravity on otoconial development and regeneration should be determined using animal models.*
- *In-flight electrophysiological recordings of otolith afferent and efferent activity and signal processing within the brain should be made in test animals. This is best accomplished by a trained physiologist serving as a payload specialist.*

SPACE MOTION SICKNESS

Space motion sickness (SMS) affects at least 70 percent of astronauts on their first flights.[96-98] An individual's susceptibility to space motion sickness seems to change little with widely separated repeat exposures. Most of the symptoms, with the exception of those characteristic of the Sopite syndrome (including extreme drowsiness, fatigue, lack of initiative, and apathy), abate within about 3 days.[99]

It is still not known whether SMS is a form of sickness fully analogous to terrestrial motion sickness.[100] Some symptoms seem different from those expressed under provocative conditions on Earth. For example, sweating and pallor are rarely reported with SMS, but episodes of sudden vomiting without clear prodromal signs can occur. However, under terrestrial conditions, the pattern and expression of signs and symptoms of motion sickness are highly dependent on the provocativeness of the test situation, and "head symptoms" tend to predominate under mildly provocative conditions.[101] Environmental factors also play a role. For example, in cool test conditions, sweating does not usually accompany motion sickness on Earth. Also, sudden vomiting can occur with terrestrial motion sickness, indicating that the lack of obvious symptoms is not equivalent to lack of sensitization.[102] In fact, common features of SMS, including lack of appetite, apathy, fatigue, difficulty in sleeping, and irritability, are characteristic features of the Sopite syndrome, which occurs with prolonged exposure to relatively low-grade stimulation on Earth. It is not yet known how long the Sopite component of SMS persists.

Reports of *mal de barquement*, the renewed symptoms of motion sickness after returning from spaceflight, tend to strengthen the analogy between SMS and terrestrial motion sickness. The incidence of symptoms is about 90 percent in cosmonauts returning from missions lasting several months.[103]

Movements of the head or of the head and body under weightless conditions are generally considered causal factors in evoking and exacerbating SMS.[104] In fact, head movements made during exposure to altered *g* levels, both for decreases and increases relative to 1 *g*, are provocative.[105] Their influence is compounded at 2 *g*, compared with 0-*g* force backgrounds. Off-axis head movements made during rotation generate unusual stimulation of the semicircular canals, known as Coriolis cross-coupling stimulation. Such stimulation occurs because as the head moves out of the plane of rotation, one set of canals also moves out of the plane of rotation and loses angular momentum and another set is brought toward the plane of rotation and gains angular momentum. This causes both sets of canals to signal rotation simultaneously but about different head axes, whereas the remaining pair of canals appropriately signals the tilting of the head. Such cross-coupling stimulation can cause profound nausea and is disorienting when head movements are made during high velocities of rotation on Earth. By contrast, head movements made at high velocities of rotation (20 to 25 rpm) after 6 days in spaceflight are not at all nausea-inducing or disorienting.[106] Unfortunately, tests have not been carried out on earlier mission days. However, in parabolic flight, susceptibility to cross-coupling stimulation changes virtually immediately as a function of *g* level, with head movements during rotation being much less provocative in 0 *g* and much more so in 1.8 *g* than in straight-and-level flight.[107] These force-dependent variations likely reflect factors related both to the altered sensorimotor control of the head and to the changes in otolith activity as a function of force level, given that input to the semicircular canals is kept constant in these studies.[108]

Subjective reports suggest that visual factors can also influence the development of SMS. For example, when astronauts are upside down in relation to the architectural ceiling of the spacecraft, or see other astronauts upside down in relation to themselves, they may develop symptoms.[109,110] This is especially true for astronauts who are visually dependent, i.e., down is where the visual architectural floor is, as opposed to astronauts who tend to perceive the "down" direction as corresponding to where their feet are located.[111]

Biofeedback control of SMS has been suggested as a means of decreasing or preventing SMS as well as terrestrial motion sickness.[112] Unfortunately, this approach has not proven successful in spaceflight. Terrestrial experiments reported to be successful in using biofeedback to control motion sickness have lacked appropriate controls for physiological changes associated with movement activity levels, initial susceptibility, and adaptation effects associated with repeated exposure conditions. The electrogastrogram (EGG) has been suggested as a potential physiological index of the presence and severity of motion sickness, because tachygastria has been reported to be related to the onset of nausea.[113] If this were consistently the case, monitoring the EGG could serve as an objective predictor of impending SMS. However, studies using controls for anxiety level that have attempted to relate symptom development and severity with changes in the EGG have failed to find a consistent correlation between EGG changes and the development of motion sickness.[114]

Progress in understanding SMS is hampered by the lack of adequate understanding of the physiological bases of motion sickness[115-118] and by the lack of adequate theories.[119-121] Current sensory-conflict theories provide useful characterizations of the condition without offering new directions to pursue for its control. One factor that correlates with severity of motion sickness in parabolic flight is the magnitude of velocity storage, as assessed by the extent of suppression of postrotary nystagmus by head tilt.[122] This is a potential link with physiological mechanisms known to be involved in eliciting motion sickness and that are altered in spaceflight, but this finding has not been pursued.

SMS can be controlled operationally by intramuscular injections of promethazine at dose levels established from parabolic flight studies.[123-125] The extreme drowsiness generally associated with promethazine during ground-testing is generally not reported in either spaceflight or parabolic flight. This may be partially attributed to the excitement inherent in being in such situations.

Summary and Recommendations

Space motion sickness is usually an operational problem for the first 72 hours of flight and can be controlled with intramuscular drug injections. Nevertheless, it is hazardous for the initial transition between different gravitational force environments. An understanding of its etiology is also important for determining human adaptability to novel environments. Moreover, the Sopite syndrome component of SMS may be a factor in generating interpersonal friction in the restrictive and stressful space environment. The use of virtual environments in space to augment training in long-duration missions and for extravehicular activities (EVAs) or for experimental purposes will likely exacerbate motion sickness. *Mal de barquement* may also occur in transitions between background force levels and is a potential operational problem for entry or reentry after long-duration missions.

- *The relationship of motion sickness to altered sensorimotor control of the head and body as a function of altered force background and effective body weight should be assessed.*
- *The possibility of maintaining dual adaptations to more than one force background simultaneously—allowing transitions between them without performance decrements—should be explored.*
- *Combined physiological and morphological studies using animal models should be performed to investigate the interaction between the vestibular system and autonomic function, including cardiovascular regulation.*
- *The time course of the Sopite syndrome component of space motion sickness should be definitively established.*
- *Testable models should be developed that integrate current knowledge of terrestrial and space motion sickness.*

CENTRAL NERVOUS SYSTEM REORGANIZATION

There is compelling evidence that cortical maps of both sensory and motor functions are highly plastic and subject to rapid reorganization.[126] Such plasticity occurs at many relay stations in the central nervous system, not just the cortex. The functional implications and consequences of such reorganizations are not understood fully under terrestrial conditions, let alone in spaceflight. Extended spaceflight must produce sensorimotor remappings. Preliminary studies should be undertaken to evaluate their implications and significance for both in-flight and postflight performance.

Recommendations

- *Pre- and postflight fMRI studies should be conducted with astronauts to determine the effects of microgravity on cortical maps. Bed-rest studies may also serve as useful models for inducing sensory-motor reorganizations.*
- *Test strategies should be developed to determine the sensorimotor and cognitive consequences of central nervous system reorganizations resulting from exposure to microgravity and their implications for reentry disturbances.*

TELEOPERATION AND TELEPRESENCE

Virtual environment technology has made it possible to create compelling artificial situations for training and operational purposes.[127] Many relevant aspects of real environments can be reproduced, creating important possibilities for teleoperation, teleexploration, and learning the layout of real environments from virtual ones. Telepresence is the sense of being physically present in a remote environment removed from an actual location. It can be induced using virtual environment technology. Virtual environments will probably be used increasingly often in preflight training and in-flight rehearsal and training for mission tasks, including familiarization with equipment and environment.

Recommendation

The extent to which telepresence is helpful in controlling equipment and robots in a remote environment should be determined.

GENERAL STRATEGIC ISSUES

NASA Life Science has been following a strategy of focusing on key problem areas using coordinated series of studies, such as in the Spacelab and Neurolab missions. This is an important approach and should be continued. The following recommendations will enhance the progress already made.

Recommendations

• *Systematic baseline data on key aspects of sensorimotor function need to be collected, preflight as well as in flight and postflight as a function of time.*

• *Data related to other changes that would affect the interpretation of sensorimotor results need to be interrelated with them (e.g., sleep cycles, hormonal and immune changes, alterations in muscle physiology, use of drugs to combat motion sickness, and so on).*

• *Sample sizes are needed that are large enough so that the range and characterization of individual differences in performance can be identified and potential age relatedness determined.*

• *The efficacy of different countermeasures against space motion sickness and postflight reentry disturbances should be validated. Those not validated should be discontinued.*

• *Identical equipment and procedures should be used as much as possible for ground-based and flight data collection in a given experiment.*

• *Neurophysiological and related experiments conducted on the International Space Station are best carried out by physiologists serving as payload specialists.*

• *Relevant animal models should be developed for exploring the physiological and morphological bases for postflight reentry disturbances.*

REFERENCES

1. Space Science Board, National Research Council. 1987. A Strategy for Space Biology and Medical Sciences for the 1980s and 1990s. National Academy Press, Washington, D.C.
2. Lackner, J.R., and DiZio, P. 1993. Multisensory, cognitive, and motor influences on human spatial orientation in weightlessness. J. Vest. Res. 3(3): 361-372.
3. Lackner, J.R., and DiZio, P. 1997. The role of reafference in recalibration of limb movement control and locomotion. J. Vest. Res. 7(2/3): 1-8.
4. Young, L.R., and Shelhamer, M. 1990. Microgravity enhances the relative contribution of visually-induced motion sensation. Aviat. Space Environ. Med. 61: 525-530.
5. Young, L.R., Oman, C.M., Merfled, D., Watt, D.G.D., Roy, S., Deluca, C., Balkwill, D., Christie, J., Groleau, N., Jackson, D.K., Law, G., Modestino, S., and Mayer, W. 1993. Spatial orientation and posture during and following weightlessness: Human experiments on Spacelab-Life Sciences-1. J. Vestib. Res. 3: 231-240.

6. Young, L.R., and Shelhamer, M. 1990. Microgravity enhances the relative contribution of visually-induced motion sensation. Aviat. Space Environ. Med. 61: 525-530.

7. Graybiel, A., and Kellogg, R.S. 1967. The inversion illusion and its probable dependence on otolith function. Aerospace Med. 38: 1099-1103.

8. Lackner, J.R., and DiZio, P. 1993. Multisensory, cognitive and motor influences on human spatial orientation in weightlessness. J. Vestib. Res. 3: 361-372.

9. Oman, C.M. 1988. The role of static visual orientation cues in the etiology of space motion sickness. Pp. 25-38 in Proceedings of the Symposium on Vestibular Organs and Altered Force Environment (M. Igarashi and K.G. Nute, eds). National Aeronautics and Space Administration, Space Biomedical Institute and Universities Space Research Association, Division of Space Biomedicine, Houston, Tex.

10. Mittelstaedt, H. 1989. The role of the pitched-up orientation of the otoliths in two recent models of the subjective vertical. Biol. Cybern. 61: 405-416.

11. Mittlestaedt, H., and Glasauer, S. 1993a. Illusions of verticality in weightlessness. Clin. Invest. 71: 732-739.

12. Mittlestaedt, H., and Glasauer, S. 1993b. Crucial effects of weightlessness on human orientation. J. Vestib. Res. 3: 307-314.

13. Lackner, J.R. 1992. Spatial orientation in weightless environments. Perception 21: 803-812.

14. Young, L.R., Oman, C.M., Watt, D.G.D., Money, K.E., Lichtenberg, B.K., Kenyon, R.V., and Arrott, A.P. 1986. M.I.T./Canadian vestibular experiments on the Spacelab-1 mission: 1. Sensory adaptation to weightlessness and readaptation to one-g: An overview. Exp. Brain Res. 64: 291-298.

15. Benson, A.J., Von Baumgarten, R., Berthoz, A., Brandt, T., Brand, U., Bruzek, W., Dichgans, J., Kass, J., Probst, T., Scherer, H., Viéville, T., Vogel, H., and Wetzig, J. 1986. Some results of the European vestibular experiments in the Spacelab-1 mission. Advisory Group for Aerosp. Res. Dev. (AGARD) Proc. 377: 1B3-1B14.

16. Arrott, A.P., and Young, L.R. 1986. M.I.T./Canadian vestibular experiments on the Spacelab-1 mission: 6. Vestibular reactions to lateral acceleration following ten days of weightlessness. Exp. Brain Res. 64: 347-357.

17. Arrott, A.P., Young, L.R., and Merfeld, D.P. 1990. Perception of linear acceleration in weightlessness. Aviat. Space Environ. Med. 61: 319-326.

18. Fisk, J., Lackner, J.R., and DiZio. P. 1993. Gravitoinertial force level influences arm movement control. J. Neurophysiol. 69(2): 504-511.

19. Young, L.R., Oman, C.M., Watt, D.G.D., Money, K.E., and Lichtenberg, B.K. 1984. Spatial orientation in weightlessness and readaptation to Earth's gravity. Science 225: 205-208.

20. Lackner, J.R., and DiZio, P. 1992. Gravitoinertial force level affects the appreciation of limb position during muscle vibration. Brain Res. 592: 175-180.

21. Lackner, J.R., DiZio, P., and Fisk, J.D. 1992. Tonic vibration reflexes and background force level. Acta Astronautica 26(2): 133-136.

22. Fisk, J., Lackner, J.R., and DiZio. P. 1993. Gravitoinertial force level influences arm movement control. J. Neurophysiol. 69(2): 504-511.

23. Schmitt, H.H., and Reid, D.J. 1985. Anecdotal information on space adaptation syndrome. National Aeronautics and Space Administration, Johnson Space Center, Houston, Tex.

24. Roll, J.P., Popov, K., Gurfinkel, V., Lipshits, M., André-Deshays, C., Gilhodes, J.C., and Quoniam, C. 1993. Sensorimotor and perceptual function of muscle proprioception in microgravity. J. Vestib. Res. 3: 259-274.

25. Ross, H.E., Brodie, E.E., and Benson, A.J. 1984. Mass discrimination during prolonged weightlessness. Science 225: 219-221.

26. Ross, H.E., Brodie, E.E., and Benson, A.J. 1986. Mass discrimination in weightlessness and readaptation to Earth's gravity. Exp. Brain Res. 64: 358-366.

27. Ross, H.E., Schwartz, E., and Emmerson, P. 1986. Mass discrimination in weightlessness improves with arm movements of higher acceleration. Naturwissenschaften 73: 453-454.

28. Clément, G., Gurfinkel, V.S., Lestienne, F., Lipshits, M.I., and Popov, K.E. 1984. Adaptation of posture control to weightlessness. Exp. Brain Res. 57: 61-72.

29. Clément, G., Gurfinkel, V.S., Lestienne, F., Lipshits, M.I., and Popov, K.E. 1985. Changes of posture during transient perturbations in microgravity. Aviat. Space Environ. Med. 56: 666-671.

30. Massion, J. 1992. Movement, posture and equilibrium: Interaction and coordination. Prog. Neurobiol. 38: 35-56.

31. Massion, J., Gurfinkel, V., Lipshits, M., Obadia, A., and Popov, K. 1993. Axial synergies under microgravity conditions. J. Vestib. Res. 3: 275-288.

32. Clément, G., Gurfinkel, V.S., Lestienne, F., Lipshits, M.I., and Popov, K.E. 1984. Adaptation of posture control to weightlessness. Exp. Brain Res. 57: 61-72.

33. Roll, J.P., Popov, K., Gurfinkel, V., Lipshits, M., André-Deshays, C., Gilhodes, J.C., and Quoniam, C. 1993. Sensorimotor and perceptual function of muscle proprioception in microgravity. J. Vestib. Res. 3: 259-274.

34. Lackner, J.R., and Graybiel, A. 1981. Illusions of postural, visual, and substrate motion elicited by deep knee bends in the increased gravitoinertial force phase of parabolic flight. Exp. Brain Res. 44: 312-316.

35. Lackner, J.R. 1992. Multimodal and motor influences on orientation: Implications for adapting to weightless and virtual environments. J. Vestib. Res. 2: 307-322.

36. Reschke, M.F., Anderson, D.J., and Homick, J.L. 1984. Vestibulospinal reflexes as a function of microgravity. Science 225: 212-214.

37. Reschke, M.F., Anderson, D.J., and Homick, J.L. 1986. Vestibulo-spinal response modification as determined with the H reflex during the Spacelab-1 flight. Exp. Brain Res. 64: 367-379.

38. Watt, D.G.D., Money, K.E., and Tomi, L.M. 1986. M.I.T./Canadian vestibular experiments on the Spacelab-1 mission: 3. Effects of prolonged weightlessness on a human otolith-spinal reflex. Exp. Brain Res. 64: 308-315.

39. Watt, D.G.D., Money, K.E., Tomi, L.M., and Better, H. 1989. Otolith-spinal reflex testing on Spacelab-1 and D-1. Physiologist 32: S45-S52.

40. Young, L.R., Oman, C.M., Watt, D.G., Money, K.E., Lichtenberg, B.K., Kenyon, R.V., and Arrott, A.P. 1986. M.I.T./Canadian vestibular experiments on the Spacelab-1 mission: 1. Sensory adaptation to weightlessness and readaptation to one-g: An overview. Exp. Brain Res. 64: 291-298.

41. Homick, J.L., and Miller, E.F. 1975. Apollo flight crew vestibular assessment. Pp. 323-340 in Biomedical Results of Apollo (R.S. Johnston, L. F. Dietlein, and C.A. Berry, eds.). NASA-SP-368. U.S. Government Printing Office, Washington, D.C.

42. Homick, J.L., and Reschke, M.F. 1977. Postural equilibrium following exposure to weightless space flight. Acta Otolaryngol. 83: 455-464.

43. Homick, J.L., Reschke, M.F., and Miller, E.F. 1977. Effects of prolonged exposure to weightlessness on postural equilibrium. Pp. 104-112 in Biomedical Results from Skylab (R.S. Johnston and L.F. Dietlein, eds.). NASA SP-377. U.S. Government Printing Office, Washington, D.C.

44. Kenyon, R.V., and Young, L.R. 1986. M.I.T./Canadian vestibular experiments on Spacelab-1 mission: 5. Postural responses following exposure to weightlessness. Exp. Brain Res. 64: 335-346.

45. Paloski, W.H., Black, F.O., Reschke, M.F., Calkins, D.S., and Shupert, C. 1993. Vestibular ataxia following shuttle flights: Effects of microgravity on otolith-mediated sensorimotor control of posture. Am. J. Otol. 14: 9-17.

46. Anderson, D.J., Reschke, M.F., Homick, J.L., and Werness, S.A.S. 1986. Dynamic posture analysis of Spacelab crew members. Exp. Brain Res. 64: 380-391.

47. Paloski, W.H., Reschke, M.F., Black, F.O., Doxey, D.D., and Harm, D.L. 1992. Recovery of postural equilibrium control following space flight. Pp. 747-754 in Sensing and Controlling Motion: Vestibular and Sensorimotor Function (B. Cohen, D.L. Tomko, and F. Guedry, eds.). New York Academy of Sciences, New York.

48. Paloski, W.H., Reschke, M.F., Doxey, D.D., and Black, F.O. 1992. Neurosensory adaptation associated with postural ataxia following space flight. Pp. 311-315 in Posture and Gait: Control Mechanisms (M. Woolacott and F. Horak, eds.). University of Oregon Press, Eugene, Ore.

49. Layne, C.S., McDonald, V.P., and Bloomberg, J.J. 1997. Neuromuscular activation patterns during treadmill walking after space flight. Exp. Brain Res. 113: 104-116.

50. Bloomberg, J.J., Peters, B.T., Smith, S.L., and Reschke, M.F. 1997. Locomotor head-trunk coordination strategies following space flight. J. Vestib. Res. 7: 161-177.

51. Reschke, M.F., and Parker, D.E. 1987. Effects of prolonged weightlessness on self-motion perception and eye movements evoked by roll and pitch. Aviat. Space Environ. Med. 58: A153-A158.

52. Reschke, M.F., Bloomberg, J.J., Harm, D.L., and Paloski, W.H. 1994. Space flight and neurovestibular adaptation. J. Clin. Pharmacol. 34: 609-617.

53. Reschke, M.F., Harm, D.L., Parker, D.E., Sandoz, G.R., Homick, J.L., and Vanderpoeg, J.M. 1994. Neurophysiologic aspects: Space motion sickness. Pp. 228-260 in Space Physiology and Medicine, 3rd ed. (A.E. Nigossian, C.L. Huntoon, and S.L. Pool, eds.). Lea and Febiger, Philadelphia.

54. Reschke, M.F., Bloomberg, J.J., Paloski, W.H., Harm, D.L., and Parker, D.E. 1994. Neurophysiologic aspects: Sensory and sensory-motor function. Pp. 261-285 in Space Physiology and Medicine, 3rd ed. (A.E. Nigossian, C.L. Huntoon, and S.L. Pool, eds.). Lea and Febiger, Philadelphia.

55. Gillingham, K.K., and Previc, F.H. 1993. Spatial orientation in flight. Technical report, AL-TR-1993-0022. Armstrong Laboratory, Brooks Air Force Base, San Antonio, Tex.

56. Parker, D.E., Arrott, A.P., Homick, J.L., and Lichtenberg, B.K. 1985. Otolith tilt-translation reinterpretation following prolonged weightlessness: Implications for pre-flight training. Aviat. Space Environ. Med. 56: 601-606.

57. Reschke, M.F., Parker, D.E., Harm, D.L., and Michaud, L. 1988. Ground-based training for the stimulus rearrangment encountered during space flight. Acta Otolaryngol. (Stockholm) 460: 87-93.

58. Young, L.R., Oman, C.M., Watt, D.G.D., Money, K.E., and Lichtenberg, B.K. 1984. Spatial orientation in weightlessness and readaptation to Earth's gravity. Science 225: 205-208.

59. Young, L.R. 1993. Space and the vestibular system: What has been learned? Guest editorial. J. Vestib. Res. 3: 203-206.

60. Scherer, H., Brandt, U., Clarke, A.H., Merbold, U., and Parker R. 1986. European vestibular experiments on the Spacelab-1 mission: 3. Caloric nystagmus in microgravity. Exp. Brain Res. 64: 255-263.

61. Von Baumgarten, R.J. 1986. European vestibular experiments on the Spacelab-1 mission: 1. Overview. Exp. Brain Res. 64: 239-246.

62. Clarke, A.H., Scherer, H., and Gundlach, P. 1988a. Caloric stimulation during short episodes of microgravity. Arch. Otorhinolaryngol. 245(3): 175-179.

63. Clarke, A.H., Scherer, H., and Schleibinger, J. 1988b. Body position and caloric nystagmus response. Acta Otolaryngol. (Stockholm) 106: 339-347.

64. Benson, A.J., and Viéville, T. 1986. European vestibular experiments on the Spacelab-1 mission: 6. Yaw axis vestibulo-ocular reflex. Exp. Brain Res. 64: 279-283.

65. DiZio, P., Lackner, J.R., and Evanoff, J.N. 1987. The influence of gravitoinertial force level on oculomotor and perceptual responses to sudden stop stimulation. Aviat. Space Environ. Med. 58: A224-A230.

66. Oman, C.M., and Kulbaski, M. 1988. Spaceflight affects the 1-g postrotatory vestibulo-ocular reflex. Adv. Otorhinolaryngol. 42: 5-8.

67. DiZio, P., and Lackner, J.R. 1988. The effects of gravitoinertial force level and head movements on post-rotational nystagmus and illusory after-rotation. Exp. Brain Res. 70: 485-495.

68. Raphan, T., Cohen, B., and Matsuo, V. 1977. A velocity storage mechanism responsible for optokinetic nystagmus (OKN), optokinetic afternystagmus (OKAN) and vestibular nystagmus. Pp. 34-37 in Control of Gaze by Brainstem Neurons (R. Baker and A. Berthoz, eds.). Elsevier, Amsterdam.

69. Cohen, B., Suzuki, J.-I., and Raphan, T. 1983. Role of the otolith organs in generation of horizontal nystagmus: Effects of selective labyrinthine lesions. Brain Res. 276: 159-164.

70. DiZio, P., and Lackner, J.R. 1992. Influence of gravito-inertial force level on vestibular and visual velocity storage in yaw and pitch. Vision Res. 32: 123-145.

71. Oman, C.M., and Balkwill, M.D. 1993. Horizontal angular VOR, nystagmus dumping, and sensation duration in Spacelab SLS-1 crewmembers. J. Vestib. Res. 3: 315-330.

72. DiZio, P., Lackner, J.R., and Evanoff, J.N. 1987. The influence of gravitoinertial force level on oculomotor and perceptual responses to sudden stop stimulation. Aviat. Space Environ. Med. 58: A224-A230.

73. DiZio, P., and Lackner, J.R. 1992. Influence of gravito-inertial force level on vestibular and visual velocity storage in yaw and pitch. Vision Res. 32: 123-145.

74. Oman, C.M., and Balkwill, M.D. 1993. Horizontal angular VOR, nystagmus dumping, and sensation duration in Spacelab SLS-1 crewmembers. J. Vestib. Res. 3: 315-330.

75. Cohen, B., Matsuo, V., and Raphan, T. 1977. Quantitative analysis of the velocity characteristics of optokinetic nystagmus and optokinetic afternystagmus. J. Physiol. (Lond.) 270: 321-344.

76. DiZio, P., and Lackner, J.R. 1992. Influence of gravito-inertial force level on vestibular and visual velocity storage in yaw and pitch. Vision Res. 32: 123-145.

77. Berthoz, A., Brandt, T., Dichgans, J., Probst, T., Bruzek, W., and Viéville, T. 1986. European vestibular experiments on the Spacelab-1 mission: 5. Contribution of the otoliths to the vertical vestibulo-ocular reflex. Exp. Brain Res. 64: 272-278.

78. Viéville, T., Clément, G., Lestienne, F., and Berthoz, A. 1986. Adaptive modifications of the optokinetic vestibulo-ocular reflexes in microgravity. Pp. 111-120 in Adaptive Processes in Visual and Oculomotor Systems (E.L. Keller and D.S. Zee, eds.). Pergamon Press, New York.

79. Young, L.R., Oman, C.M., Watt, D.G.D., Money, K.E., and Lichtenberg, B.K. 1984. Spatial orientation in weightlessness and readaptation to Earth's gravity. Science 225: 205-208.

80. Vogel, H., and Kass, J.R. 1986. European vestibular experiments on Spacelab-1 mission: 7. Ocular counterrolling measurements pre- and post-flight. Exp. Brain Res. 64: 284-290.

81. Hofstetter-Degen, K., Wetzig, J., and Von Baumgarten, R.J. 1993. Oculovestibular interactions under microgravity. Clin. Invest. 71: 749-756.

82. Arrott, A.P., and Young, L.R. 1986. M.I.T./Canadian vestibular experiments on the Spacelab-1 mission: 6. Vestibular reactions to lateral acceleration following ten days of weightlessness. Exp. Brain Res. 64: 347-357.

83. Dai, M., Cohen, B., Raphan, T., McGarvie, I., and Kozlovskaya, I. 1993. Reduction of ocular counterrolling (OCR) after space flight. Soc. Neurosci. Abst. 19: 343.

84. Clarke, A.H., Scherer, H., and Schleibinger, J. 1993. Evaluation of the torsional VOR in weightlessness. J. Vestib. Res. 3: 207-218.

85. Clément, G., and Berthoz, A. 1988. Vestibulo-ocular reflex and optokinetic nystagmus in microgravity. Adv. Otorhinolaryngol. 42: 1-4.

86. Clément, G., and Berthoz, A. 1990. Cross-coupling between horizontal and vertical eye movements during optokinetic nystagmus and optokinetic after-nystagmus elicited in microgravity. Acta Otolaryngol. (Stockholm) 109: 179-187.

87. Clément, G., Popov, K.E., and Berthoz, A. 1993. Effects of prolonged weightlessness on human horizontal and vertical optokinetic nystagmus and optokinetic after-nystagmus. Exp. Brain Res. 94: 456-462.

88. Clément, G., Reschke, M.F., Verrett, C.M., and Wood, S.J. 1992. Effects of gravitoinertial force variations on optokinetic nystagmus and on perception of visual stimulus orientation. Aviat. Space Environ. Med. 63: 771-777.

89. Clément, G., Viéville, T., Lestienne, F., and Berthoz, A. 1986. Modifications of gain asymmetry and beating field of vertical optokinetic nystagmus in microgravity. Neurosci. Lett. 63: 271-274.

90. Clément, G., Wood, S.J., and Reschke, M.F. 1992. Effects of microgravity on the interaction of vestibular and optokinetic nystagmus in the vertical plane. Aviat. Space Environ. Med. 63: 778-784.

91. Kornilova, L.N., Grigorova, V., and Bod, G. 1993. Vestibular function and sensory interaction in space flight. J. Vestib. Res. 3: 219-230.

92. André-Deshays, C., Israël, I., Charade, O., Berthoz, A., Popov, K., and Lipshits, M. 1993. Gaze control in microgravity: 1. Saccades, pursuit, eye-head coordination. J. Vestib. Res. 3: 331-344.

93. Ross, M.D. 1992. A study of the effects of space travel on mammalian gravity receptors. Space Life Sciences-1 180 Day Experimental Reports. National Aeronautics and Space Administration, Washington, D.C.

94. Ross, M.D. 1993. Morphological changes in rats' vestibular system following weightlessness. J. Vestib. Res. 3: 241-251.

95. Correia, M.J., Perachio, A.A., and Dickman, J.D. 1992. Changes in monkey horizontal semicircular afferent responses following space flight. J. Appl. Physiol. 73 (Suppl. 2): 112S-120S.

96. Davis, J.R., Vanderpoeg, J.M., Santy, P.A., Jennings, R.T., and Stewart, D.F. 1988. Space motion sickness during 24 flights of the Space Shuttle. Aviat. Space Environ. Med. 59: 1185-1189.

97. Reschke, M.F., Harm, D.L., Parker, D.E., Sandoz, G.R., Homick, J.L., and Vanderpoeg, J.M. 1994. Neurophysiologic aspects: Space motion sickness. Pp. 228-260 in Space Physiology and Medicine, 3rd ed. (A.E. Nicogossian, C.L. Huntoon, and S.L. Pool, eds.). Lea and Febiger, Philadelphia.

98. Jennings, R.T. 1997. Managing space motion sickness. J. Vestib. Res. 8: 67-70.

99. Graybiel, A., and Knepton, J. 1976. Sopite syndrome: A sometimes sole manifestation of motion sickness. Aviat. Space Environ. Med. 47: 1096-1100.

100. Graybiel, A. 1980. Space motion sickness: Skylab revisited. Aviat. Space Environ. Med. 51: 814-822.

101. Reason, J.T., and Brand, J.J. 1975. Motion Sickness. Academic Press, London.

102. Lackner, J.R., and Graybiel, A. 1986. Sudden emesis following parabolic flight maneuvers: Implications for space motion sickness. Aviat. Space Environ. Med. 57: 343-347.

103. Jennings, R.T. 1997. Managing space motion sickness. J. Vestib. Res. 8: 67-70

104. Lackner, J.R., and Graybiel, A. 1984. Elicitation of motion sickness by head movements in the microgravity phase of parabolic flight maneuvers. Aviat. Space Environ. Med. 55: 513-520.

105. Lackner, J.R., and Graybiel, A. 1986. Head movements made in non-terrestrial force environments elicit symptoms of motion sickness: Implications for the etiology of space motion sickness. Aviat. Space Environ. Med. 57: 443-448.

106. Graybiel, A., Miller, E.F., and Homick, J.L. 1974. Experiment M-131: Human vestibular function. 1. Susceptibility to motion sickness. Pp. 169-198 in Proceedings of the Skylab Life Sciences Symposium, Vol. 1. TMX-58154. NASA Johnson Space Center, Houston, Tex.

107. DiZio, P., Lackner, J.R., and Evanoff, J.N. 1987. The influence of gravitointertial force level on oculomotor and perceptual responses to coriolis, cross-coupling stimulation. Aviat. Space Environ. Med. 58: A218-A223.

108. Lackner, J.R., and DiZio, P. 1992. Gravitational, inertial, and coriolis force influences on nystagmus, motion sickness, and perceived head trajectory. Pp. 216-222 in The Head-Neck Sensory-Motor Symposium (A. Berthoz, W. Graf, and P.P. Vidal, eds.). Oxford University Press, New York.

109. Oman, C.M., Lichtenberg, B.K., and Money, K.E. 1990. Space motion sickness monitoring experiment: Spacelab 1. Pp. 217-246 in Motion and Space Sickness (G.H. Crampton, ed.). CRC Press, Boca Raton, Fla.

110. Harm, D.L., and Parker, D.E. 1993. Perceived self-orientation and self-motion in microgravity after landing and during preflight adaptation training. J. Vestib. Res. 3: 297-305.

111. Oman, C.M. 1997. Sensory conflict theory and space sickness: Our changing perspective. J. Vestib. Res. 8: 51-56.

112. Cowings, P.S., Suter, S., Toscano, W.B., Kamiya, J., and Naifeh, K. 1986. General autonomic components of motion sickness. Psychophysiology 23: 542-551.

113. Stern, R.M., Koch, K.L., Leibowitz, H.W., Inger, M., Lindblad, I.M., Schupert, C.L., and Stewart, M.S. 1985. Tachygastria and motion sickness. Aviat. Space Environ. Med. 56: 1074-1077.

114. Lawson, B.D. 1993. Human physiological and subjective responses during motion sickness induced by unusual visual and vestibular stimulation. Ph.D. dissertation, Brandeis University, Waltham, Mass.

115. Yates, B.J., Sklare, D.A., and Frey, M.A. 1997. Vestibular autonomic regulation: Overview and conclusions of a recent workshop at the University of Pittsburgh. J. Vestib. Res. 8: 1-5.

116. Balaban, C.D., and Porter, J.D. 1997. Neuroanatomical substrates for vestibulo-autonomic interactions. J. Vestib. Res. 8: 7-16.

117. Yates, B.J., and Miller, A.D. 1997. Physiological evidence that the vestibular system partcipates in autonomic and respiratory control. J. Vestib. Res. 8: 17-25.

118. Money, K.E., Lackner, J.R., and Cheung, R.S.K. 1996. The autonomic nervous system and motion sickness. Vestibular Autonomic Regulation (B.J. Yates and A.D. Millers, eds.). CRC Press, Boca Raton, Fla.

119. Reschke, M.F., Harm, D.L., Parker, D.E., Sandoz, G.R., Homick, J.L., and Vanderpoeg, J.M. 1994. Neurophysiologic aspects: Space motion sickness. Pp. 228-260 in Space Physiology and Medicine, 3rd ed. (A.E. Nicogossian, C.L. Huntoon, and S.L. Pool, eds.). Lea and Febiger, Philadelphia.

120. Oman, C.M. 1997. Sensory conflict theory and space sickness: Our changing perspective. J. Vestib. Res. 8: 51-56.

121. Oman, C.M. 1990. Motion sickness: A synthesis and evaluation of the sensory conflict theory. Can. J. Physiol. Pharmacol. 68: 294-303.

122. DiZio, P., and Lackner, J.R. 1991. Motion sickness susceptibility in parabolic flight and velocity storage activity. Aviat. Space Environ. Med. 62: 300-307.

123. Bagian, J.P. 1991. First intramuscular administration in the U.S. Space Program. J. Clin. Pharmacol. 31: 920.

124. Graybiel, A., and Lackner, J.R. 1987. Treatment of severe motion sickness with antimotion sickness drug injections. Aviat. Space Environ. Med. 58: 773-776.

125. Jennings, R.T. 1997. Managing space motion sickness. J. Vestib. Res. 8: 67-70.

126. Kass, J. 1995. The reorganization of sensory and motor maps in adult mammals. Pp. 51-72 in The Cognitive Neurosciences (M.S. Gazzaniga, ed.). MIT Press, Cambridge, Mass.

127. Durlach, N.I., and Mavor, A.S. 1995. Virtual Reality—Scientific and Technological Challenges. National Academy Press, Washington, D.C.

6

Bone Physiology

INTRODUCTION

One of the best-documented pathophysiological changes associated with microgravity and the spaceflight environment is bone loss. The reduction in bone mass and its effect following reentry could substantially limit long-term human exploration of space. The development of effective countermeasures through better scientific understanding of this phenomenon is therefore essential for future crewed flights.

This chapter very briefly reviews the functions of bone, bone growth and development, the process of bone remodeling (which underlies its physiological function), and the effects of hormones on this process. It summarizes current information on mechanical effects on bone, effects that may be the basis for the changes observed in microgravity or space environment conditions, briefly reviewing clinical observations; experiments on humans, animals, and cells; and putative mechanisms. It also summarizes spaceflight effects on the skeleton in humans and animals and gives caveats for these data. It then presents open questions and directions for future research aimed at continuing to characterize and understand, at a fundamental level, microgravity or space-environment-related bone loss and at developing effective countermeasures.

BONE FUNCTIONS, GROWTH AND DEVELOPMENT, AND REMODELING

Functions of Bone

Bone has four major functions: (1) mechanical, including support of soft tissues and locomotion; (2) storage of ions and ion homeostasis; (3) housing of the bone marrow and support of hemopoiesis; and (4) protection of the central nervous system. By fulfilling these functions, the mineralized skeleton played a central role in the evolution of terrestrial vertebrates.

1. Bones act as levers for muscles, and all aspects of locomotion (walking, running, climbing, flying) were tied to the evolution of the skeleton. Moreover, breathing (which requires expansion of the rib cage), brachiation, erect posture, and the use of tools are all skeletal functions. Maintenance of a healthy musculoskeletal system is essential for well-being, and its deterioration is often associated with aging, manifesting itself as a reduction in muscle mass and osteoporosis, probably due in part to reduced usage.

2. Calcium is a major ion recruited from bone for homeostatic needs. The marine environment provides a constant adequate supply of calcium; however, on land, organisms rely on dietary calcium supplied intermittently during feeding. Between meals, the organism maintains a stringently controlled steady-state calcium concentration in the extracellular fluid by withdrawing it from bone through bone resorption (degradation). Steady calcium levels are needed for normal neural, muscular, and endocrine functions, as well as for blood clotting and for cellular adhesion, migration, and proliferation. Other ions stored in the skeleton are phosphate and hydrogen, both of which may affect the rate of mineralization and demineralization. In addition, the skeleton stores potassium and magnesium, for which no homeostatic mechanisms have been documented. Important in the context of this discussion is the fact that calcium can only be withdrawn by destroying the skeletal structure that contains it, weakening the skeleton at that site. Furthermore, calcium homeostasis seems to have priority over other skeletal functions, so that calcium deprivation leads to thinning and weakening of the skeleton. However, the bones exposed to maximum mechanical loads are destroyed least and last.

3. Adult hemopoiesis in humans is totally confined to the bone marrow. Bone marrow cells with osteogenic potential may support hemopoiesis that, at least in tissue culture, requires a feeder layer of supportive cells. The bone-resorbing cells, the osteoclasts, are of hemopoietic origin. They are related to macrophages and originate from GM-CFUs (granulocyte macrophage colony forming units). In hypoxia, due to high altitude, for example, which causes active erythropoiesis, there is an expansion of the bone marrow cavity at the expense of bone. Whenever bone is formed, as at an ectopic site in muscle, it is always populated by bone marrow. With aging, active bone marrow is replaced by fatty marrow. It is presumed that stromal cells with dual capability (osteogenic and adipogenic) become adipocytes.[1] It has been suggested that there is a gradual exhaustion with age of osteogenic cells in the marrow; however, fracture repair occurs at all ages.

4. Housing of the brain and spinal column has survival advantages. Interestingly, unlike the rest of the skeleton, the skull does not seem to be subject to estrogen deficiency or to immobilization or microgravity-related bone loss.

It is important to recognize that to fulfill most of its functions, bone has to be destroyed and eventually rebuilt. For mechanical function, bone is continuously rebuilding itself to optimize its structure and architecture. For calcium mobilization, bone packets are destroyed, potentially to be rebuilt when calcium is again available. There are fewer fluctuations in bone mass related to the other two functions, except during very active physiological or pathological hemopoiesis, when the marrow cavity expands at the expense of bone. The process of bone destruction and rebuilding is called remodeling.

Bone Growth and Development

Embryologically there are two types of bone: the membranous bone of the face and skull and endochondral bone. For flat membranous bone, bone develops directly in connective tissue, formed by osteoblasts (bone-forming cells) that differentiate locally from mesenchymal cells. The rest of the skeleton appears first as a cartilagenous anlage. Starting at ossification centers, the cartilage is calcified and is invaded by blood vessels. Osteoprogenitor cells start depositing bone on the calcified cartilage that is subsequently removed by osteoclast-like cells. This mixture of calcified cartilage and early bone is subsequently replaced by lamellar bone, the dense type found in the mature organism. At the periphery of all bones, there is bone deposition in the periosteum, which resembles membranous bone formation. This process of bone growth according to the genetic template is called modeling, whereas the process of replacement of existing bone (for carrying out its functions) is remodeling. Bones, as well as other structures, are apparently unaffected when embryonic development of chicks occurs during microgravity in shuttle flights.[2]

Bones grow in length only at the cartilagenous epiphyses, where chondrocytes proliferate, differentiate, undergo hypertrophy, and mineralize the surrounding matrix on which the initial bone deposits, as described above for endochondral bone formation. To maintain its shape, the new bone is actively remodeled. Bones grow in diameter by deposition of external bone at the periosteum, a process that continues throughout life and is subject to mechanical regulation. Most of the load in long bones is sustained by the cylindrical bone of the shaft, which is formed in that manner by the periosteum.

It should be noted for experimental studies in rodents (especially mice) that the epiphyses never close and there is continuous longitudinal growth. In rats, the growth rate is significantly attenuated at about 12 to 14 months of age. In humans, epiphyses close around the age of 18 and longitudinal growth ceases. However, peak bone mass (the maximum amount of bone before age-related loss starts) is only reached toward the middle of the third decade.

Multiple humoral, local, and systemic factors affect skeletal growth and development and remodeling in the adult.

Bone Remodeling: Hormonal Effects

As mentioned above, bone remodeling is the process of replacement and rebuilding of packets of bone as part of bone's normal functions. Architecturally, bone structure is either cancellous (trabecular), shaped like honeycomb plates in the interior of each bone; or cortical (compact), denser bone at the boundary. The skeleton contains 80 percent cortical bone (by mass, not volume) and 20 percent cancellous bone. In humans, the turnover rate is 20 to 30 percent per year for cancellous bone, depending on the site, and about 3 percent per year for cortical bone. The cancellous bone is thus more readily affected by conditions that increase bone remodeling.

This section is a summary of the action of hormones that affect the skeleton by playing a role in calcium homeostasis or bone metabolism. Changes in the level of these hormones due to gravity and/or the space environment could play a role in the bone loss observed. Systemic hormonal changes that stimulate bone resorption (destruction) include a rise in parathyroid hormone (from calcium deficiency, for example), a rise in thyroid hormone, or a decrease in sex steroids (estrogens in women and androgens in men). Both sex steroids have profound effects on the skeleton in both genders. For instance, males usually have a larger skeleton than females. Congenital mutations in the estrogen or androgen receptor in humans suggest that estrogen is required for the closure of the epiphyses and both hormones are needed to produce a skeleton with normal density. In androgen-receptor-deficient individuals, who therefore have a female phenotype despite their XY chromosomes, there is reduced bone density that could not be corrected by estrogen administration. In a male patient who was estrogen-receptor deficient, the epiphyses were not closed at the age of 28 and the bone density was significantly decreased. In mature females, estrogen deficiency causes an increase in bone resorption leading to osteoporosis. However, androgens seem to be required for normal bone as well, since the addition of androgens to estrogen in hormone replacement therapy further increased bone mineral density.[3]   In males, testosterone deficiency has effects similar to those of estrogen deficiency in women (increasing bone resorption and causing bone loss), which can be reversed by testosterone administration.   These hormonal effects could be important during prolonged spaceflights, which could affect the level of sex steroids due to stress or other causes.[4]

A reduction in estrogen (along with amenorrhea) can be induced by sustained stress, vigorous exercise (in marathon runners and gymnasts), or dietary deprivations (in people with anorexia).   Exogenous administration of glucocorticoids ("steroids") also suppresses sex hormone production and has additional effects that reduce bone mass: inhibition of calcium absorption that causes (via feedback) elevations in parathyroid hormone (PTH), and direct suppression of bone formation.   However, these effects are seen at relatively high glucocorticoid levels (e.g., 7.5 mg prednisolone/day) that are not reached during physiological stress.

Bone resorption as well as defective mineralization is also caused by deficiency of $1,25(OH)_2$ vitamin D_3. The major action of $1,25(OH)_2D_3$ is to facilitate calcium absorption in the gut.   Its absence reduces circulating calcium levels and raises PTH.   High doses of vitamin D, reached by exogenous administration, stimulate bone resorption.   Endogenous $1,25(OH)_2D_3$ levels are controlled by PTH through its induction of the 1-hydroxylase in the kidney, part of the calcium homeostasis feedback loop.[5]

The only systemic hormone that inhibits bone resorption is calcitonin, a 32-amino-acid peptide produced by the thyroid clear cells.   Calcitonin is released in response to elevated calcium concentrations.[6]

A systemic hormone needed for skeletal development is growth hormone (GH).   It is released by the pituitary and increases liver production of insulin-like growth factor (IGF-1), which acts on the epiphyses to stimulate longitudinal growth in endochondral bone formation.   GH may also act directly on cartilage or bone, which has been shown to contain GH receptors.   In adults, GH or IGF-1 stimulate both bone formation and bone resorption without increasing bone mass.   In individuals with a GH deficiency, bone and muscle mass are reduced and have been reported to increase following GH administration.[7] IGF-1 and IGF-2 have been extracted from bone, are made by bone cells, and were reported in experimental studies to stimulate osteoblast proliferation and collagen synthesis in vitro.   In rat bone explants and cells, osteogenic agents, such as PTH and prostaglandin E, stimulate IGF-1 production.   Taken together, these observations suggest that IGF-1 could be important for maintaining normal bone.   The effects of IGF-1 on its target tissues depend on the level of the IGF binding proteins, both present in the circulation and locally produced.[8]   Changes in GH have been reported in rats exposed to the space environment (see Chapter 9, "Endocrinology").   Thyroid hormones also play a role in the development and maintenance of the skeleton, either indirectly, through effects on GH production, or through direct action on osteoblasts.   Thyroid hormone deficiency during development causes short stature and skeletal malformations, known as cretinism.   Thyroid hormone excess in adults, usually caused by hormone replacement therapy, increases bone turnover and causes bone loss.   Insulin also acts on the skeleton; anomalies in bone metabolism have been associated with diabetes, but the link is not clearly established.[9]

A large number of local factors increase bone resorption and/or formation, and some have been shown to participate in the mechanical effects on bone.   Skeletal remodeling resembles inflammation, and most inflammatory cytokines have been shown to affect osteoblasts and/or osteoclasts.

Prostaglandin E (PGE), like PTH, stimulates both bone resorption and bone formation.   Prostaglandins E and I2 have been implicated as mediators of mechanical effects on bone remodeling (see below), as well as in the bone loss associated with inflammation in periodontal disease and rheumatoid arthritis. Prostaglandins are produced by many cell types, including osteoblast lineage cells and macrophages.

Potent local stimulators of bone resorption include Interleukin-1 (IL-1) and tumor necrosis factor α (TNFα), which are produced by macrophages.[10] Other cytokines are IL-4, reported to suppress osteoclast formation in vitro, and IL-6, implicated in osteoclast formation and estrogen-deficiency bone loss, at least in mice. IL-11 is also a potent stimulator of osteoclast formation in vitro.[11]

Locally produced growth factors that may play a role in skeletal remodeling are the bone morphogenetic proteins (BMPs), a large family of growth and differentiation factors related to transforming growth factor β (TGFβ). Initially identified in bone extracts found to induce bone formation when injected into muscle or dermis, BMPs also play a role in limb development and stimulate differentiation of pluripotent mesenchymal cells into the osteogenic lineage. They have therefore been implicated in osteogenesis and possibly in bone remodeling. BMPs also play a role in the development of other organs. Evidence for the involvement of specific BMPs in physiological or pathological bone remodeling or bone formation in the adult or in response to mechanical or gravitation changes is not yet available.

TGFβ, like BMPs, stimulates bone formation when applied directly to bone. In vitro, it inhibits osteoclast formation. TGFβ is produced by platelets, as well as osteoblasts and other cells, and is abundant in bone in the precursor form that is activated by proteolytic cleavage.[12]

There are at least two growth factors that may participate in the link between blood flow or vascular changes and bone changes and thus play a role in the microgravity-induced pathophysiology. Fibroblast growth factor-2 (basic FGF) is produced by osteoblasts as well as endothelial cells and stimulates bone formation when administered to rats in vivo, either locally or systemically.[13] Vascular endothelial growth factor is produced in bone and by osteoblasts in vitro in response to PGE. It is a potent stimulator of angiogenesis, which is required for osteogenesis, and may participate in a positive feedback loop that connects the two processes.[14]

Bone cells have receptors for many other factors and hormones, such as endothelin, enkephalin, thrombin, epidermal growth factor, epinephrine, and norepinephrine, but their role in bone metabolism is uncertain.

Bone metabolism is thus subject to a large number of systemic and local factors, many of which are part of other physiological processes that could be affected by microgravity, stress, nutrition, fluid, and electrolyte balance.

MECHANICAL EFFECTS ON BONE REMODELING

Clinical Observations and Human Experimentation

Clinical observations and human experimentation show that human bone mass and structure are ideally suited to sustain the loads exerted upon them. There is well-documented bone loss after the removal of mechanical loading and somewhat more limited documentation of increased bone mass in response to mechanical stimulation. This information is highly relevant to gravitational effects on the skeleton.

The effect of mechanical loads on the architecture of the human skeleton has been recognized for some time and was scientifically described at the end of the 19th century by Roux and Wolff. The ability of bone to adapt to mechanical forces has been extensively used in orthopedics and orthodontics. Briefly stated, bone will change its mass and architecture to adapt to the forces exerted upon it, providing maximum strength for minimum material. This is achieved through an intricate structure where appropriately oriented collagen fibers provide tensile strength and the minerals embedded in the matrix provide compressive strength.[15] This mechanical adaptation is responsible for the reshaping of bone during fracture healing to resume the prefracture anatomy (if properly set).

Bone loss, caused by paralysis and immobilization, is well documented and selectively affects the immobilized bones (for example, lower extremities in paraplegia, or a single arm kept in a cast). This bone loss is always associated with the loss of muscle mass and strength and is a result (at least in part) of the loss of muscular tension continuously exerted on bones under 1 *g*.[16] Bone degradation can be evaluated by monitoring bone collagen degradation products in urine and blood (C-terminal or N-terminal pyridinoline peptides), for which commercial ELISA assays are available. Bone formation rates can be estimated by measuring the osteoblast products, bone alkaline phosphatase, or osteocalcin in the plasma.

In human volunteers under forced bed-rest conditions, calcium balance and the excretion of bone collagen degradation products showed that immobilization causes bone destruction that was detectable within days.[17] The bone loss caused by microgravity (see below) is consistent with these observations. There is no question that the lack of mechanical stimulation causes bone loss.

The increase in bone mass in response to mechanical stimulation is less well documented. Extensive controlled exercise programs in postmenopausal women who are estrogen deficient produced inconclusive results or had limited effects.[18,19]

It had been noted that the bone mass is much larger in the dominant arm of professional tennis players. Data from bone mass measurements collected over the last couple of years on gymnasts and other athletes show that mechanical stimulation produced by high-impact loading increases bone mass.[20,21] Weight lifters and other athletes also had higher bone density in their extremities and vertebral column during the years they practiced the sport, but bone mass returned to average levels later in life.[22] Mechanical stimulation seems to be more effective in increasing bone mass before peak bone mass is reached at approximately age 25.[23,24]

Animal Studies

Observations regarding bone remodeling in humans have all been reproduced in animal studies, in which some of the mechanisms have been elucidated at the tissue level. Hindlimb paralysis caused in rats by severance of the sciatic nerve or of the knee tendons leads to significant bone loss in the affected limb, due to an initial increase in bone resorption and a sustained decrease in bone formation.[25] When immobilization is produced in rats by putting a limb in a cast, most of the bone loss is reversible after remobilization.

An extensively used model that aims to mimic microgravity is hindlimb unloading, often by tail suspension or a body sling in rats. Within 2 weeks, these animals lose about 25 percent of their cancellous bone in the proximal tibia and show a 30 percent reduction in the mechanical strength of the shaft.[26,27] Interestingly, recent studies have shown that hindlimb unloading causes a 40 percent reduction in blood flow in the unloaded extremities.[28]

Immobilization experiments conducted in dogs, whose cortical bone is more similar to that in humans, also showed bone loss, including significant reduction in the cancellous bone of the proximal radius of the immobilized dogs. There was significant recovery of bone following remobilization, except for the bone mineral density in the central radius.[29] An interesting model is the disarticulated ulna of a turkey, which maintains its vascular supply but is totally unloaded.[30] This bone is undergoing rapid resorption, which can be prevented with the amazingly limited mechanical stimulation of four loading cycles per day. Further mechanical loading stimulates periosteal bone formation that depends on loading strain and frequency.[31,32]

Other experimental protocols have shown that mechanical stimulation increases bone formation in the rat caudal vertebra or the rat tibia. Short stimulation cycles (30 per day) with loads that produce strains approximately equivalent to those generated by walking were effective (700 $\mu\varepsilon$; a strain of 1 ε produces a deformation of 0.1 percent).[33] There is also evidence of interaction between hormonal factors and mechanical stimuli on bone. It was recently shown that bone loss caused by estrogen deficiency in rats is influenced by running treadmill exercises and functional unloading.[34] Whether these findings extrapolate to humans remains to be seen.

Putative Mechanisms

There is no question that bone cells respond to and therefore perceive mechanical changes that cause deformation of the matrix and shear stress due to fluid flow. The proposed sensors (whose function has yet to be proven in bone) include ion channels and integrins. Initial stimuli are followed by signal transduction and amplification, as well as secondary responses that include secretion of prostaglandins and possibly nitrous oxide and growth factors, which act in an autocrine and paracrine fashion. However, many details of this general scheme have not yet been elucidated.

To explain the adaptation of the skeleton and its response to mechanical loads, Frost proposed the existence of a mechanostat or sensor, set at about 1,200 to 1,400 $\mu\varepsilon$, that would initiate bone loss if the strain in the tissue is below that and bone gain if it is above.[35] Such a mechanism is consistent with the physiological evidence and predicts a system of feedback-controlled bone mass homeostasis responsible for adjusting the amount of bone to its mechanical function.[36]

The proposed cellular and molecular basis of this feedback-regulated system is a controlling effect of mechanical strain on osteoblasts (the bone-forming cells) and osteoclasts (the bone-resorbing cells). Strong experimental evidence points to a decrease in osteoblastic bone formation during immobilization (lack of mechanical load), primarily in rats. In several species, there is fairly good evidence for increased osteoblastic bone formation with increase in mechanical loads. There is less evidence for mechanical regulation of osteoclast activity, which is usually indirectly controlled through osteoblast-lineage and other cells.

Several types of mechanical perturbations are believed to be perceived by the cells. Mechanical loading increases the strain in the matrix, and that strain could be transmitted to the cells through their attachments. In addition, deformation of bone or cartilage causes the displacement of fluid, which produces shear stress.

The current hypothesis for mechanical sensors is that specific structures in bone cells act as mechanochemical transducers that generate biochemical signals, which are part of the known signal transduction pathways. One of the outcomes is the production of cytokines that act in an autocrine and paracrine manner to propagate the signal to neighboring cells. Based on information from other systems, three types of sensors have been proposed, with functions that are not mutually exclusive.

Stretch-activated cation channels show increased calcium permeability in membranes stretched in patch clamp experiments and were proposed as mechanosensors.[37] Such channels have also been described in osteoblasts.[38] Similar ion channels act as mechanosensors in *C. elegans*.[39] It is not clear if the strain in the bone matrix (1,500 $\mu\varepsilon$ to 5,000 $\mu\varepsilon$) can cause sufficient distortion of the membranes in attached cells to activate these channels. Calcium channel inhibitors were shown to block mechanical responses in rat bone.[40] A recent mathematical model attempted to quantify the relationship between fluid shear stress on the extracellular surface and the increased calcium permeability of stress-gated ion channels in endothelial cells.[41]

The second class of possible sensors are the integrins through which cells attach to the extracellular matrix. Integrin ligation by antibodies or ligand occupancy activates the sodium proton exchanger or increases intracellular calcium. In addition, integrins can activate signal transduction pathways that involve kinases, similar to those activated by growth factors and other extracellular signals.[42] Mechanical effects on smooth muscle cells can be blocked by arginine-glycine-aspartic acid peptides and antibodies against β3 and αvβ5 integrins (cell adhesion receptors).[43] Integrins were shown to be mechanically connected to cytoskeletal filaments and the nucleoplasm.[44] Cyclical strain (10 percent, 60 Hz) of human umbilical vein endothelial cells caused redistribution of β1 integrins and tyrosine phosphorylation of focal adhesion kinase (pp125[FAK]).[45]

Several secondary responses can participate in the propagation of mechanical effects to neighboring cells. The best documented is upregulation of prostaglandin release.[46,47] Inhibitors of prostaglandin synthesis block responses to mechanical stimulation;[48,49] prostaglandin E2 and I2 have been implicated.[50] Other possible secondary responses include TGFβ and IGF1, which increase bone formation and bone mass in unloaded bones in rats.[51,52] Mechanical strain promotes release of FGF2 from vascular smooth muscle cells,[53] and shear stress induces cyclooxygenase 2, as well as endothelial cell nitric oxide synthase and manganese superoxide dismutase in endothelial cells.[54] In osteoblasts, shear stress also causes nitric oxide release.[55] In addition, pulsating fluid flow was shown to stimulate prostaglandin release and cyclooxygenase 2 in mouse bone cells.[56] G proteins and nitric oxide were also shown to mediate the response of bovine articular chondrocytes to shear stress.[57] Another recently reported secondary response is the induction of the gene for the glutamate transporter,[58] which is of interest because the glutamate receptor in *C. elegans* has been implicated in tactile responses.[59]

MICROGRAVITY EFFECTS ON THE SKELETON

Caveats

More than 30 years of microgravity research has clearly established that the skeleton is one of the organs at risk (for a recent review see van Loon et al., 1993[60]). However, the available data have several limitations. For human studies, the sample size is always small and data collection often incomplete. Collection of urine and feces is difficult in outer space. The high-precision bone density measurement technology (with a 1 to 2 percent coefficient of variation) is relatively new and was too bulky in the past to be used on spacecrafts to monitor the rate of bone loss during flight. Countermeasures were not always rigorously observed. Confounding variables, which may affect the skeleton in addition to microgravity, have not always been recorded.

Animal studies have similar limitations to those enumerated for human studies. The sample sizes are small. The most frequently used species are rodents. These animals grow throughout their lives, and their cortical bone is not similar to that of humans. Because of payload limitations, animals were often very young and actively growing. Flights were mostly short (less than 2 weeks). During spaceflight, animals were usually unattended, and no samples were collected or experiments conducted. Some, but not all, experiments had 1-*g* controls on board; otherwise, the references were "synchronous" ground-based and vivarium controls. Confounding conditions included the acceleration and stress associated with launching and landing, and occasionally flight conditions not precisely mimicked on land, such as higher temperature, crowding, and so on, plus delays between landing and sample collection.

Human Studies

The comprehensive experience of spaceflight effects on the human skeleton has recently been reviewed by van Loon et al. [61] and is summarized here. Major recurrent findings were a negative calcium balance and a decrease in bone mineral density (BMD). BMD estimates the amount of bone by measuring the attenuation of two narrow beams of x rays of different wavelengths, by the calcium in the x-ray path, with correction for soft tissue. A precision of 1 to 2 percent coefficient of variation can be achieved with this instrumentation, which is relatively bulky (the size of a large table plus console). However, portable smaller-size equipment for measuring bone mineral density in some bones is currently available and is being adapted for in-flight measurements. Following are several illustrations. A reduction of 3 to 9 percent in the bone mineral content of os calcis was observed following a 4- to 14-day orbital spaceflight.[62] A loss of 0.2 percent total body calcium was observed in astronauts on Apollo 17 after a 14-day flight.[63] A reduction in the BMD of os calcis and radius was also observed in astronauts on the Skylab 3 and 4 missions, and there was no return to preflight levels after 97 days.[64,65] A negative calcium balance was recorded in the Skylab 2 crew (50 mg/day).[66]

The most compelling data have recently been compiled from the experience on the Mir space station. BMD was measured on various bones in 18 crew members who spent 4 to 14.4 months on the station. The data show an average loss of up to 1 percent BMD per month (Table 6.1). For comparison, the rapid bone loss in some women in early menopause is 2 to 4 percent per year. The in-flight exercise program was clearly not sufficient to prevent the bone loss caused by weightlessness and flight condition. Other changes include a relative reduction in lean mass (primarily muscle) and relative increase in fat mass.

It should be emphasized (see caveats) that the flight database is still very small (by comparison, the natural history of age-related and postmenopausal bone loss is based on thousands of subjects). There was significant variability in spaceflight-related bone loss among subjects as well as in postflight recovery, on which there is relatively limited information. Different bones were affected to varying degrees. There have been few female astronauts so far, and there are no reported data on the effect of microgravity on skeletal changes, specifically in women. Some women astronauts could also be estrogen deficient, an additional risk for bone loss. Evaluating the effect of microgravity on the female skeleton and the possible interaction (additive, synergistic, and so on) between microgravity and hor-

TABLE 6.1 Bone Loss on Mir Space Station (percent bone mineral density lost/month)

Variable	Number of Crew Members	Mean Loss (percent)	Standard Deviation
Spine	18	1.07*	0.63
Neck of femur	18	1.16*	0.85
Trochanter	18	1.58*	0.98
Total body	17	0.35*	0.25
Pelvis	17	1.35*	0.54
Arm	17	0.04	0.88
Leg	16	0.34*	0.33

*$p < 0.01$.

SOURCE: LeBlanc, A., Schneider, V., Shackelford, L., West, S., Ogavov, V., Bakulin, A., and Veronin, L. 1996. Bone mineral and lean tissue loss after long duration spaceflight. J. Bone Miner. Res. 11: S323.

monal status will be essential. This knowledge could be used to help develop appropriate preventive countermeasures.

Genetic Variability

Usually the number of subjects in a given database is small and there is considerable variation in individual responses. Some of this might be the result of experimental "noise," but it is likely that humans will show considerable individual variability in how they respond to microgravity. With the prospect of relatively large numbers of individuals residing in space for extended periods, it is time to begin investigating the genetic variability in the response to spaceflight for those situations where ground-based studies suggest a genetic component.

From the perspective of spaceflight, one of the most important effects on human health is bone loss. It is well established from identical twin studies that osteoporosis has a strong genetic component, accounting for about 60 percent of the variance in bone mineral density.[67] However, no single gene has been convincingly proven to be a risk factor for osteoporosis. The genetic component has been attributed instead to the cumulative effects of a number of genes with small individual effects. There is some evidence from a twins study suggesting that the vitamin D receptor gene may correlate with bone mineral density.[68,69] Other polymorphisms that reportedly correlate with bone density or osteoporosis but have not been as extensively studied include collagen type I, estrogen receptor, and interleukin 1. There is also a report on familial high bone density, localized by linkage analysis to chromosome 11. Osteoporosis, like atherosclerosis and other similar diseases, is complex and multifactorial, dependent on several genes that act as risk factors. Identification of such genes is currently under way. The point is that there is human polymorphism with respect to osteoporosis and bone mineral density; this may imply that the degree of bone loss found during spaceflight may involve a genetic component. Locating and identifying the genes involved in osteoporosis is currently being pursued. To quote a recent review on the topic, this goal is "certainly attainable within the foreseeable future, given proper application of newer advances in molecular genetics and genetic epidemiology."[70]

Although NASA should not be involved in this task, the information and methodology for identifying those individuals most vulnerable to osteoporosis is extremely important to NASA. Countermeasures for bone mineral loss are important; selecting crew persons for long-term missions with consideration of their susceptibility to osteoporosis would be a complementary approach.

Several effective therapies are currently available for the treatment and/or prevention of osteoporosis. These include hormone replacement therapy for estrogen-deficient women, bisphosphonates, and calcitonin. Other therapies are currently being developed. The efficacy of these agents in preventing microgravity bone loss should be explored. Well-designed studies (including randomization, placebo groups, and/or crossover) should be conducted. Although the number of astronauts is small, the magnitude of the bone loss (about 1 percent per month) may make it possible to obtain conclusive data for therapeutic effects and identify pharmacological means for the prevention of microgravity bone loss. Studies in space should be preceded by evaluation of therapeutic intervention in the bed-rest model.

Renal Stone Formation

One of the consequences of bone calcium loss is an increase in susceptibility to renal stone formation. Microgravity changes urine composition to favor supersaturation of stone-forming salts.[71] Specifically, the combination of increased calcium excretion in the urine secondary to the bone calcium loss, with decreased fluid intake, high salt intake,[72] and a reduction in urine pH, will predispose a person to renal stone formation. This is a potentially serious problem. Dietary factors (especially fluid intake and pharmacological interventions) can significantly influence urinary chemical composition, and their use as countermeasures should be further explored.

Recommendations

To address the questions related to microgravity effects on bone loss in humans, the following studies should be conducted:

1. Obtain a comprehensive and detailed description of the phenomenon.
   —A careful record of skeletal changes occurring during microgravity and postflight should be generated for each astronaut using dual-beam x-ray absorptiometry (DXA). Monitoring should compare weight-bearing bones of the lower extremities and vertebral column to those of the upper extremities and the cranium.
   —A database should be established to correlate skeletal changes with age and gender, muscle changes, hormonal changes during flight, diet, and genetic factors (e.g., susceptibility to osteoporosis) if and when these genetic factors become known.
   —The course of microgravity-related bone loss should be thoroughly documented, with attention paid to the rate of bone loss as a function of flight duration. Researchers should establish if bone loss levels off and bone mass stabilizes at a new steady-state level and to what extent the bone loss is reversible after reentry.

2. The mechanism of bone loss should be determined, with emphasis on useful information for developing countermeasures.
   —Determine if bone loss is the result of increased bone destruction (resorption), decreased bone formation, or both. Markers of bone formation and resorption are currently in clinical use. Given the variability of these parameters and the small sample size, astronauts should serve as their own controls.
   —Further validate the bed-rest model as representative for microgravity-induced bone loss and use it to test countermeasures.

3. Effective countermeasures for preventing bone loss should be developed.
   —Instrumentation that can be used for different types of mechanical stimulation should be developed and different types of exercise should be evaluated. Ground experiments suggest that impact loading is most effective in maintaining or increasing bone mass, and animal experiments suggest that short-duration mechanical stimulation may be sufficient to maintain bone mass in immobilized bones.
   —The use of pharmacological means should be examined.

Animal Studies

Findings on the skeletal changes observed in rats exposed to microgravity have been recently reviewed[73] and are briefly summarized here. It should be reiterated that the applicability of the rat data to the human skeleton is limited due to the use of young animals, the continuous growth of the rodent skeleton, and the differences in cortical bone, which lacks vascular canals in rodents. However, there is a vast database on rodent bone biology, and gravitation-related changes in rats or mice could be explored for mechanistic studies. In the Cosmos flights, the major positive finding was a reduction in the rate of periosteal bone formation, which returned to control levels approximately 4 weeks postflight. Decreases in the mechanical properties of weight-bearing bones were also observed.[74-81] These findings were not always reproducible.[82-86]

In Spacelab 3, defective mineralization was observed.[87-90] In some experiments, there was evidence of increased bone resorption.[91] There were no differences in the biochemistry of collagen or proteoglycans. Thus, examination of animal bones, primarily from rats exposed to microgravity for short durations (4 to 15 days) showed either no effect or a reduction in bone formation, relative to controls that were not always perfectly matched. A limited number of experiments have been conducted using bone cells cultured under microgravity conditions. Considering the difficulty in controlling experimental variables in space, such as convection, media changes, partial pressure of gases, and the like, no definitive conclusions could yet be reached regarding microgravity's effects on isolated cells in petri dishes. However, similar experiments in the space station, conducted side-by-side with controls in a 1-g centrifuge, could generate relevant data on cultured cells. Reproducible changes, if observed, should provide a suitable system for exploring the mechanism for microgravity perception by, and effects on, isolated bone cells.

Well-documented ground-based in vitro models for studying the interaction between osteoblasts and osteoclasts are currently available. Most recently, RANK ligand (TRANCE), which is expressed on membranes of osteoblast lineage cells, was shown to promote osteoclast differentiation in macrophage or osteoclast precursors. Other agents active in this coculture system include prostaglandins, interleukins, osteotropic hormones, and hematopoietic growth factors. This experimental system could be used under microgravity conditions with 1-g centrifuge controls. If differences are observed, their basis can be investigated further.

Recommendations

The two crucial questions are, What is the validity of animal models for mimicking the changes produced by microgravity in the human skeleton? and, What experiments can be conducted under ground-based conditions? Both in space and on the ground, mostly young rodents with an actively growing skeleton have been used. When an effect was observed, it was primarily a decrease in bone formation. The following recommendations are listed in order of priority.

1. Determine if changes produced by microgravity in animal bones are similar to those in humans, and if they have a similar basis—that is, increased bone resorption and/or decreased formation.

—Older animals (rats or mice) should be used (the growth in rats is significantly attenuated at about 12 to 14 months). Mice are preferable, because they are smaller and include genetically homogenous strains, mutants, and transgenic animals ("knockouts," and so on).

—Dual x-ray absorptiometry, histomorphometry, and biochemical markers (described under human studies) should be used.

2. When an animal model is identified that mimics human changes in spaceflight, it should be compared to ground-based models, such as hindlimb unloading. If appropriate, it should be used to study mechanisms that can then be corroborated in space. Emphasis should be on investigations that can help develop countermeasures and shed light on human pathology (e.g., osteoporosis). Studies should be conducted that will do the following:

—Evaluate the contribution of changes in muscle function and blood flow. Muscle, bone, and blood flow, which strongly affect both, should be studied side by side. Given the dynamic nature of

the vascular system, in-flight experiments should include rapid tissue fixation to best address this question.

—Evaluate to what extent the bone loss is secondary to systemic effects on hormones, growth factors, and cytokines, which should be measured in the same experiments. Identify putative mediators involved in the tissue response: prostaglandins (PGE2, PGI2), growth factors (FGF, TGFβ), and so on. Elegant pharmacological studies were used in the past. Recombinant proteins and mutated cells and animals can supplement the pharmacological approach.

—Identify the cells responsive to gravity changes (osteoblasts, osteocytes, osteoclasts) and determine at what level the cellular processes are altered: cell recruitment, cell activity, cell survival (apoptosis). These questions can be addressed by the combination of in vivo (histology) and in vitro (cell biology) methods.

—Evaluate the interaction between osteoblast lineage cells and osteoclast precursors under microgravity conditions with 1-g centrifuge controls, using the well-established ground-based experimental models. Examine if production of factors known to mediate the communication between osteoblasts and osteoclasts (such as prostaglandins, inflammatory cytokines, hemopoietic growth factors, matrix molecules, and others) are altered under microgravity conditions.

—Examine whether bone cells, or cells that convey the information to bone cells, respond to (a) strain in the matrix to which they are attached, (b) shear stress produced by fluid flow, and/or (c) electrical fields produced by deformation or fluid flow. This should be investigated using primarily ground-based experiments, taking advantage of the potential of current cellular and molecular biology methods (transgenic animals and the like). Identify the sensors for these perturbations (e.g., integrins, ion channels), and determine the pathways for their signal transduction and amplification.

3. Determine if there are molecular alterations in the structure of bone matrix or bone mineral.

—Examine the composition of the matrix and the presence of posttranslational modifications (e.g., collagen cross-links). The discovery of new proteins and their secondary modifications warrants reexamination of this question in the context of mechanical and gravitational changes.

—Determine if there are changes in mineral structure (crystal size and composition), the degree of mineralization, or relationship of the bone mineral to the bone matrix.

Many of these issues have been addressed in the past, but few conclusive answers are available at this point. It is also clear from this list that many experiments could be conducted in ground-based research and that solid results obtained from such investigations, especially regarding mechanisms, could increase the payoff from experiments conducted in microgravity.

Equipment Needs

Specific

1. Bone mineral density measurement instrument suited for use in flight, for humans and animals.
2. Exercise instrumentation that provides different types of mechanical stimulation (impact-loading, muscle-mediated bone strain) for humans and animals.

General

1. Animal facilities with in-flight accommodations for at least 30 adult rats or mice.

2. In-flight centrifuge that generates 1-*g* control conditions for rats or mice.

3. Equipment needed for in-flight animal handling: feeding, injections, blood withdrawal, sacrifice, dissection, and tissue fixation.

REFERENCES

1. Prockop, D.J. 1997. Marrow stromal cells as stem cells for nonhematopoietic tissues. Science 276: 71-74.

2. Suda, T., Abe, E., Shinki, T., Katagiri, T., Yamaguchi, A., Yokose, S., Yoshiki, S., Horikawa, H., Cohen, G.W., Yasugi, S., et al. 1994. The role of gravity in chick embryogenesis. FEBS Lett. 340: 34-38.

3. Schmidt, Z., Harada, S., Rodan, G.A. 1996. Anabolic steroid effects on bone in women. Pp. 1125-1134 in Principles of Bone Biology (J.P. Bilezikian, L.G. Raisz, and G.A. Rodan, eds.). Academic Press, San Diego.

4. Orwoll, E.S. 1996. Androgens. Pp. 563-580 in Principles of Bone Biology (J.P. Bilezikian, L.G. Raisz, and G.A. Rodan, eds.). Academic Press, San Diego.

5. Norman, A.W., and Collins, E.D. 1996. Vitamin D receptor structure, expression, and nongenomic effects. Pp. 419-434 in Principles of Bone Biology (J.P. Bilezikian, L.G. Raisz, and G.A. Rodan, eds.). Academic Press, San Diego.

6. Becker, K.L., Nylen, E.S., Cohen, R., and Snider, R.H., Jr. 1996. Calcitonin: Structure, molecular biology, and actions. Pp. 471-494 in Principles of Bone Biology (J.P. Bilezikian, L.G. Raisz, and G.A. Rodan, eds.). Academic Press, San Diego.

7. Burman, P., Johansson, A.G., Siegbahn, A., Vessby, B., and Karlsson, F.A. 1996. Growth hormone (GH)-deficient men are more responsive to GH replacement therapy than women. J. Clin. Endocrinol. Metab. 82: 550-555.

8. Conover, C.A. 1996. The role of insulin-like growth factors and binding proteins in bone cell biology. Pp. 607-618 in Principles of Bone Biology (J.P. Bilezikian, L.G. Raisz, and G.A. Rodan, eds.). Academic Press, San Diego.

9. Verhaegh, J., and Bouillon, R. 1996. Effects of diabetes and insulin on bone metabolism. Pp. 549-561 in Principles of Bone Biology (J.P. Bilezikian, L.G. Raisz, and G.A. Rodan, eds.). Academic Press, San Diego.

10. Yoneda, T. 1996. Local regulators of bone: Epidermal growth factor-transforming growth factor-α. Pp. 729-738 in Principles of Bone Biology (J.P. Bilezikian, L.G. Raisz, and G.A. Rodan, eds.). Academic Press, San Diego.

11. Horowitz, M.C., and Lorenzo, J.A. 1996. Local regulators of bone: IL-1, TNF, lymphotoxin, interferon-g, IL-8, IL-10, IL-4, the LIF/IL-6 family, and additional cytokines. Pp. 687-700 in Principles of Bone Biology (J.P. Bilezikian, L.G. Raisz, and G.A. Rodan, eds.). Academic Press, San Diego.

12. Bonewald, L.F. 1996. Transforming growth factor-β. Pp. 647-659 in Principles of Bone Biology (J.P. Bilezikian, L.G. Raisz, and G.A. Rodan, eds.). Academic Press, San Diego.

13. Hurley, M.M., and Florkiewicz, R.Z. 1996. Fibroblast growth factor and vascular endothelial cell growth factor families. Pp. 627-645 in Principles of Bone Biology (J.P. Bilezikian, L.G. Raisz, and G.A. Rodan, eds.). Academic Press, San Diego.

14. Hurley, M.M., and Florkiewicz, R.Z. 1996. Fibroblast growth factor and vascular endothelial cell growth factor families. Pp. 627-645 in Principles of Bone Biology (J.P. Bilezikian, L.G. Raisz, and G.A. Rodan, eds.). Academic Press, San Diego.

15. Einhorn, T.A. 1996. Biomechanics of bone. Pp. 25-37 in Principles of Bone Biology (J.P. Bilezikian, L.G. Raisz, and G.A. Rodan, eds.). Academic Press, San Diego.

16. Marcus, R. 1996. Mechanisms of exercise effects on bone. Pp. 1135-1146 in Principles of Bone Biology (J.P. Bilezikian, L.G. Raisz, and G.A. Rodan, eds.). Academic Press, San Diego.

17. Ruml, L.A., Dubois, S.K., Roberts, M.L., and Pak, C.Y.C. 1995. Prevention of hypercalciuria and stone-forming propensity during prolonged bedrest by alendronate. J. Bone Miner. Res. 10: 655-662.

18. Bassey, E.J., and Ramsdale, S.J. 1995. Weight-bearing exercise and ground reaction forces: A 12-month randomized controlled trial of effects on bone mineral density in healthy postmenopausal women. Bone 16: 469-476.

19. Pruitt, L.A., Taaffe, D.R., and Marcus, R. 1996. Effects of a one-year high-intensity versus low-intensity resistance training program on bone mineral density in older women. J. Bone Miner. Res. 10: 1788-1795.

20. Heinonen, A., Kannus, P., Sievanen, H., Oja, P., Pasanen, M., Rinne, M., Uusi-Rasi, K., and Vuori, I. 1996. Randomised controlled trial of effect of high-impact exercise on selected risk factors for osteoporotic fractures. Lancet 348: 1343-1347.

21. Taaffe, D.R., Robinson, T.L., Snow, C.M., and Marcus, R. 1997. High-impact exercise promotes bone gain in well-trained female athletes. J. Bone Miner. Res. 12: 255-260.

22. Karlsson, M.K., Hasserius, R., and Obrant, K.J. 1996. Bone mineral density in athletes during and after career: A comparison between loaded and unloaded skeletal regions. Calcif. Tissue Int. 59: 245-248.

23. Nordstrom, P., Nordstrom, G., and Lorentzon, R. 1997. Correlation of bone density to strength and physical activity in young men with a low or moderate level of physical activity. Calcif. Tissue Int. 60: 332-337.

24. Friedlander, A.L., Genant, H.K., Sadowsky, S., Byl, N.N., and Gluer, C.-C. 1995. A two-year program of aerobics and weight training enhances bone mineral density of young women. J. Bone Miner. Res. 10: 574-593.

25. Weinreb, M., Rodan, G.A., and Thompson, D.D. 1989. Osteopenia in the immobilized rat hind limb is associated with increased bone resorption and decreased bone formation. Bone 10: 187-194.

26. Morey, E.R. 1979. Spaceflight and bone turnover: Correlation with a new rat model of weightlessness. BioScience 29: 168-172.

27. Vico, L., Novikov, V.E., Very, J.M., and Alexandre, C. 1991. Bone histomorphometric comparison of rat tibial metaphysis after 7 day hindlimb unloading vs. 7 day spaceflight. Aviat. Space Environ. Med. 62: 26-31.

28. Roer, R.D., and Dillaman, R.M. 1994. Decreased femoralarterial flow during simulated microgravity in the rat. J. Appl. Physiol. 76: 2125-2129.

29. Lane, N.E., Kaneps, A.J., Stover, S.M., Modin, G., and Kimmel, D.B. 1996. Bone mineral density and turnover following forelimb immobilization and recovery in young adult dogs. Calcif. Tissue Int. 59: 401-409.

30. Gross, T.S., and Rubin, C.T. 1995. Uniformity of resorptive bone loss induced by disuse. J. Orthop. Res. 13: 708-714.

31. Rubin, C.T., and Lanyon, L.E. 1984. Regulation of bone formation by applied dynamic loads. J. Bone Jt. Surg. Am. 66: 397-402.

32. Rubin, C.T., and McLeod, K.J. 1994. Promotion of bony ingrowth by frequency-specific, low-amplitude mechanical strain. Clin. Orthop. 298: 165-174.

33. Chow, J.W.M., Jagger, C.J., and Chambers, T.J. 1993. Characterization of osteogenic response to mechanical stimulation in cancellous bone of rat caudal vertebrae. Am. J. Physiol. 265: E340-E347.

34. Westerlind, K.C., Wronski, T.J., Ritman, E.L., Luo, Z.-P., An, K.-N., Bell, N.H., and Turner, R.T. 1997. Estrogen regulates the rate of bone turnover but bone balance in ovariectomized rats is modulated by prevailing mechanical strain. Proc. Natl. Acad. Sci. U.S.A. 94: 4199-4204.

35. Frost, H.M. 1996. Perspectives: A proposed general model of the "mechanostat." Anat. Rec. 244: 139-147.

36. Rodan, G.A. 1997. Bone mass homeostasis and bisphosphonate action. Bone 20: 1-4.

37. Hoyer, J., Kohler, R., Haase, W., and Distler, A. 1996. Up-regulation of pressure-activated Ca^{2+}-permeable cation channel in intact vascular endothelium of hypertensive rats. Proc. Natl. Acad. Sci. U.S.A. 93: 11253-11258.

38. Kizer, N., Guo, X.-L., and Hruska, K. 1997. Reconstitution of stretch-activated cation channels by expression of the a-subunit of the epithelial sodium channel cloned from osteoblasts. Proc. Natl. Acad. Sci. U.S.A. 94: 1013-1018.

39. Corey, D.P., and Garcia-Anoveros, J. 1996. Mechanosensation and the DEG/ENaC ion channels. Science 273: 323-324.

40. Rawlinson, S.C.F., Pitsillides, A.A., and Lanyon, L.E. 1996. Involvement of different ion channels in osteoblasts' and osteocytes' early responses to mechanical strain. Bone 19: 609-614.

41. Wiesner, T.F., Berk, B.C., and Nerem, R.M. 1997. A mathematical model of the cytosolic-free calcium response in endothelial cells to fluid shear stress. Proc. Natl. Acad. Sci. U.S.A. 94: 3726-3731.

42. Hynes, R.O. 1992. Integrins: Versatility, modulation and signaling in cell adhesion. Cell 69: 11-25.

43. Wilson, E., Sudhir K., and Ives, H.E. 1995. Mechanical strain of rat vascular smooth muscle cells is sensed by specific extracellular matrix/integrin interactions. J. Clin. Invest. 96: 2364-2372.

44. Maniotis, A.J., Chen, C.S., and Ingber, D.E. 1997. Demonstration of mechanical connections between integrins, cytoskeletal filaments, and nucleoplasm that stabilize nuclear structure. Proc. Natl. Acad. Sci. U.S.A. 94: 849-854.

45. Yano, Y., Geibel, J., and Sumpio, B.E. 1997. Cyclic strain induces reorganization of integrin a5b1 and a2b1 in human umbilical vein endothelial cells. J. Cell Biochem. 64: 505-513.

46. Somjen, D., Binderman, I., Berger, E., and Harell, A. 1980. Bone remodeling induced by physical stress is prostaglandin E2 mediated. Biochim. Biophys. Acta 627: 91-100.

47. Yeh, C., and Rodan, A. 1984. Tensile forces enhance PGE synthesis in osteoblasts grown on collagen ribbon. Calcif. Tissue Int. 36: S67-S71.

48. Pead, M.J., and Lanyon, L.E. 1989. Indomethacin modulation of load-related stimulation of new bone formation in vivo. Calcif. Tissue Int. 45: 34-40.

49. Forwood, M.R. 1996. Inducible cyclo-oxygenase (COX-2) mediates the induction of bone formation by mechanical loading in vivo. J. Bone Miner. Res. 11: 1688-1693.

50. Rawlinson, S.C.F., Pitsillides, A.A., and Lanyon, L.E. 1996. Involvement of different ion channels in osteoblasts' and osteocytes' early responses to mechanical strain. Bone 19: 609-614.

51. Machwate, M., Zerath, E., Holy, X., Pastoureau, P., and Marie, P.J. 1994. Insulin-like growth factor-I increases trabecular bone formation and osteoblastic cell proliferation in unloaded rats. Endocrinology 134: 1031-1038.

52. Machwate, M., Zerath, E., Holy, X., Hott, M., Godet, D., Lomri, A., and Marie, P.J. 1995. Systemic administration of transforming growth factor-β2 prevents the impaired bone formation and osteopenia induced by unloading in rats. J. Clin. Invest. 96: 1245-1253.

53. Cheng, G.C., Briggs, W.H., Gerson, D.S., Liby, P., Grodzinsky, A.J., Gray, M.L., and Lee, R.T. 1997. Mechanical strain tightly controls fibroblast growth factor-2 release from cultured human vascular smooth muscle cells. Circ. Res. 80: 28-36.

54. Topper, J.N., Cai, J., Falb, D., and Gimbrone, M.A. Jr. 1996. Identification of vascular endothelial genes differentially responsive to fluid mechanical stimuli: Cyclooxygenase-2, manganese superoxide dismutase, and endothelial cell nitric oxide synthase are selectively up-regulated by steady laminar shear stress. Proc. Natl. Acad. Sci. U.S.A. 93: 10417-10422.

55. Johnson, D.L., McAllister, T.N., and Frangos, J.A. 1996. Fluid flow stimulates rapid and continuous release of nitric oxide in osteoblasts. Am. J. Physiol. 271: E205-E208.

56. Klein-Nulend, J., Burger, E.H., Semeins, C.M., Raisz, L.G., and Pilbeam, C.C. 1997. Pulsating fluid flow stimulates prostaglandin release and inducible prostaglandin G/H synthase mRNA expression in primary mouse bone cells. J. Bone Miner. Res. 12: 45-51.

57. Das, P., Schurman, D.J., and Smith, R.L. 1997. Nitric oxide and G proteins mediate the response of bovine articular chondrocytes to fluid-induced shear. J. Orthop. Res. 15: 87-93.

58. Mason, D.J., Suva, L.J., Genever, P.G., Patton, A.J., Steucklem S., Hillam, R.A., and Skerry, T.M. 1997. Mechanically regulated expression of a neural glutamate transporter in bone: A role for excitatory amino acids as osteotropic agents? Bone 20: 199-205.

59. Maricq, A.V., Peckol, E., Driscoll, M., and Bargmann, C.I. 1995. Mechanosensory signalling in C. elegans mediated by the GLR-1 glutamate receptor. Nature 378: 78-81.

60. van Loon, J.J., van den Bergh, L.C., Schelling, R., Veldhuijzen, J.P., and Huijser, R.H. 1993. Development of a centrifuge for acceleration research in cell and development biology. Space Safety and Rescue, 1993 [proceedings of a symposium of the International Academy of Astronautics held in conjunction with the 44th International Astronautical Federation Congress, October 16-22, 1993, Graz, Austria] (Gloria W. Heath, ed.). IAF/IAA-93-G.4-166. American Astronautical Society, Springfield, Va.

61. van Loon, J.J., van den Bergh, L.C., Schelling, R., Veldhuijzen, J.P., and Huijser, R.H. 1993. Development of a centrifuge for acceleration research in cell and development biology. Space Safety and Rescue, 1993 [proceedings of a symposium of the International Academy of Astronautics held in conjunction with the 44th International Astronautical Federation Congress, October 16-22, 1993, Graz, Austria] (Gloria W. Heath, ed.). IAF/IAA-93-G.4-166. American Astronautical Society, Springfield, Va

62. Vose, G.P. 1974. Review of roentgenographic bone demineralization studies of the Gemini spaceflights. Am. J. Roentgenol., Radium Ther. Nucl. Med. 121: 1-4.

63. Rambaut, P.C., Leach, C.S., and Johnson, P.C. 1975. Calcium and phosphorus change of the Apollo 17 crew members. Nutr. Metab. 18: 62-69.

64. Vogel, J.M., and Whittle, M.W. 1976. Bone mineral changes: The second manned Skylab mission. Aviat. Space Environ. Med. 47: 396-400.

65. Tilton, F.E., Degioanni, J.J., and Schneider, V.S. 1980. Long-term follow-up of Skylab bone demineralization. Aviat. Space Environ. Med. 11: 1209-1213.

66. Whedon, G.D., Lutwak, L., Reid, J., Bambaut, P., Whittle, M., Smith, M., and Leach, C. 1975. Mineral and nitrogen balance study, results of metabolic observations on Skylab II 28-day orbital mission. Acta Astronautica 2: 297-309.

67. Rogers, J., Mahaney, M.C., Beamer, W.G., Donahue, L.R., and Rosen, C.J. 1997. Beyond one gene-one disease: Alternative strategies for deciphering genetic determinants of osteoporosis. Calcif. Tissue Int. 60: 225-228.

68. Morrison, N.A., Qi, J.C., Tokita, A., Kelley, P.J., Crofts, L., Nguen, T.V., Sambrook, P.N., and Eisman, J.A. 1994. Prediction of bone density from vitamin D receptor alleles. Nature 367: 284-287.

69. Peacock, M. 1955. Vitamin D receptor gene alleles and osteoporosis: A contrasting view. J. Bone Miner. Res. 10: 1294-1297.

70. Tipton, C.M. 1996. Animal models and their importance to human physiological responses in microgravity. Med. Sci. Sports Exercise. 28: S94-S100.

71. Whitson, P.A., Pietrzyk, R.A., and Pack, C.Y.C. 1997. Renal stone assessment during space shuttle flights. J. Urol. 158: 2305-2310.

72. Navidi, M., Wolinsky, I., Fung, P., and Arnaud, S.B. 1995. Effect of excess dietary salt on calcium metabolism and bone mineral in a spaceflight rat model. J. Appl. Physiol. 78: 70-75.

73. van Loon, J.J., van den Bergh, L.C., Schelling, R., Veldhuijzen, J.P., and Huijser, R.H. 1993. Development of a centrifuge for acceleration research in cell and development biology. Space Safety and Rescue, 1993 [proceedings of a symposium of the International Academy of Astronautics held in conjunction with the 44th International Astronautical Federation Congress, October 16-22, 1993, Graz, Austria] (Gloria W. Heath, ed.). IAF/IAA-93-G.4-166. American Astronautical Society, Springfield, Va.

74. Turner, R.T., Bell, N.H., Duvall, P., Bobyn, J.D., Spector, M., Morey-Holton, E., and Baylink, D.J. 1985. Spaceflight results in formation of defective bone (42215). Proc. Soc. Exp. Biol. Med. 180: 544-549.

75. Spengler, D.M., Morey, E.R., Carter, D.R., Turner, R.T., and Baylink, D.J. 1983. Effects of spaceflight on structural and material strength of growing bone. Proc. Soc. Exp. Biol. Med. 174: 224-228.

76. Spector, M., Turner, R.T., Morey-Holton, E., Baylink, D.J., and Bell, N.H. 1983. Arrested bone formation during spaceflight results in a hypomineralized skeletal defect. Physiologist 26: S110-S111.

77. France, E.P., and Oloff, C.M. 1982. Bone mineral analysis of rat vertebrae following spaceflight COSMOS 1129. Physiologist 25: S147-S148.

78. Eurell, J.A., and Kazarian, L.E. 1983. Quantitative histochemistry of rat lumbar vertebrae following spaceflight. Am. J. Physiol. 244: R315-R318.

79. Jee, W.S.S., Wronski, T.J., Morey, E.R., and Kimmel, D.B. 1983. Effects of spaceflight on trabecular bone in rats. Am. J. Physiol. 244: R310-R314.

80. Bakulin, A.V., Ilyan, E.A., Organov, V.S., and Lebedev, V.I. 1985. The state of bones of pregnant rats during an acute stage of adaptation to weightlessness. Pp. 225-259 in Proceedings, 2nd International Conference on Space Physiology, Toulouse, France, Nov. 20-22, 1985 (J.J. Hunt, ed.). ESA-SP-237. European Space Agency, Paris.

81. Vico, L., Chappard, D., Alexandre, C., Palle, S., Minaire, P., Riffat, G., Novikov, V.E., and Bakulin, A.V. 1987. Effects of weightlessness on bone mass and osteoclast number in pregnant rats after a five-day spaceflight (Cosmos 1514). Bone 8: 95-103.

82. Vico, L., Chappard, D., Palle, S., Bakulin, A.V., and Alexandre, C. 1988. Trabecular bone remodeling after seven days of weightlessness exposure (Biocosmos 1667). Am. J. Physiol. 255: R243-R247.

83. Vico, L., Novikov, V.E., Very, J.M., and Alexandre, C. 1991. Bone histomorphometric comparison of rat tibial metaphysis after 7 day hindlimb unloading vs. 7 day spaceflight. Aviat. Space Environ. Med. 62: 26-31.

84. Vico, L., Bourrin, J.M., Very, D., Chappard, D., and Alexandre, C. 1990. Bone adaptation to real and simulated microgravity. Pp. 359-361 in Proceedings of the 4th European Symposium on Life Sciences Research in Space (David V. Paris, ed.). European Space Agency, Paris.

85. Vico, L., Bourrin, S., Genty, C., Palle, S., and Alexandre, C. 1993. Histomorphometric analyses of cancellous bone from Cosmos 2044 rats. J. Appl. Physiol. 75: 2203-2208.

86. Vailas, A.C., Vanderby, R., Martinez, D.A., Ashman, R.B., Ulm, M.J., Grindeland, R.E., Durnova, G.N., and Kaplansky, A.S. 1992. Adaptations of young adult rat cortical bone to 14 days of spaceflight. J. Appl. Physiol. 73: 4S-9S.

87. Patterson-Buckendahl, P.E., Arnaud, S.B., Mechanic, G.L., Martin, R.B., Grindeland, R.E., and Can, C.E. 1987. Fragility and composition of growing rat bone after one week in spaceflight. Am. J. Physiol. 252: R240-R246.

88. Simmons, D.J., Russell, J.E., and Grynpas, M.D. 1986. Bone maturation and quality of bone material in rats flown on the space shuttle "Spacelab 3 mission." Bone Miner. 1: 485-493.

89. Duke, J., Janer, L., Campbell, M., and Morrow, J. 1985. Microprobe analyses of epiphyseal plates from Spacelab 3 rats. Physiologist 28: S217-S218.

90. Wronski, T.J., Morey-Holton, E.R., Doty, S.B., Maese, A.C., and Walsh, C.C. 1987. Histomorphometric analysis of rat skeleton following spaceflight. Am. J. Physiol. 252: R252-R255.

91. Kaplansky, A.S., Durnova, G.N., Burkovskaya, T.E., and Vorotnikova, E.V. 1991. The effect of microgravity on bone fracture healing in rats flown on Cosmos 2044. Physiologist 34: S196-S199.

7

Skeletal Muscle

INTRODUCTION

Even though humans have not been reported to suffer permanent neuromuscular deficits from working in space for periods of a year or more, significant changes have been observed both in flight and after return to Earth. These changes include muscle weakness, fatigue, incoordination, and delayed-onset muscle soreness. Loaded treadmill exercises reduce but do not prevent the loss of strength in the lower limbs, and bicycle ergometry is ineffective for preserving muscle mass. The increased fatigability, incoordination, and susceptibility to reloading injury are not remedied. Postflight recovery has not been analyzed adequately in either humans or animal models to determine the mechanisms, efficacy, and temporal progress of the rehabilitation. Muscle deterioration remains a major concern that warrants continued flight and ground studies to prepare for longer-duration missions. The goal is to maintain neuromuscular structure and function while minimizing the time required for countermeasures and enhancing the productivity of mission tasks. A consensus on the most appropriate types of countermeasures for prolonged spaceflight is far from being reached.[1-3] There is general agreement that a multipronged approach, employing exercise plus other synergistic measures, is necessary. The other measures may include strategies such as the concomitant utilization of hormones, growth factors, drugs, and lower-body negative-pressure devices. To date, microgravity has been only minimally exploited as a unique tool for understanding the fundamental mechanisms that underlie its effects on neuromuscular function and provide the basis for development of effective countermeasures. The vast majority of spaceflight studies have been conducted pre- and postflight. Completion of the International Space Station and maintenance of a continuous human presence in space promise in-flight studies that will generate a wealth of new information on neuromuscular biology to advance basic knowledge and benefit humans in space and on Earth.

As currently understood, many effects on muscle can only be described as spaceflight-induced. Few changes can be confidently defined as resulting solely from microgravity. Humans continue to be undernourished during flight, and this has global effects on the body—most especially muscle, which constitutes more than 30 percent of the body mass. In-flight studies so far have been inadequate to permit assignment of changes found during postflight sampling as resulting either from primary in-flight effects or secondary reentry or reloading alterations postflight. Dramatic secondary changes introduced during reentry and reexposure to unit gravity may alter or mask microgravity's primary effects. Despite this limitation, there are sufficient data from in-flight and simulated microgravity ground studies of humans and rats to put forth a plausible scenario of the adaptation of muscle to microgravity and readaptation to terrestrial gravity.

This chapter discusses skeletal muscle biology in the context of whether or not we currently understand muscle physiology in sufficient detail to proceed with sending crews on long-term space missions, such as to Mars, and returning them to gravity without fear of serious clinical consequences. The answer is that we do not, but we can be ready in the next decade, given adequate ground-based and flight research opportunities. There are a number of key areas requiring attention. For example, muscle atrophy is induced by reduced contractile activity, unloaded contractions, and a shortened working range. Contractile proteins are selectively targeted for degradation through ubiquitination and the multicatalytic proteasome pathway. Stress hormones and undernutrition exacerbate a negative nitrogen balance due to loss of muscle protein. Lowered levels of bioactive growth hormone and diminished growth factors may be additional contributing factors (see Chapter 9). Changes in autonomic vasomotor regulation and blood volume affect muscle microcirculation. Musculovenous pumping is curtailed in quiescent muscles. The energy metabolism of muscle shifts from oxidative and lipid utilization toward reliance on glucose consumption. Novel patterns of central nervous system motor and reflex control of muscle are acquired. Most of these changes are appropriate adaptations to microgravity but are undesirable and debilitating in normal gravity.

Most spaceflight investigations have focused on growing or mature animals. Preliminary spaceflight results and hindlimb unloading studies involving developing animals suggest that gravity loading may not only be necessary for the maintenance of mature skeletal muscle but may also be essential for differentiation of the developing neuromuscular system in a direction appropriate for functioning efficiently in terrestrial gravity. A period of unloading during spaceflight renders mature muscle susceptible to reloading injury. A major area of concern is that, while adaptation to microgravity is well tolerated, the stress of returning to gravity loading reveals serious impairments to normal functioning. The issue of neuromuscular deterioration involves multiple systems, which necessitates multifaceted solutions.

BACKGROUND

Research Done on Muscle Biology

Significant progress has been made in understanding the molecular basis of embryonic muscle cell differentiation. Sequential expression of regulatory factors of the MyoD family of myogenic proteins (MyoD, Myf5, MRF4, and myogenin) commits mesodermal cells to a myocyte lineage.[4] These proteins are members of the basic helix-loop-helix superfamily of transcription factors that, after forming heterodimers, bind to E-box sites on muscle-specific genes to activate transcription.[5] Cells, programmed as myoblasts, migrate from somites to seed the body musculature; eventually, they cease proliferation and differentiate by coalescing to generate multinucleated myofibers that synthesize large quantities of muscle-specific proteins. Myocyte enhancer factor-2 (MEF2) transcription factors are necessary and interact synergistically with myogenic factors to maintain synthesis of muscle proteins.

Growth factors regulate plasticity of the differentiating muscle. Insulin-like growth factors (IGFs) promote differentiation of myoblasts in response to injury, possibly through stimulating myogenin expression. Basic fibroblast growth factor represses differentiation and keeps myoblasts in the proliferative state. The mature fiber adapts its phenotype to optimize performance under conditions of altered workload, length, patterns of use, and humoral factors.

Regulatory factors partition myofibers into multiple fast and slow fiber types. Primary slow fibers change from expression of embryonic to slow myosin heavy chain;[6,7] secondary slow fibers produce embryonic, neonatal, fast, and slow myosins. [8,9] There are several other proteins, such as the troponins and myosin light chains, which have fast and slow isoforms.[10] DNA sequences with E-box and MEF2-like motifs upstream from the promoters (slow upstream regulatory elements) and intronic enhancers (fast intronic regulatory elements) are candidate factors for conferring fiber-type-specific expression. The phenotype of the secondary fibers is modulated by impulse patterns from the motor innervation.[11] Thyroid hormone levels can also profoundly affect expression of myosin heavy and light chain isoforms.[12,13] Mounting evidence indicates that *cis-* and *trans*-acting regulatory factors, as yet uncharacterized, control specific fiber-type differentiation at the transcriptional level.[14] Myogenin mRNA levels are higher in slow fibers, and MyoD transcript levels are higher in fast fibers.[15] The characterization of *cis-* and *trans*-acting factors that regulate transcription of fiber-type-specific protein isoforms is an important area to pursue in understanding muscle plasticity.

Whole-muscle physiological and biochemical studies have yielded insights into muscle adaptation but have stopped short of elucidating the unique repertoire of characteristics and functions at the single-fiber (cell) level. Transgenic animal studies have shown that c-ski proto-oncogene product induces hypertrophy of a subset of fast muscle fibers.[16] The fast and slow phenotypes are altered by motor nerve impulse patterns and neurotrophic interactions, a cadre of hormones, and muscle cell working load and length.

Previous Space- and Ground-based Research

With the advent of the International Space Station and a permanent presence of humans in space, astronauts will spend longer periods in reduced gravity before returning to terrestrial gravity. Before this step can be accomplished with confidence, the deleterious effects that spaceflight and reloading upon return to Earth have on skeletal muscle must be better understood to ensure performance and prevent injury. The fact that many humans have successfully sojourned into space and returned without apparent persistent debilitation does not adequately diffuse the concern for ensuring healthy long-term space travel. A recent NASA panel on countermeasures concluded that many questions still have to be investigated so that researchers can identify efficacious countermeasure protocols. Exercise protocols appear to have a positive effect because the degree of muscle wasting correlates inversely with the amount of exercise, rather than directly with the duration of space travel.[17] In addition, crew members increased their upper arm strength in microgravity by working against elastic cords. However, the tremendous compensatory and regenerative capacities of the neuromuscular system could have repaired and masked pathological changes during their recovery periods. If this is true, relying too heavily on such reserves may cause their capacity to be exceeded and culminate in permanent disabilities. This may be what happens in the so-called postpolio syndrome, in which individuals who have recovered from polio experience unexplained dramatic reductions in motor performance some 30 years later.[18]

To date, there have only been three studies of pre- and postflight muscle biopsies from astronaut crews of relatively short missions lasting 5, 11, and 17 days. Physiological, biochemical, and structural changes were studied at the cellular and molecular levels.[19-22] Interpretation of the findings is complicated by factors that have continually plagued spaceflight studies and could have influenced the outcomes: the number of human subjects was small, and their preflight levels of exercise conditioning varied greatly. In addition to possible gender-unique responses, the test subjects participated in a wide range of in-flight payload activities that required variable and undocumented muscle use. Given the lack of specific details and high cost of time and effort, replaying the mission as a postflight ground control study in 1 *g* would be virtually impossible. This argues for better-controlled flight studies that may focus intensively on neuromuscular investigations for some future missions.

Current understanding of the changes in skeletal muscles is derived from pioneering contributions of over two decades of pre- and postflight studies of rodents that were orbited 1 to 3 weeks in Russian biosatellites and U.S. space shuttles. A single in-flight tissue acquisition from adult rats was achieved as part of the 1993 Spacelab Life Sciences-2 (SLS-2) mission.[23] A second on-orbit procurement of tissues from adult rats as well as developing postnatal rats flew in early 1998 as part of the Neurolab (Neurosciences Spacelab) mission. Data on ground-based rat hindlimb unloading and preliminary flight (NIH.R3 mission on STS-72) results indicate that gravity loading is required for normal development of the neuromuscular system of the weight-bearing soleus muscle, whereas loading is not critical for the non-weight-bearing extensor digitorum longus muscle. In the absence of weight-bearing, soleus muscle fibers failed to grow in size and differentiate normally into slow fibers, and elaboration of the motor nerve terminals was retarded. In addition, myoblast (satellite cell) proliferation was inhibited, producing a possible permanent deficit in the number of myonuclei per fiber.[24] In a ground-based study, postnatal day-13 rats that were suspension unloaded until postnatal day 31 developed an abnormal hindlimb gait that persisted into adulthood.[25] These findings suggest that neuromuscular development of terrestrial mammals requires timed environmental cues from gravity loading.

A picture of microgravity-induced primary changes and reloading-induced secondary changes can be deduced from the collection of human spaceflight studies, ground-based simulations of spaceflight (including prolonged bed rest), and complementary investigations of rodents subjected to spaceflight or simulated microgravity unloading by hindlimb unloading (HU). HU involves harnessing rats to elevate the hindquarters and remove weight bearing (loading) from the muscles of the hindlimbs.[26]

PRIMARY IN-FLIGHT CHANGES

Simple Deconditioning and Adaptation

The primary effects of spaceflight and HU on rodent skeletal muscles have been established by in-flight dissection and by taking tissues in ground-based models before the affected muscles have reexperienced weight bearing.[27] These changes are distinguished from secondary alterations that appear in muscle tissue obtained hours to days after return to Earth or let down from HU.[28-50] Broadly speaking, leg extensor muscles such as soleus and adductor longus lift the body against gravity (antigravity muscles). Antigravity slow-twitch muscles generally show the greatest deterioration following spaceflight and HU. These fibers have low-myosin ATPase activity and slow shortening contractions, and they are specialized for oxidative metabolism that provides fatigue resistance. In contrast, flexor muscles such as tibialis anterior and extensor digitorum longus, which contain mostly fast-twitch fibers, lack a weight-bearing function, are rapidly contracting, and are enriched with enzymes for anaerobic glycolysis. For human extensor muscles with a high proportion of fast fibers, spaceflight-induced atrophy of fast fiber types can be greater than that of slow fibers in the same muscle.[51] In microgravity, novel motor patterns evolve as both humans and rats relegate the lower (hind) limbs to perching-type activities and rely on the forelimbs for translocation. Chronic electromyographic recordings of soleus muscles of HU adult male rats show an immediate and persistent 75 percent reduction in contractile activity.[52,53] Suspended female rats exhibit a transient decrease in contractile activity, but in both sexes there is continuous unloading of soleus muscles.[54] Neglect of the lower limbs during spaceflight correlates with the loss of proprioception in weightlessness and reliance on visual feedback of position. Early in flight, humans lose anticipatory extensor muscle (electromyographic) activity in response to unannounced sudden falling.[55] Rats cease reaching for the floor when being set down. Both reflexes return quickly when the subjects are returned to gravity.

Most of the primary changes represent simple deconditioning without pathology. These changes are appropriate adaptations for efficient functioning in a low-workload microgravity environment. Slow to fast transformation of muscle fiber type and decreased fiber size would be tolerable if astronauts did not have to return, rather abruptly, to gravity and use microgravity-weakened muscles to deal with heavy workloads. Thus, effective in-flight countermeasures should have the goal of maintaining the readiness of the skeletal muscle system to handle transitions without undue injury while delivering a high level of performance.

Pathological Alteration and Metabolic Adaptation

Not all of the expected primary changes in the muscles of humans exposed to microgravity will be free of pathology. Hindlimb unloading of adult rats for 10 days caused ischemic-like necrosis of fast-twitch, oxidative glycolytic fibers in the soleus.[56] One interpretation is that the marked reduction in blood flow that occurs when a tonically active muscle becomes quiescent deprives highly oxidative fibers of sufficient blood-borne metabolites.[57] This metabolic vulnerability may be transient, because during unloading, soleus fibers gradually acquire an increased capacity for glycolytic metabolism. The metabolically adapted cells are better equipped biochemically to derive energy anaerobically and tolerate ischemia.[58-60] The reduced blood flow in and dehydration of the leg muscles cause enhanced heart rate responses.[61] The adaptation toward glycolysis is as yet unexplained and is accompanied by a compromised ability to oxidize long-chain fatty acids.[62] This shift has the downside of rendering the muscle more fatigable, even though the capacity to transport glucose is enhanced.[63,64] The shift in metabolism is evident in the increased content of glycolytic energy-deriving enzymes, elevated storage of glycogen, and disappearance of peripheral mitochondria.[65-69] Following bed rest, glycogen storage increased within the I bands and occupied spaces within myofibrils vacated by thin filaments lost during fiber atrophy.[70] Astronauts and rats returning from a short-duration spaceflight of 1 to 2 weeks may experience muscle fatigue, weakness, incoordination, and delayed-onset muscle soreness.[71-73] The greater reliance on glycolysis contributes to the reduced endurance and increased fatigability. Clearly the spaceflight-induced shift from fatty acid metabolism to glycolysis is an important issue in muscle metabolism that warrants continued investigation. Additional muscle biopsy studies are necessary to distinguish between simple deconditioning adaptive changes and pathological disruptions that invoke regenerative processes.

Contractile Physiology, Contractile Proteins, and Myofilaments

Muscle weakness following spaceflight and HU is consistent with the reported 20 to 50 percent decrease in muscle fiber cross-sectional area (CSA) and preferential loss of contractile proteins relative to cytoplasmic proteins.[74-83] Surprisingly, significant atrophy was evident in human muscles after only 5 days in space.[84] It has not been determined whether muscle deterioration reaches a plateau during long-duration spaceflight. Reduced output of muscle fiber force/CSA (specific tension) was found to be directly proportional to or greater than the reduction in CSA.[85-89] In human bed-rest subjects, specific tension was not significantly less for soleus fibers after 17 days but was down by 40 percent for the quadriceps muscle by 6 weeks.[90,91] Some reduction is expected when slow fibers shift toward fast ones because fast fibers reportedly generate lower specific tensions. More direct measures of single fiber tensions and diameters in isolated physiological preparations indicate no difference in slow and fast fiber-specific tensions, which range from 110 to 160 kN/m^2. These physiological measurements may also be distorted by the correction factor of 20 percent swelling used for skinned fibers.[92-94] Regardless, the longer-duration bed rest may have achieved a more complete fiber-type transformation. Fast fibers have more sarcoplasmic reticulum and thinner myofibrils to permit expedient and uniform exchange of Ca^{+2} ions during cycles of rapid contraction/relaxation.[95,96] The soleus muscles of 7-day-HU rats exhibited dramatic (54 percent) increases in the velocity of shortening and a 17 percent decrease in specific tension.[97] One explanation for the reduced tension is the disproportionate loss of thick (myosin-containing) filaments as documented ultrastructurally.[98] Accelerated loss of thick filaments is postulated to be the consequence of the foot-drop posture in the HU rat, which chronically shortens the working range of the soleus by ~20 percent.[99] Reorganization of sarcomeres in shortened muscles is necessary to reestablish optimal overlap (cross-bridge interaction) of thick and thin filaments for the midpoint of the abbreviated working range. Adjustment in the number of sarcomeres in series in fibers operates throughout the human's lifetime. The process is especially important for growth of fibers in length as the skeleton elongates. A second consequence of reduced packing density of contractile filaments is increased shortening velocity.[100] This may account for single-fiber physiology and biochemistry measurements, which demonstrate speeding up of slow fibers without an associated elevation of fast myosin (heavy and light chain) expression.[101,102] There is morphological and physiological evidence that the 20 percent reduction in thin filaments following a 17-day bed rest contributes to elevated shortening velocities of soleus fibers.[103] A similar reduction was detected after a 17-day spaceflight.[104] When floating in microgravity, humans are prone to foot-drop posture (ankle plantar flexion). This shortens the extensor compartment and appears to accelerate thick filament loss. The removal of weight bearing (unloading) appears to diminish thin filament concentration. Astronauts who exercise on bicycle ergometers and treadmills to preserve muscle strength and endurance also are counteracting the shortening adaptation. During these exercises, dorsiflexion of the foot stretches the soleus through its full range. Even the strength-testing sessions conducted during bed rest and in orbit, involving around 300 voluntary contractions of the foot against a strain gauge (dynamometer), may have blunted degenerative changes in soleus muscle fibers.[105-107] The health benefits of stretching muscles through their full range are well known to physiotherapists. Brief exercise can be beneficial.[108] Preparation of the musculoskeletal system for reentry to a gravity environment through appropriate exercise and mechanical stimulation should be developed.

Preservation of Function During Atrophy

Muscle tissue is truly amazing in its ability to ameliorate the loss of function during atrophy. The elevated speed of shortening, resulting from decreased contractile-filament packing density and increased fast myosin expression, compensates for the reduced force by diminishing the loss in power output (power is the product of velocity times force).[109-113] Another example is the increased capillary density and cross-sectional area of tissue that occurs when muscle fibers shrink more rapidly than the downsizing of the microvascular network.[114,115] This theoretically compensates for lower blood flow and greater susceptibility to fatigue because the average diffusion distance is decreased from the capillary to the centers of muscle fibers. Muscle fatigability is also forestalled by the slower reduction in mitochondrial content relative to contractile protein loss. This conserves the normal concentration of intermyofibrillar mitochondria and associated oxidative enzymes.[116-120]

REENTRY- AND RELOADING-INDUCED SECONDARY CHANGES

Movements in Space and Upon Return to Earth

In weightlessness, bipedal humans and quadrupedal rats both move about using the upper limbs or forelimbs.[121,122] Because of the loss of proprioception in weightlessness and the dependence on visualizing limbs for positional information, it is not surprising that the less visible hindlimbs or lower limbs are used less frequently. A novel pattern of locomotion evolves that is appropriate and sufficient for directed movements in microgravity. The weightless astronaut soon ceases reflex stepping out with a foot when moving forward. Rats pull themselves around with their forelimbs, and the hindlimbs trail outstretched behind.[123] Upon return to Earth, astronauts must reactivate 1-g motor skills such as recalling to step out to prevent falling when moving forward. Immediately upon their return to Earth, they are very unstable from a combination of orthostatic intolerance, altered otolith-spinal reflexes, reliance on weakened atrophic muscles, and inappropriate motor patterns.[124,125] In the first few hours after landing, spaceflown rats do not extend their limbs and reach for the ground when lowered to the ground. This reflex, beneficial in 1 g, returns in a day.[126] Spaceflown rats walked significantly more slowly than normal the first 2 days, but they moved as rapidly as ground controls by the third day. The jerky, stilted stepping of the hindlimbs quickly evolved to the smooth walking pattern of a terrestrially readapted rat.[127] Unloading studies with immature rats indicate that gravity loading during the third and fourth weeks after birth is essential for normal development of 1-g locomotion.[128] Early in flight, humans subjected to sudden "drop tests" ceased anticipatory contractile (electromyographic) activity in extensor muscles.[129] This reflex returned to normal within a day after landing. Thus, terrestrial motor skills are rapidly restored and performed forcefully. This occurs well before muscle-fiber regrowth in cross-sectional areas and during the period of slow muscle-fiber necrosis. It appears that the central nervous system undergoes significant reprogramming (plasticity) and performs compensatory activation of motor units that masks the deteriorated state of the neuromuscular system.[130-133] There may also be rapid changes in the neuromuscular junctions. The physiological and morphological adaptations at the nerve/muscle synapse under conditions of reduced or elevated use have not been adequately examined.[134]

Compromised Microcirculation

The headward fluid shift in microgravity and reduced muscle contractions (musculovenous pumping) during unloading result in reduced blood flow in the lower (hind) limbs.[135-137] This is associated with the movement of blood proteins, such as albumin, into the interstitium. Without musculovenous pumping, extravasated proteins are recovered less well into the vascular system via postcapillary venules and lymphatics.[138-140] At the shortest time examined after shuttle landing (0.5 hour after reentry plus 2 hours after wheel stop), the slow adductor longus muscles of rats already showed simple (noninflammatory) interstitial edema that was not evident in flight.[141] By 2 days postflight, the edema had advanced to inflammatory myopathy with more severe edema. The scenario of reloading-induced edema is mimicked by HU for 12.5 days and subsequent reloading of rat antigravity slow muscles.[142] Postflight pooling of blood in the lower limbs was not present in the legs of astronauts during quiet standing initiated about 4 hours after landing.[143,144] This indicates that the pull of gravity (hydrostatic pressure) alone is not sufficient in the short term to cause fluid accumulation in relatively quiescent limbs. The onset and severity of interstitial edema in rats appears directly related to the intensity of postflight muscle contractile activity.[145-150] The osmotic pressure of the extravascular proteins is thought to pull water into the interstitium when the microvascular network is reperfused with blood at high pressure and flow in response to resumption of strong muscle contractions.[151,152] If the muscle activity is sufficiently strenuous (a level that is undefined), interstitial edema increases and leads to ischemic-like tissue necrosis with mast cell degranulation and increased vascular permeability.[153] At this stage, muscles exhibit inflammatory-like myopathy with infiltration of mononuclear cells.[154-156] Neutrophils may exacerbate damage by activating complement-mediated membrane disruption and releasing reactive oxygen radicals.[157]

Muscles adapted to 1 *g* can also become edematous during intense exercise.[158,159] Perhaps adaptation to microgravity lowers a threshold for the onset of edema. It appears that underused microvessels adapt to low flow and pressure during spaceflight or HU and may become inherently "more leaky" during the rapid onset of high blood flow and pressure when gravity-loaded muscle contractions are resumed.[160-162] Furthermore, the unloading-induced shift to glycolytic metabolism away from oxidative metabolism results in a more robust stimulation of blood pressure during muscle contraction.[163] The results suggest that avoiding strenuous muscle contractions during reacclimatization to gravity and possibly medicating before landing with drugs that block mast cell degranulation may minimize edema in returning astronauts. Possible countermeasures that should be investigated to minimize reentry and reloading-induced edema and ischemic tissue necrosis include flushing the interstitium of excess proteins by exercise-induced muscle contractions, combined with lower body negative pressure (LBNP) induction of microcirculation filling. LBNP is achieved by sealing the lower half of the body in a chamber in which air pressure is lowered to draw blood into the lower extremities. LBNP regimens and loaded treadmill exercises routinely performed by Russian cosmonauts are believed by them to be essential for successful readaptation to gravity after a year in space.[164] However, these promising countermeasures, including stationary bicycle and rowing exercises used on other missions, require further controlled studies to verify their efficacy. This circulation-related problem emphasizes the point that muscle adaptation during spaceflight goes beyond changes restricted to muscle fibers (cells). In fact, nervous and cardiovascular system involvement in muscle performance serves as a reminder that we are sending organisms, not isolated organ systems, into space. Effective countermeasures will need to target multiple systems.

Increased Susceptibility to Structural Damage

Atrophic muscle fibers resulting from spaceflight and hindlimb unloading are structurally weaker and more susceptible to eccentric-like (lengthening) contraction-induced tearing of the contractile elements, the fiber cell membrane (sarcolemma), and associated connective tissue.[165-169] The severity of the damage appears directly correlated to the magnitude of the reloading workload. These tissue changes are reminiscent of those associated with delayed-onset muscle soreness in 1-*g*-adapted human muscles after unaccustomed strenuous exercise (especially muscle lengthening) and in rat muscles electrically stimulated to generate forceful eccentric contractions.[170-173] Some astronauts are aware that minimizing activities that eccentrically load their leg muscles, such as walking down stairs during the first days back on Earth, reduces the severity of delayed-onset soreness and stiffness.[174] Adaptation to the lower workload history of microgravity or HU appears to render muscle tissue more prone to structural failure when reloaded, especially by lengthening contractions.[175-180] This is partly explained by the relatively greater workload on the antigravity muscles because of fiber atrophy.[181,182] A

50 percent loss of soleus mass is equivalent to increasing muscle loading by doubling the body weight. Further studies on Earth subjecting 1-*g*-adapted animals to hypergravity on a variable force centrifuge would be useful.

In addition to the relative increase in workload, however, there still may be a lowering of a threshold for structural failure so that muscle fibers are damaged at a level of specific tension previously tolerated without injury.[183,184] The disproportionate loss of actin thin filaments and increased tension per filament in human soleus muscles, after spaceflight and chronic bed rest, is one possible molecular mechanism of lowering the injury threshold. Another possibility is a disproportionate decrease in structural proteins (harnessing tension) relative to the contractile proteins (generating tension).[185] There is evidence for unlinking of the cytoskeleton (actin and intermediate filaments) and myofibrils to costameres.[186] Postflight damage, sports-related injuries, and degeneration in muscular dystrophies share structural similarities.[187-189] Skeletal muscles are capable of generating more force than the connective tissue (interstitium and myotendinous junctions) can tolerate without structural failure.[190] For both spaceflight and HU, the atrophic degenerative changes that have been documented at the myotendinous junctions and along the length of the fibers are consistent with a reduced margin of safety for structural integrity during weight-bearing contractions.[191-194]

Structural proteins can fail within sarcomeres. This process has been thoroughly reviewed for 1-*g*-adapted muscles.[195] Null allele mutations to eliminate selected sarcomeric proteins provide valuable insights into the mechanism of myofibril assembly.[196,197] The sarcolemma (cell membrane and basal lamina) is another potentially weakened component. At costameres and myotendinous junctions, cytoskeletal actin and intermediate filaments normally transmit contractile force through linking proteins to integral membrane glycoproteins that bind to extracellular matrix (integrins to fibronectin and dystrophin/dystroglycan to laminin-2).[198-202] These sites may also function as receptors for sensing workload and generate intracellular signals through enzymes like focal adhesion kinase that affect the synthesis of myofibrillar proteins.[203] Mechanosensing of workload by heart myocytes (likely to be analogous for skeletal muscle) involves autocrine/paracrine growth factors signal transduction, immediate-early genes, and multiple second messengers (phospholipase, eicosinoids, tyrosine kinases, RAF-1, MAP kinases, stretch-activated ion channels, and protein translation regulators).[204] The in vitro model of applying stretch to skeletal muscle cells has shown a convincing correlation between active tension and prostaglandin release; these autocrine factors modulate protein degradation and synthesis.[205] Absence of a single component protein of the dystrophin glycoprotein complex can result in greater susceptibility to contraction-induced sarcolemma tearing.[206-208] This is seen in human dystrophies and mouse dystrophy mutants.[209] Muscles of the mdx dystrophic mouse are more easily torn during contraction than the same muscles in normal animals.[210] HU renders the atrophic soleus muscle more susceptible to contraction-induced muscle tearing.[211] The nonatrophic extensor digitorum longus muscle in the same animals did not exhibit increased vulnerability.[212] Fortunately, genetically normal muscle fibers rapidly repair sarcomere lesions by Z-line-like patching and restore segmental necrosis by membrane sealing and satellite cell regeneration.[213-217] In human muscles, the number of satellite cell divisions normally may be limited. Dystrophic muscles are compromised in these reparative abilities and therefore undergo more extensive and persistent degeneration. Spaceflight-induced muscle atrophy may also compromise the healing capacity of injured muscles.[218] Investigations pursuing means of promoting repair processes with growth factors and pharmacological agents are needed to minimize the negative impact of muscle injury on astronaut performance.[219] Extravehicular activities (space walks) require significant muscular effort, making astronauts vulnerable to muscular injury. This is an important consideration because repeated space walks are obvious requirements for space station construction. During manned exploration of the surface of Mars, performance could be negatively affected by such injuries, even those as commonplace as painful pulled back muscles or stiff neck muscles. Muscle soreness in astronauts needs to be examined rigorously.

Recommendations

• *The efficacy of muscle repair in microgravity should be more thoroughly examined at the sarcomere, myofiber, motor neuron, and tissue levels.*

• *Myotendinous junctions as structural failure sites warrant further study to prevent tendon rupture during transition to gravity reloading.*

• *Development of improved countermeasures is a high-priority goal, and should include the following:*

—*Consideration of the potential value of hormones, growth factors, second messengers, and drugs in the design of novel, more effective countermeasures; and*

—*Improved understanding of mechanisms of microcirculation failure, which is required for design of in-flight countermeasures for preventing reloading damage.*

• *The process of transitioning from microgravity to higher gravity without undue damage and loss of performance needs to be explored by examination of microgravity-adapted individuals transitioning to 1 g and 1-g-adapted individuals exposed to hypergravity by centrifugation.*

• *NASA should actively work with the scientific community and industry to develop equipment to reduce in-flight muscle stress and fatigue as well as instrumentation necessary for in-flight experimentation and monitoring of muscle function and health. Examples include the following:*

—*Rapid freezing apparatus to preserve biospecimens obtained in flight for later biochemical and molecular assays;*

—*An ergonomic space suit for hand movements to reduce fatigue; and*

—*Dual-beam x ray and ultrasonography for noninvasive monitoring of muscle deterioration and real-time assessment of the efficacy of countermeasures.*

• *An astronaut's history of muscle use and condition needs to be better monitored before, during, and after flight to reduce uncontrolled parameters that introduce excessive intersubject variation for statistical detection of changes.*

CELLULAR AND MOLECULAR MECHANISMS

Up to now, researchers have only minimally exploited microgravity for advancing the understanding of muscle biology. Microgravity has proven an excellent tool for noninvasively perturbing the synthesis of muscle proteins in the search for molecular signals and gene regulatory factors influencing differentiation, growth, maintenance, plasticity, and atrophy of muscle. The relationships between blood flow history and interstitial edema and between workload history and structural failure are but two of the important problems that require serious attention. The roles of hormones and growth factors as well as of fiber-type-specific *cis*- and *trans*-acting factors in regulating gene expression, and the microgravity-induced alterations in their production, are other urgent issues. These types of studies will yield information that will advance basic knowledge of muscle biology and offer insights into countermeasure design. This knowledge is also likely to assist in the rehabilitation of diseased or injured muscles in humans on Earth, especially individuals in the more vulnerable aging population or those participating in strenuous sports. Microgravity can be exploited as a unique tool to perturb the normal adult and developing neuromuscular systems to derive insights into fundamental muscle function on Earth, as well as to define the epigenetic requirement of gravity for the maintenance and differentiation of the neuromuscular system. The past two decades of spaceflight research have yielded a description of the physiological effects of microgravity on multiple systems. The next generation of experiments must define the cellular and molecular mechanisms of microgravity-induced changes.

The fundamental questions regarding control of muscle function, causes of atrophic changes, regulation of muscle phenotype, the mechanism of muscle plasticity in response to altered demands, and identification of optimal countermeasures will require rigorous ground-based research as well as space-flight studies. The bed-rest paradigm for humans and the hindlimb unloading model for rats have provided high-fidelity ground-based models for many but not all aspects of spaceflight. These models can continue to produce significant information. Multidisciplinary approaches should be encouraged that probe the molecular changes in regulation of transcription and translation in individual muscle cells and the mechanisms whereby other organ systems (vascular, hormonal, neural) interact to bring about changes in gene expression and cellular function. However, no Earth-based model of microgravity is sufficiently faithful to escape some influences of the gravity vector. For example, gravity-induced postural distortion of the hindleg in hindlimb-suspension unloaded rats introduces a shortened working range for the soleus with associated changes in muscle structure and physiology. It will therefore be important to conduct in-flight experiments to test molecular and cellular mechanisms of response identified in ground-based models.

Studies on regulation of gene expression will be increasingly important. Molecular tools are available to probe properties specific to a particular muscle fiber type; sensing of muscle length and load; and influence of autocrine, paracrine, and humoral factors. Transgenic mice with over- and underexpression of specific molecules thought to be involved in the adaptation process are invaluable research animals to incorporate into the ground and flight studies.[220,221] The roles played by membrane receptors or ligands and cytoskeletal elements (costameres, dystrophin glycoprotein complex, integrins) in load bearing and length sensing need to be understood.[222] Degradation of muscle proteins by ubiquitination and the proteolytic proteasome complex is a fundamental process of degrading contractile proteins during atrophy that requires further investigation. Factors responsible for the shift from lipid oxidation to glucose utilization need to be understood from the viewpoint of control of energy metabolism. Development from conception to the mature adult and the F1 generations are needed to determine whether critical periods require gravitation influences for normal gene expression.[223] As described in other chapters of this report, a key question is how cells detect the effects of gravity. The myotendinous junctions and costameres are the most likely sites at which loading imparts stresses on the connective tissues and tendons and the basement membrane, which signals through integral membrane proteins to the cytoskeleton and contractile proteins.[224-226] These pathways will involve multiple second messengers, such as signaling through kinase activation and eicosinoids.[227] Mechanisms of sensing working length remain undefined.

Recommendations

1. The highest priority for research projects should be those investigations designed to elucidate the cellular and molecular mechanisms underlying muscle weakness, fatigue, incoordination, and delayed-onset muscle soreness. These deficits are associated with muscle cell atrophy, greater susceptibility of muscle fibers to contraction-induced destruction of fibers, and a compromised microcirculation leading to ischemic necrosis and secondary changes in muscle tissue.

2. Ground-based models such as bed rest for humans and hindlimb unloading for rats should continue to be used for testing and refining hypotheses seeking to understand the fundamental mechanisms of how workload is transduced into molecular signals regulating muscle mass and protein isoform expression, as well as the cellular and molecular control myofilament assembly and length. Integrated studies employing multiple intra- and interdisciplinary approaches are encouraged.

3. Fundamental work is needed on myogenesis, fiber type differentiation, neuromuscular development, and spinal development using neonatal rodents as model systems. The mechanisms should be determined whereby muscle cells sense working length and the mechanical stress of gravity. Signal transduction pathways for growth factors, stretch-activated ion channels, regulators of protein synthesis, and interactions of extracellular matrix and membrane proteins with cytoskeleton should be investigated.

4. In-flight and hindlimb unloading testing of genetically engineered murine models with gain and loss of function, up regulation and down regulation of key proteins, and release of candidate hormones and growth factors and their respective receptors should be exploited for probing basic mechanisms of neuromuscular function during adaptation to unique hypo- and hypergravity environments.

REFERENCES

1. Baldwin, K.M., White, T.P., Arnaud, S.B., Edgerton, V.R., Kraemer, W.J., Kram, R., Raab-Cullen, D., and Snow, C.M. 1996. Musculoskeletal adaptations to weightlessness and development of effective countermeasures. Med. Sci. Sports Exercise 10: 1247-1253.

2. Convertino, V.A. 1991. Neuromuscular aspects in development of exercise countermeasures. Physiologist 34: S125-S128.

3. Ferrando, A.A., Tipton, K.D., Bamman, M.M., and Wolfe, R.R. 1997. Resistance exercise maintains skeletal muscle protein synthesis during bed rest. J. Appl. Physiol. 82: 807-810.

4. Kablar, B., Krastel, K., Ying, C., Asakura, A., Tapscott, S., and Rudnicki, M. 1997. MyoD and Myf-5 differentially regulate the development of limb versus trunk skeletal muscle. Dev. Suppl. 124: 4729-4738.

5. Buonanno, A., and Rosenthal, N. 1996. Molecular control of muscle diversity and plasticity. Dev. Genet. 19: 95-107.

6. Hoh, J.F., and Hughes, S. 1989. Immunocytochemical analysis of the perinatal development of cat masseter muscle using anti-myosin antibodies. J. Muscle Res. Cell Motil. 10: 312-325.

7. Wright, C., Haddad, F., Qin, A.X., and Baldwin, K.M. 1997. Analysis of myosin heavy chain mRNA expression by RT-PCR. J. Appl. Physiol. 83: 1389-1396.

8. Hoh, J.F., and Hughes, S. 1989. Immunocytochemical analysis of the perinatal development of cat masseter muscle using anti-myosin antibodies. J. Muscle Res. Cell Motil. 10: 312-325.

9. Wright, C., Haddad, F., Qin, A.X., and Baldwin, K.M. 1997. Analysis of myosin heavy chain mRNA expression by RT-PCR. J. Appl. Physiol. 83: 1389-1396.

10. Schiaffino, S., and Salviati, F. 1998. Molecular diversity of myofibrillar proteins: isoforms analysis at the protein and mRNA level. Methods Cell Biol. 52: 349-369.

11. Hoh, J.F., and Hughes, S. 1989. Immunocytochemical analysis of the perinatal development of cat masseter muscle using anti-myosin antibodies. J. Muscle Res. Cell Motil. 10: 312-325.

12. Caiozzo, V.J. 1996. Thyroid hormone: Modulation of muscle structure, function, and adaptive responses to mechanical loading. Exercise Sport Sci. Rev. 24: 321-361.

13. Devor, S.T., and White, T.P. 1996. Myosin heavy chain of immature soleus muscle grafts adapts to hyperthyroidism more than to physical activity. J. Appl. Physiol. 80: 789-794.

14. Buonanno, A., and Rosenthal, N. 1996. Molecular control of muscle diversity and plasticity. Dev. Genet. 19: 95-107.

15. Buonanno, A., and Rosenthal, N. 1996. Molecular control of muscle diversity and plasticity. Dev. Genet. 19: 95-107.

16. Engert, J.C., Servaes, S., Sutrave, P., Hughes, S.H., and Rosenthal, N. 1995. Activation of a muscle-specific enhancer by the Ski proto-oncogene. Nucleic Acids Res. 23: 2988-2994.

17. Koslovskaya, I.B., Barmin, V.A., Stepantsov, V.I., and Kharitinov, N.M. 1990. Results of studies of motor functions in long-term spaceflights. Physiologist 33: S1-S3.

18. Thorteinsson, G. 1997. Management of postpolio syndrome. Mayo Clin. Proc. 72: 627-638.

19. Edgerton, V.R., Zhou, M.-Y., Ohira, Y., Klitgaard, H., Jiang, B., Bell, G., Harris, B., Saltin, B., Gollnick, P.D., Roy, R.R., Day, M.K., and Greenisen, M. 1995. Human fiber size and enzymatic properties after 5 and 11 days of spaceflight. J. Appl. Physiol. 78: 1733-1739.

20. Fitts, R.H., Widrick, J.J., Knuth, S.T., Blaser, C.A., Karhanek, M., Trappe, S.W., Trappe, T.A., and Costill, D.L. 1997. Force-velocity and force-power properties of human muscle fiber after spaceflight. Med. Sci. Sports Exercise 29: S190.

21. Trappe, S.W., Trappe, T.A., Costill, D.L., Lee, G.A., Widrick, J.J., and Fitts, R.H. 1997. Effect of spaceflight on human calf muscle morphology and function. Med. Sci. Sports Exercise 29: S190.

22. Zhou, M.-Y., Klitgaard, H., Saltin, B., Roy, R.R., Edgerton, V.R., and Gollnick, P.D. 1995. Myosin heavy chain isoforms of human muscle after short-term spaceflight. J. Appl. Physiol. 78: 1740-1744.

23. Riley, D.A., Thompson, J.L., Krippendorf, B.B., and Slocum, G.R. 1995. Review of spaceflight and hindlimb unloading induced sarcomere damage and repair. Basic Appl. Myology 5: 139-145.

24. Darr, K.C., and Schultz, E. 1989. Hindlimb suspension suppresses muscle growth and satellite cell proliferation. J. Appl. Physiol. 67: 1827-1834.

25. Walton, K., Heffernan, C., Sluice, D., and Benavides, L. 1997. Changes in gravity influenced rat postnatal motor system development: From simulation to spaceflight. Gravit. Space Biol. Bull. 10: 111-118.

26. Fitts, R.H., Metzger, J.M., Riley, D.A., and Unsworth, B.R. 1986. Models of skeletal muscle disuse: A comparison of suspension hypokinesia and hindlimb immobilization. J. Appl. Physiol. 60: 1946-1953.

27. Krippendorf, B.B., and Riley, D.A. 1993. Distinguishing unloading- versus reloading-induced changes in rat soleus muscle. Muscle Nerve 16: 99-108.

28. Edgerton, V.R., Zhou, M.-Y., Ohira, Y., Klitgaard, H., Jiang, B., Bell, G., Harris, B., Saltin, B., Gollnick, P.D., Roy, R.R., Day, M.K., and Greenisen, M. 1995. Human fiber size and enzymatic properties after 5 and 11 days of spaceflight. J. Appl. Physiol. 78: 1733-1739.

29. Riley, D.A., Slocum, G.R., Bain, J.L.W., Sedlak, F.R., Sowa, T.E., and Mellender, J.W. 1990. Rat hindlimb unloading: Soleus histochemistry, ultrastructure, and electromyography. J. Appl. Physiol. 69: 58-66.

30. Fitts, R.H., Widrick, J.J., Knuth, S.T., Blaser, C.A., Karhanek, M., Trappe, S.W., Trappe, T.A., and Costill, D.L. 1997. Force-velocity and force-power properties of human muscle fiber after spaceflight. Med. Sci. Sports Exercise 29: S190.

31. Trappe, S.W., Trappe, T.A., Costill, D.L., Lee, G.A., Widrick, J.J., and Fitts, R.H. 1997. Effect of spaceflight on human calf muscle morphology and function. Med. Sci. Sports Exercise 29: S190.

32. Zhou, M.-Y., Klitgaard, H., Saltin, B., Roy, R.R., Edgerton, V.R., and Gollnick, P.D. 1995. Myosin heavy chain isoforms of human muscle after short-term spaceflight. J. Appl. Physiol. 78: 1740-1744.

33. Baldwin, K.M., Herrick, R.E., and McCue, S.A. 1993. Substrate oxidation capacity in rodent skeletal muscle: Effects of exposure to zero gravity. J. Appl. Physiol. 75: 2466-2470.

34. Baranski, S., Baranska, W., Marciniak, M., and Ilyina-Kakueva, E.I. 1979. Ultrasonic (ultrastructural) investigations of the soleus muscle after spaceflight on the Biosputnik 936. Aviat. Space Environ. Med. 50: 930-934.

35. Ilyina-Kakueva, E.I., Portugalov, V.V., and Kirvenkova, N.P. 1976. Spaceflight effects on the skeletal muscle of rats. Aviat. Space Environ. Med. 47: 700,703.

36. Martin, I.P., Edgerton, V.R., and Grindeland, R.E. 1988. Influence of spaceflight on rat skeletal muscle. J. Appl. Physiol. 65: 2318-2325.

37. Riley, D.A., Ellis, S., Slocum, G.R., Satyanarayana, T., Bain, J.L.W., and Sedlak, F.R. 1987. Hypogravity-induced atrophy of rat soleus and extensor digitorum longus muscles. Muscle Nerve 10: 560-568.

38. Riley, D.A., Ilyina-Kakueva, E.I., Ellis, S., Bain, J.L.W., Slocum, G.R., and Sedlak, F.R. 1990. Skeletal muscle fiber, nerve, and blood vessel breakdown in space-flown rats. FASEB J. 4: 84-91.

39. Riley, D.A., Ellis, S., Giometti, C.S., Hoh, J.F.Y., Ilyina-Kakueva, E.I., Oganov, V., Slocum, G.R., Bain, J.L.W., and Sedlak, F.R. 1992. Muscle sarcomere lesions and thrombosis after spaceflight and suspension unloading. J. Appl. Physiol. 73: S33-S43.

40. Riley, D.A., Thompson, J.L., Krippendorf, B.B., and Slocum, G.R. 1995. Review of spaceflight and hindlimb suspension unloading induced sarcomere damage and repair. Basic Appl. Myology 5: 139-145.

41. Riley, D.A., Ellis, S., Slocum, G.R., Sedlak, F.R., Bain, J.L.W., Krippendorf, B.B., Lehman, C.T., Macias, M.Y., Thompson, J.L., Vijayan, K., and DeBruin, J.A. 1996. In-flight and postflight changes in skeletal muscles of SLS-1 and SLS-2 spaceflown rats. J. Appl. Physiol. 81: 133-144.

42. Tischler, M.E., Henriksen, E.J., Munoz, K.A., Stump, C.S., Woodman, C.R., and Kirby, C.R. 1993. Spaceflight on STS-48 and Earth-based unweighting produce similar effects on skeletal muscle of young rats. J. Appl. Physiol. 74: 2161-2165.

43. Caiozzo, V.J., Haddad, F., Baker, M.J., Herrick, R.E., Prietto, N., and Baldwin, K.M. 1996. Microgravity-induced transformations of myosin isoforms and contractile properties of skeletal muscle. J. Appl. Physiol. 81: 123-132.

44. Diffee, G.M., Caiozzo, V.J., Herrick, R.E., and Baldwin, K.M. 1991. Contractile and biochemical properties of rat soleus and plantaris after hindlimb suspension. Am. J. Physiol. 260: C528-C534.

45. Diffee, G.M., Haddad, F., Herrick, R.E., and Baldwin, K.M. 1991. Control of myosin heavy chain expression: Interaction of hypothyroidism and hindlimb suspension. Am. J. Physiol. 261: C1099-C1106.

46. Krippendorf, B.B., and Riley, D.A. 1993. Distinguishing unloading- versus reloading-induced changes in rat soleus muscle. Muscle Nerve 16: 99-108.

47. Krippendorf, B.B., and Riley, D.A. 1994. Temporal changes in sarcomere lesions of rat adductor longus muscles during hindlimb reloading. Anat. Rec. 238: 304-310.

48. McDonald, K.S., Blaser, C.A., and Fitts, R.H. 1994. Force-velocity and power characteristics of rat soleus muscle fibers after hindlimb suspension. J. Appl. Physiol. 77: 1609-1616.

49. McDonald, K.S., Delp, M.D., and Fitts, R.H. 1992. Effect of hindlimb unweighting on tissue blood flow in the rat. J. Appl. Physiol. 72: 2210-2218.

50. Riley, D.A., Slocum, G.R., Bain, J.L.W., Sedlak, F.R., Sowa, T.E., and Mellender, J.W. 1992. Rat hindlimb unloading: Soleus histochemistry, ultrastructure and electromyography. J. Appl. Physiol. 69: 58-66.

51. Edgerton, V.R., Zhou, M.-Y., Ohira, Y., Klitgaard, H., Jiang, B., Bell, G., Harris, B., Saltin, B., Gollnick, P.D., Roy, R.R., Day, M.K., and Greenisen, M. 1995. Human fiber size and enzymatic properties after 5 and 11 days of spaceflight. J. Appl. Physiol. 78: 1733-1739.

52. Riley, D.A., Slocum, G.R., Bain J.L.W., Sedlak, F.R., Sowa, T.E., and Mellender, J.W. 1990. Rat hindlimb unloading: Soleus histochemistry, ultrastructure, and electromyography. J. Appl. Physiol. 69: 58-66.

53. Blewett, C., and Elder, G.C. 1993. Quantitative EMG analysis in soleus and plantaris during hindlimb suspension and recovery. J. Appl. Physiol. 74: 2057-2066.

54. Alford, E.K., Roy, R.R., Hodgson, J.A., and Edgerton, V.R. 1987. Electromyography of rat soleus, medial gastrocnemius, and tibialis anterior during hind limb suspension. Exp. Neurol. 96: 635-649.

55. Young, L.R., Oman, C.M., Watt, D.G.D., Money, K.E., Lichtenberg, B.K., Kenyon, R.V., and Arrott, A.P. 1986. M.I.T./Canadian vestibular experiments on the Spacelab-1 mission: 1. Sensory adaptation to weightlessness and readaptation to one-g: An overview. Exp. Brain Res. 64: 291-298.

56. Riley, D.A., Slocum, G.R., Bain, J.L.W., Sedlak, F.R., Sowa, T.E., and Mellender, J.W. 1990. Rat hindlimb unloading: Soleus histochemistry, ultrastructure, and electromyography. J. Appl. Physiol. 69: 58-66.

57. McDonald, K.S., Delp, M.D., and Fitts, R.H. 1992. Effect of hindlimb unweighting on tissue blood flow in the rat. J. Appl. Physiol. 72: 2210-2218.

58. Baldwin, K.M., Herrick, R.E., and McCue, S.A. 1993. Substrate oxidation capacity in rodent skeletal muscle: Effects of exposure to zero gravity. J. Appl. Physiol. 75: 2466-2470.

59. Baranski, S., Baranska, W., Marciniak, M., and Ilyina-Kakueva, E.I. 1979. Ultrasonic (ultrastructural) investigations of the soleus muscle after spaceflight on the Biosputnik 936. Aviat. Space Environ. Med. 50: 930-934.

60. Riley, D.A., Ellis, S., Slocum, G.R., Satyanarayana, T., Bain, J.L.W., and Sedlak, F.R. 1987. Hypogravity-induced atrophy of rat soleus and extensor digitorum longus muscles. Muscle Nerve 10: 560-568.

61. Wilson, L.B., Dyke, C.K., Parsons, D., Wall, P.T., Pawelczyk, J.A., Williams, R.S., and Mitchell, J.H. 1995. Effect of skeletal muscle fiber type on the pressor response evoked by static contraction in rabbits. J. Appl. Physiol. 79: 1744-1752.

62. Baldwin, K.M., Herrick, R.E., and McCue, S.A. 1993. Substrate oxidation capacity in rodent skeletal muscle: Effects of exposure to zero gravity. J. Appl. Physiol. 75: 2466-2470.

63. Baldwin, K.M., Herrick, R.E., and McCue, S.A. 1993. Substrate oxidation capacity in rodent skeletal muscle: Effects of exposure to zero gravity. J. Appl. Physiol. 75: 2466-2470.

64. Tischler, M.E., Henriksen, E.J., Munoz, K.A., Stump, C.S., Woodman, C.R., and Kirby, C.R. 1993. Spaceflight on STS-48 and Earth-based unweighting produce similar effects on skeletal muscle of young rats. J. Appl. Physiol. 74: 2161-2165.

65. Edgerton, V.R., Zhou, M.-Y., Ohira, Y., Klitgaard, H., Jiang, B., Bell, G., Harris, B., Saltin, B., Gollnick, P.D., Roy, R.R., Day, M.K., and Greenisen, M. 1995. Human fiber size and enzymatic properties after 5 and 11 days of spaceflight. J. Appl. Physiol. 78: 1733-1739.

66. Martin, I.P., Edgerton, V.R., and Grindeland, R.E. 1988. Influence of spaceflight on rat skeletal muscle. J. Appl. Physiol. 65: 2318-2325.

67. Riley, D.A., Ellis, S., Slocum, G.R., Satyanarayana, T., Bain, J.L.W., and Sedlak, F.R. 1987. Hypogravity-induced atrophy of rat soleus and extensor digitorum longus muscles. Muscle Nerve 10: 560-568.

68. Riley, D.A., Ilyina-Kakueva, E.I., Ellis, S., Bain, J.L.W., Slocum, G.R., and Sedlak, F.R. 1990. Skeletal muscle fiber, nerve, and blood vessel breakdown in space-flown rats. FASEB J. 4: 84-91.

69. Riley, D.A., Ellis, S., Giometti, C.S., Hoh, J.F.Y., Ilyina-Kakueva, E.I., Oganov, V., Slocum, G.R., Bain, J.L.W., and Sedlak, F.R. 1992. Muscle sarcomere lesions and thrombosis after spaceflight and suspension unloading. J. Appl. Physiol. 73: S33-S43.

70. Widrick, J.J., Romatowski, J.G., Bain, J.L.W., Trappe, S.W., Trappe, T.A., Thompson, J.L., Costill, D.L., Riley, D.A., and Fitts, R.H. 1997. Effect of 17 days of bed rest on peak isometric force and unloaded shortening velocity of human soleus fibers. Am. J. Physiol. 273: C1690-C1699.

71. Edgerton, V.R., and Roy, R.R. 1994. Neuromuscular adaptation to actual and simulated weightlessness. Pp. 33-67 in Advances in Space Biology and Medicine, Vol. 4 (S.L. Bonting, ed.). JAI Press, Greenwich, Conn.

72. Riley, D.A., Thompson, J.L., Krippendorf, B.B., and Slocum, G.R. 1995. Review of spaceflight and hindlimb suspension unloading induced sarcomere damage and repair. Basic Appl. Myology 5: 139-145.

73. Stauber, W.T., Clarkson, P.M., Fritz, V.K., and Evans, W.J. 1990. Extracellular matrix disruption and pain after eccentric muscle action. J. Appl. Physiol. 69: 868-874.

74. Edgerton, V.R., and Roy, R.R. 1994. Neuromuscular adaptation to actual and simulated weightlessness. Pp. 33-67 in Advances in Space Biology and Medicine, Vol. 4 (S.L. Bonting, ed.). JAI Press, Greenwich, Conn.

75. Koslovskaya, I.B., Barmin, V.A., Stepantsov, V.I., and Kharitinov, N.M. 1990. Results of studies of motor functions in long-term spaceflights. Physiologist 33: S1-S3.

76. Thornton, W.E., and Rummel, J.A. 1977. Muscular deconditioning and its prevention in spaceflight. Pp. 191-197 in Biomedical Results from Skylab (R.S. Johnston and L.F. Dietlein, eds.). NASA-SP-377. National Aeronautics and Space Administration, Houston, Tex.

77. Baranski, S., Baranska, W., Marciniak, M., and Ilyina-Kakueva, E.I. 1979. Ultrasonic (ultrastructural) investigations of the soleus muscle after spaceflight on the Biosputnik 936. Aviat. Space Environ. Med. 50: 930-934.

78. Ilyina-Kakueva, E.I., Portugalov, V.V., and Kirvenkova, N.P. 1976. Spaceflight effects on the skeletal muscle of rats. Aviat. Space Environ. Med. 47: 700,703.

79. Martin, I.P., Edgerton, V.R., and Grindeland, R.E. 1988. Influence of spaceflight on rat skeletal muscle. J. Appl. Physiol. 65: 2318-2325.

80. Riley, D.A., Ilyina-Kakueva, E.I., Ellis, S., Bain, J.L.W., Slocum, G.R., and Sedlak, F.R. 1990. Skeletal muscle fiber, nerve, and blood vessel breakdown in space-flown rats. FASEB J. 4: 84-91.

81. Riley, D.A., Ellis, S., Giometti, C.S., Hoh, J.F.Y., Ilyina-Kakueva, E.I., Oganov, V., Slocum, G.R., Bain, J.L.W., and Sedlak, F.R. 1992. Muscle sarcomere lesions and thrombosis after spaceflight and suspension unloading. J. Appl. Physiol. 73: S33-S43.

82. Riley, D.A., Thompson, J.L., Krippendorf, B.B., Slocum, G.R. 1995. Review of spaceflight and hindlimb suspension unloading induced sarcomere damage and repair. Basic Appl. Myology 5: 139-145.

83. Riley, D.A., Ellis, S., Slocum, G.R., Sedlak, F.R., Bain, J.L.W., Krippendorf, B.B., Lehman, C.T., Macias, M.Y., Thompson, J.L., Vijayan, K., and DeBruin, J.A. 1996. In-flight and postflight changes in skeletal muscles of SLS-1 and SLS-2 spaceflown rats. J. Appl. Physiol. 81: 133-144.

84. Edgerton, V.R., Zhou, M.-Y., Ohira, Y., Klitgaard, H., Jiang, B., Bell, G., Harris, B., Saltin, B., Gollnick, P.D., Roy, R.R., Day, M.K., and Greenisen, M. 1995. Human fiber size and enzymatic properties after 5 and 11 days of spaceflight. J. Appl. Physiol. 78: 1733-1739.

85. Diffee, G.M., Caiozzo, V.J., Herrick, R.E., and Baldwin, K.M. 1991. Contractile and biochemical properties of rat soleus and plantaris after hindlimb suspension. Am. J. Physiol. 260: C528-C534.

86. McDonald, K.S., Blaser, C.A., and Fitts, R.H. 1994. Force-velocity and power characteristics of rat soleus muscle fibers after hindlimb suspension. J. Appl. Physiol. 77: 1609-1616.

87. Widrick, J.J., and Fitts, R.H. 1997. Peak force and maximal shortening velocity of soleus fibers after non-weight-bearing and resistance exercise. J. Appl. Physiol. 82: 189-195.

88. Widrick, J.J., Romatowski, J.G., Bain, J.L.W., Trappe, S.W., Trappe, T.A., Thompson, J.L., Costill, D.L., Riley, D.A., and Fitts, R.H. 1997. Effect of 17 days bedrest on peak isometric force and maximal shortening velocity of human soleus fibers. Am. J. Physiol. 273: C1690-C1699.

89. Larsson, L., Li, X., Berg, H.E., and Frontera, W.R. 1996. Effects of removal of weight-bearing function on contractility and myosin isoform composition in single human skeletal muscle cells. Pfluegers Arch. 432: 320-328.

90. Widrick, J.J., Romatowski, J.G., Bain, J.L.W., Trappe, S.W., Trappe, T.A., Thompson, J.L., Costill, D.L., Riley, D.A., and Fitts, R.H. 1997. Effect of 17 days bedrest on peak isometric force and maximal shortening velocity of human soleus fibers. Am. J. Physiol. 273: C1690-C1699.

91. Larsson, L., Li, X., Berg, H.E., and Frontera, W.R. 1996. Effects of removal of weight-bearing function on contractility and myosin isoform composition in single human skeletal muscle cells. Pfluegers Arch. 432: 320-328.

92. Widrick, J.J., Romatowski, J.G., Bain, J.L.W., Trappe, S.W., Trappe, T.A., Thompson, J.L., Costill, D.L., Riley, D.A., and Fitts, R.H. 1997. Effect of 17 days of bed rest on peak isometric force and unloaded shortening velocity of human soleus fibers. Am. J. Physiol. 273: C1690-C1699.

93. Larsson, L., Li, X., Berg, H.E., and Frontera, W.R. 1996. Effects of removal of weight-bearing function on contractility and myosin isoform composition in single human skeletal muscle cells. Pfluegers Arch. 432: 320-328.

94. Metzger, J.M., and Moss, R.L. 1987. Shortening velocity in skinned single muscle fibers—Influence of filament lattice spacing. Biophys. J. 52: 127-131.

95. Howells, K.F., Jordan, T.C., and Howells, J.D. 1978. Myofibril content of histochemical fibre types in rat skeletal muscle. Acta Histochem. 63: 177-182.

96. Kugelberg, E., and Thornell, L.E. 1983. Contraction time, histochemical type, and terminal cisternae volume of rat motor units. Muscle Nerve 6: 149-153.

97. McDonald, K.S., Blaser, C.A., and Fitts, R.H. 1994. Force-velocity and power characteristics of rat soleus muscle fibers after hindlimb suspension. J. Appl. Physiol. 77: 1609-1616.

98. Riley, D.A., Slocum, G.R., Bain, J.L.W., Sedlak, F.R., Sowa, T.E., and Mellender, J.W. 1990. Rat hindlimb unloading: Soleus histochemistry, ultrastructure, and electromyography. J. Appl. Physiol. 69: 58-66.

99. Riley, D.A., Slocum, G.R., Bain, J.L.W., Sedlak, F.R., Sowa, T.E., and Mellender, J.W. 1990. Rat hindlimb unloading: Soleus histochemistry, ultrastructure, and electromyography. J. Appl. Physiol. 69: 58-66.

100. Metzger, J.M., and Moss, R.L. 1987. Shortening velocity in skinned single muscle fibers—Influence of filament lattice spacing. Biophys. J. 52: 127-131.

101. McDonald, K.S., Blaser, C.A., and Fitts, R.H. 1994. Force-velocity and power characteristics of rat soleus muscle fibers after hindlimb suspension. J. Appl. Physiol. 77: 1609-1616.

102. Widrick, J.J., Romatowski, J.G., Bain, J.L.W., Trappe, S.W., Trappe, T.A., Thompson, J.L., Costill, D.L., Riley, D.A., and Fitts, R.H. 1997. Effect of 17 days bedrest on peak isometric force and maximal shortening velocity of human soleus fibers. Am. J. Physiol. 273: C1690-C1699.

103. Widrick, J.J., Romatowski, J.G., Bain, J.L.W., Trappe, S.W., Trappe, T.A., Thompson, J.L., Costill, D.L., Riley, D.A., and Fitts, R.H. 1997. Effect of 17 days bedrest on peak isometric force and maximal shortening velocity of human soleus fibers. Am. J. Physiol. 273: C1690-1699.

104. Fitts, R.H., Widrick, J.J., Knuth, S.T., Blaser, C.A., Karhanek, M., Trappe, S.W., Trappe, T.A., and Costill, D.L. 1997. Force-velocity and force-power properties of human muscle fiber after spaceflight. Med. Sci. Sports Exercise 29: S190.

105. Edgerton, V.R., Zhou, M.-Y., Ohira, Y., Klitgaard, H., Jiang, B., Bell, G., Harris, B., Saltin, B., Gollnick, P.D., Roy, R.R., Day, M.K., and Greenisen, M. 1995. Human fiber size and enzymatic properties after 5 and 11 days of spaceflight. J. Appl. Physiol. 78: 1733-1739.

106. Zhou, M.-Y., Klitgaard, H., Saltin, B., Roy, R.R., Edgerton, V.R., and Gollnick, P.D. 1995. Myosin heavy chain isoforms of human muscle after short-term spaceflight. J. Appl. Physiol. 78: 1740-1744.

107. Widrick, J.J., Romatowski, J.G., Bain, J.L.W., Trappe, S.W., Trappe, T.A., Thompson, J.L., Costill, D.L., Riley, D.A., and Fitts, R.H. 1997. Effect of 17 days bedrest on peak isometric force and maximal shortening velocity of human soleus fibers. Am. J. Physiol. 273: C1690-C1699.

108. Pierotti, D.J., Roy, R.R., Flores, V. and Edgerton, V.R. 1990. Influence of 7 days of hindlimb suspension and intermittent weight support on rat muscle mechanical properties. Aviat. Space Environ. Med. 61:205-210.

109. Fitts, R.H., Widrick, J.J., Knuth, S.T., Blaser, C.A., Karhanek, M., Trappe, S.W., Trappe, T.A., and Costill, D.L. 1997. Force-velocity and force-power properties of human muscle fiber after spaceflight. Med. Sci. Sports Exercise 29: S190.

110. Caiozzo, V.J., Haddad, F., Baker, M.J., Herrick, R.E., Prietto, N., and Baldwin, K.M. 1996. Microgravity-induced transformations of myosin isoforms and contractile properties of skeletal muscle. J. Appl. Physiol. 81: 123-132.

111. Diffee, G.M., Caiozzo, V.J., Herrick, R.E., and Baldwin, K.M. 1991. Contractile and biochemical properties of rat soleus and plantaris after hindlimb suspension. Am. J. Physiol. 260: C528-C534.

112. McDonald, K.S., Blaser, C.A., and Fitts, R.H. 1994. Force-velocity and power characteristics of rat soleus muscle fibers after hindlimb suspension. J. Appl. Physiol. 77: 1609-1616.

113. Widrick, J.J., Romatowski, J.G., Bain, J.L.W., Trappe, S.W., Trappe, T.A., Thompson, J.L., Costill, D.L., Riley, D.A., and Fitts, R.H. 1997. Effect of 17 days bedrest on peak isometric force and maximal shortening velocity of human soleus fibers. Am. J. Physiol. 273: C1690-C1699.

114. Edgerton, V.R., Zhou, M.-Y., Ohira, Y., Klitgaard, H., Jiang, B., Bell, G., Harris, B., Saltin, B., Gollnick, P.D., Roy, R.R., Day, M.K., and Greenisen, M. 1995. Human fiber size and enzymatic properties after 5 and 11 days of spaceflight. J. Appl. Physiol. 78: 1733-1739.

115. Hudlicka, O., Brown, M., and Egginton, S. 1992. Angiogenesis in skeletal and cardiac muscle. Physiol. Rev. 72: 369-417.

116. Edgerton, V.R., and Roy, R.R. 1994. Neuromuscular adaptation to actual and simulated weightlessness. Pp. 33-67 in Advances in Space Biology and Medicine, Vol. 4 (S.L. Bonting, ed.). JAI Press, Greenwich, Conn.

117. Zhou, M.-Y., Klitgaard, H., Saltin, B., Roy, R.R., Edgerton, V.R., and Gollnick, P.D. 1995. Myosin heavy chain isoforms of human muscle after short-term spaceflight. J. Appl. Physiol. 78: 1740-1744.

118. Riley, D.A., Ellis, S., Slocum, G.R., Satyanarayana, T., Bain, J.L.W., and Sedlak, F.R. 1987. Hypogravity-induced atrophy of rat soleus and extensor digitorum longus muscles. Muscle Nerve 10: 560-568.

119. Riley, D.A., Ellis, S., Giometti, C.S., Hoh, J.F.Y., Ilyina-Kakueva, E.I., Oganov, V., Slocum, G.R., Bain, J.L.W., and Sedlak, F.R. 1992. Muscle sarcomere lesions and thrombosis after spaceflight and suspension unloading. J. Appl. Physiol. 73: S33-S43.

120. Riley, D.A., Slocum, G.R., Bain, J.L.W., Sedlak, F.R., Sowa, T.E., and Mellender, J.W. 1990. Rat hindlimb unloading: Soleus histochemistry, ultrastructure, and electromyography. J. Appl. Physiol. 69: 58-66.

121. Edgerton, V.R., and Roy, R.R. 1994. Neuromuscular adaptation to actual and simulated weightlessness. Pp. 33-67 in Advances in Space Biology and Medicine, Vol. 4 (S.L. Bonting, ed.). JAI Press, Greenwich, Conn.

122. Riley, D.A., Ellis, S., Slocum, G.R., Sedlak, F.R., Bain, J.L.W., Krippendorf, B.B., Lehman, C.T., Macias, M.Y., Thompson, J.L., Vijayan, K., and DeBruin, J.A. 1996. In-flight and postflight changes in skeletal muscles of SLS-1 and SLS-2 spaceflown rats. J. Appl. Physiol. 81: 133-144.

123. Riley, D.A., Ellis, S., Slocum, G.R., Sedlak, F.R., Bain, J.L.W., Krippendorf, B.B., Lehman, C.T., Macias, M.Y., Thompson, J.L., Vijayan, K., and DeBruin, J.A. 1996. In-flight and postflight changes in skeletal muscles of SLS-1 and SLS-2 spaceflown rats. J. Appl. Physiol. 81: 133-144.

124. Edgerton, V.R., and Roy, R.R. 1994. Neuromuscular adaptation to actual and simulated weightlessness. Pp. 33-67 in Advances in Space Biology and Medicine, Vol. 4 (S.L. Bonting, ed.). JAI Press, Greenwich, Conn.

125. Riley, D.A., Ellis, S., Slocum, G.R., Sedlak, F.R., Bain, J.L.W., Krippendorf, B.B., Lehman, C.T., Macias, M.Y., Thompson, J.L., Vijayan, K., and DeBruin, 1996. J.A. In-flight and postflight changes in skeletal muscles of SLS-1 and SLS-2 spaceflown rats. J. Appl. Physiol. 81: 133-144.

126. Riley, D.A., Ellis, S., Slocum, G.R., Sedlak, F.R., Bain, J.L.W., Krippendorf, B.B., Lehman, C.T., Macias, M.Y., Thompson, J.L., Vijayan, K., and DeBruin, J.A. 1996. In-flight and postflight changes in skeletal muscles of SLS-1 and SLS-2 spaceflown rats. J. Appl. Physiol. 81: 133-144.

127. Riley, D.A., Ellis, S., Slocum, G.R., Sedlak, F.R., Bain, J.L.W., Krippendorf, B.B., Lehman, C.T., Macias, M.Y., Thompson, J.L., Vijayan, K., and DeBruin, J.A. 1996. In-flight and postflight changes in skeletal muscles of SLS-1 and SLS-2 spaceflown rats. J. Appl. Physiol. 81: 133-144.

128. Walton, K., Heffernan, C., Sulica, D., and Benavides, L. 1997. Changes in gravity influence rat postnatal motor system development: From simulation to spaceflight. Gravit. Space Biol. Bull. 10: 111-118.

129. Young, L.R., Oman, C.M., Watt, D.G.D., Money, K.E., Lichtenberg, B.K., Kenyon, R.V., and Arrott, A.P. 1986. M.I.T./Canadian vestibular experiments on the Spacelab-1 mission: 1. Sensory adaptation to weightlessness and readaptation to one-g: An overview. Exp. Brain Res. 64: 291-298.

130. Edgerton, V.R., and Roy, R.R. 1994. Neuromuscular adaptation to actual and simulated weightlessness. Pp. 33-67 in Advances in Space Biology and Medicine, Vol. 4 (S.L. Bonting, ed.). JAI Press, Greenwich, Conn.

131. Young, L.R., Oman, C.M., Watt, D.G.D., Money, K.E., Lichtenberg, B.K., Kenyon, R.V., and Arrott, A.P. 1986. M.I.T./Canadian vestibular experiments on the Spacelab-1 mission: 1. Sensory adaptation to weightlessness and readaptation to one-g: An overview. Exp. Brain Res. 64: 291-298.

132. Riley, D.A., Ilyina-Kakueva, E.I., Ellis, S., Bain, J.L.W., Slocum, G.R., and Sedlak, F.R. 1990. Skeletal muscle fiber, nerve, and blood vessel breakdown in space-flown rats. FASEB J. 4: 84-91.

133. Riley, D.A., Ellis, S., Slocum, G.R., Sedlak, F.R., Bain, J.L.W., Krippendorf, B.B., Lehman, C.T., Macias, M.Y., Thompson, J.L., Vijayan, K., and DeBruin, J.A. 1996. In-flight and postflight changes in skeletal muscles of SLS-1 and SLS-2 spaceflown rats. J. Appl. Physiol. 81: 133-144.

134. Deschenes, M.R., Covault, J., Kraemer, W.J., and Maresh, C.M. 1994. The neuromuscular junction. Muscle fibre type differences, plasticity and adaptability to increased and decreased activity. Sports Med. 17: 358-372.

135. Buckey, J.C., Jr., Lane, L.D., Levine, B.D., Watenpaugh, D.E., Wright, S.J., Moore, W.E., Gaffney, F.A., and Blomqvist, C.G. 1996. Orthostatic intolerance after spaceflight. J. Appl. Physiol. 81: 7-18.

136. McDonald, K.S., Delp, M.D., and Fitts, R.H. 1992. Effect of hindlimb unweighting on tissue blood flow in the rat. J. Appl. Physiol. 72: 2210-2218.

137. Stick, C., Grau, H., and Witzleb, E. 1989. On the edema-preventing effect of the calf muscle pump. Eur. J. Appl. Physiol. 59: 39-47.

138. Stick, C., Grau, H., and Witzleb, E. 1989. On the edema-preventing effect of the calf muscle pump. Eur. J. Appl. Physiol. 59: 39-47.

139. Olszewski, W., Engeset, A., Jaeger, P.M., Sokolowski, J., and Theodorsen, L. 1977. Flow and composition of leg lymph in normal men during venous stasis, muscular activity and local hyperthermia. Acta Physiol. Scand. 99: 149-155.

140. Stauber, W.T., and Ong, S.-H. 1981. Extravascular localization of albumin and IgG in normal tonic and phasic muscles of the rat. Histochem. J. 13: 1009-1015.

141. Riley, D.A., Ellis, S., Slocum, G.R., Sedlak, F.R., Bain, J.L.W., Krippendorf, B.B., Lehman, C.T., Macias, M.Y., Thompson, J.L., Vijayan, K., and DeBruin, J.A. 1996. In-flight and postflight changes in skeletal muscles of SLS-1 and SLS-2 spaceflown rats. J. Appl. Physiol. 81: 133-144.

142. Krippendorf, B.B., and Riley, D.A. 1993. Distinguishing unloading- versus reloading-induced changes in rat soleus muscle. Muscle Nerve 16: 99-108.

143. Buckey, J.C., Jr., Lane, L.D., Levine, B.D., Watenpaugh, D.E., Wright, S.J., Moore, W.E., Gaffney, F.A., and Blomqvist, C.G. 1996. Orthostatic intolerance after spaceflight. J. Appl. Physiol. 81: 7-18.

144. Buonanno, A., and Rosenthal, N. 1996. Molecular control of muscle diversity and plasticity. Dev. Genet. 19: 95-107.

145. Riley, D.A., Ellis, S., Slocum, G.R., Satyanarayana, T., Bain, J.L.W., and Sedlak, F.R. 1987. Hypogravity-induced atrophy of rat soleus and extensor digitorum longus muscles. Muscle Nerve 10: 560-568.

146. Riley, D.A., Ilyina-Kakueva, E.I., Ellis, S., Bain, J.L.W., Slocum, G.R., and Sedlak, F.R. 1990. Skeletal muscle fiber, nerve, and blood vessel breakdown in space-flown rats. FASEB J. 4: 84-91.

147. Riley, D.A., Ellis, S., Giometti, C.S., Hoh, J.F.Y., Ilyina-Kakueva, E.I., Oganov, V., Slocum, G.R., Bain, J.L.W., and Sedlak, F.R. 1992. Muscle sarcomere lesions and thrombosis after spaceflight and suspension unloading. J. Appl. Physiol. 73: S33-S43.

148. Riley, D.A., Thompson, J.L., Krippendorf, B.B., and Slocum, G.R. 1995. Review of spaceflight and hindlimb suspension unloading induced sarcomere damage and repair. Basic Appl. Myology 5: 139-145.

149. Riley, D.A., Ellis, S., Slocum, G.R., Sedlak, F.R., Bain, J.L.W., Krippendorf, B.B., Lehman, C.T., Macias, M.Y., Thompson, J.L., Vijayan, K., and DeBruin, J.A. 1996. In-flight and postflight changes in skeletal muscles of SLS-1 and SLS-2 spaceflown rats. J. Appl. Physiol. 81: 133-144.

150. Krippendorf, B.B., and Riley, D.A. 1993. Distinguishing unloading- versus reloading-induced changes in rat soleus muscle. Muscle Nerve 16: 99-108.

151. Riley, D.A., Ellis, S., Giometti, C.S., Hoh, J.F.Y., Ilyina-Kakueva, E.I., Oganov, V., Slocum, G.R., Bain, J.L.W., and Sedlak, F.R. 1992. Muscle sarcomere lesions and thrombosis after spaceflight and suspension unloading. J. Appl. Physiol. 73: S33-S43.

152. Riley, D.A., Ellis, S., Slocum, G.R., Sedlak, F.R., Bain, J.L.W., Krippendorf, B.B., Lehman, C.T., Macias, M.Y., Thompson, J.L., Vijayan, K., and DeBruin, J.A. 1996. In-flight and postflight changes in skeletal muscles of SLS-1 and SLS-2 spaceflown rats. J. Appl. Physiol. 81: 133-144.

153. Riley, D.A., Ellis, S., Slocum, G.R., Sedlak, F.R., Bain, J.L.W., Krippendorf, B.B., Lehman, C.T., Macias, M.Y., Thompson, J.L., Vijayan, K., and DeBruin, J.A. 1996. In-flight and postflight changes in skeletal muscles of SLS-1 and SLS-2 spaceflown rats. J. Appl. Physiol. 81: 133-144.

154. Riley, D.A., Ilyina-Kakueva, E.I., Ellis, S., Bain, J.L.W., Slocum, G.R., and Sedlak, F.R. 1990. Skeletal muscle fiber, nerve, and blood vessel breakdown in space-flown rats. FASEB J. 4: 84-91.

155. Riley, D.A., Ellis, S., Slocum, G.R., Sedlak, F.R., Bain, J.L.W., Krippendorf, B.B., Lehman, C.T., Macias, M.Y., Thompson, J.L., Vijayan, K., and DeBruin, J.A. 1996. In-flight and postflight changes in skeletal muscles of SLS-1 and SLS-2 spaceflown rats. J. Appl. Physiol. 81: 133-144.

156. Krippendorf, B.B., Riley, D.A. Distinguishing unloading- versus reloading-induced changes in rat soleus muscle. Muscle Nerve 16: 99-108, 1993.

157. Tidball, J.G. 1995. Inflammatory cell response to acute muscle injury. Med. Sci. Sports Exercise 27: 1022-1032.

158. Stick, C., Grau, H., and Witzleb, E. 1989. On the edema-preventing effect of the calf muscle pump. Eur. J. Appl. Physiol. 59: 39-47.

159. Olszewski, W., Engeset, A., Jaeger, P.M., Sokolowski, J., and Theodorsen, L. 1977. Flow and composition of leg lymph in normal men during venous stasis, muscular activity and local hyperthermia. Acta Physiol. Scand. 99: 149-155.

160. McDonald, K.S., Delp, M.D., and Fitts, R.H. 1992. Effect of hindlimb unweighting on tissue blood flow in the rat. J. Appl. Physiol. 72: 2210-2218.

161. Stick, C., Grau, H., and Witzleb, E. 1989. On the edema-preventing effect of the calf muscle pump. Eur. J. Appl. Physiol. 59: 39-47.

162. Olszewski, W., Engeset, A., Jaeger, P.M., Sokolowski, J., and Theodorsen, L. 1977. Flow and composition of leg lymph in normal men during venous stasis, muscular activity and local hyperthermia. Acta Physiol. Scand. 99: 149-155.

163. Wilson, L.B., Dyke, C.K., Parsons, D., Wall, P.T., Pawelczyk, J.A., Williams, R.S., and Mitchell, J.H. 1995. Effect of skeletal muscle fiber type on the pressor response evoked by static contraction in rabbits. J. Appl. Physiol. 79: 1744-1752.

164. Koslovskaya, I.B., Barmin, V.A., Stepantsov, V.I., and Kharitinov, N.M. 1990. Results of studies of motor functions in long-term spaceflights. Physiologist 33: S1-S3.

165. Riley, D.A., Ellis, S., Slocum, G.R., Satyanarayana, T., Bain, J.L.W., and Sedlak, F.R. 1987. Hypogravity-induced atrophy of rat soleus and extensor digitorum longus muscles. Muscle Nerve 10: 560-568.

166. Riley, D.A., Ilyina-Kakueva, E.I., Ellis, S., Bain, J.L.W., Slocum, G.R., and Sedlak, F.R. 1990. Skeletal muscle fiber, nerve, and blood vessel breakdown in space-flown rats. FASEB J. 4: 84-91.

167. Riley, D.A., Ellis, S., Giometti, C.S., Hoh, J.F.Y., Ilyina-Kakueva, E.I., Oganov, V., Slocum, G.R., Bain, J.L.W., and Sedlak, F.R. 1992. Muscle sarcomere lesions and thrombosis after spaceflight and suspension unloading. J. Appl. Physiol. 73: S33-S43.

168. Riley, D.A., Thompson, J.L., Krippendorf, B.B., and Slocum, G.R. 1995. Review of spaceflight and hindlimb suspension unloading induced sarcomere damage and repair. Basic Appl. Myology 5: 139-145.

169. Warren, G.L., Hayes, D.A., Lowe, D.A., Williams, J.H., and Armstrong, R.B. 1994. Eccentric contraction-induced injury in normal and hindlimb-suspended mouse soleus and EDL muscles. J. Appl. Physiol. 77: 1421-1430.

170. Warren, G.L., Hayes, D.A., Lowe, D.A., Williams, J.H., and Armstrong, R.B. 1994. Eccentric contraction-induced injury in normal and hindlimb-suspended mouse soleus and EDL muscles. J. Appl. Physiol. 77: 1421-1430.

171. Russell, B., Dix, D.J., Haller, D.L., and Jacobs-El, J. 1992. Repair of injured skeletal muscle: A molecular approach. Med. Sci. Sports Exercise 24: 189-196.

172. Gibala, M.J., MacDougall, J.D., Tarnopolsky, M.A., Stauber, W.T., and Elorriaga, A. 1995. Changes in human skeletal muscle ultrastructure and force production after acute resistance exercise. J. Appl. Physiol. 78: 702-708.

173. Macpherson, P.C.D., Schork, M.A., and Faulkner, J.A. 1996. Contraction-induced injury to single fiber segments from fast and slow muscles of rats by single stretches. Am. J. Physiol. 271: C1438-C1446.

174. Stauber, W.T., Clarkson, P.M., Fritz, V.K., and Evans, W.J. 1990. Extracellular matrix disruption and pain after eccentric muscle action. J. Appl. Physiol. 69: 868-874.

175. Riley, D.A., Ellis, S., Slocum, G.R., Satyanarayana, T., Bain, J.L.W., and Sedlak, F.R. 1987. Hypogravity-induced atrophy of rat soleus and extensor digitorum longus muscles. Muscle Nerve 10: 560-568.

176. Riley, D.A., Ilyina-Kakueva, E.I., Ellis, S., Bain, J.L.W., Slocum, G.R., and Sedlak, F.R. 1990. Skeletal muscle fiber, nerve, and blood vessel breakdown in space-flown rats. FASEB J. 4: 84-91.

177. Riley, D.A., Ellis, S., Giometti, C.S., Hoh, J.F.Y., Ilyina-Kakueva, E.I., Oganov, V., Slocum, G.R., Bain, J.L.W., and Sedlak, F.R. 1992. Muscle sarcomere lesions and thrombosis after spaceflight and suspension unloading. J. Appl. Physiol. 73: S33-S43.

178. Riley, D.A., Thompson, J.L., Krippendorf, B.B., and Slocum, G.R. 1995. Review of spaceflight and hindlimb suspension unloading induced sarcomere damage and repair. Basic Appl. Myology 5: 139-145.

179. Riley, D.A., Ellis, S., Slocum, G.R., Sedlak, F.R., Bain, J.L.W., Krippendorf, B.B., Lehman, C.T., Macias, M.Y., Thompson, J.L., Vijayan, K., and DeBruin, J.A. 1996. In-flight and postflight changes in skeletal muscles of SLS-1 and SLS-2 spaceflown rats. J. Appl. Physiol. 81: 133-144.

180. Warren, G.L., Hayes, D.A., Lowe, D.A., Williams, J.H., and Armstrong, R.B. 1994. Eccentric contraction-induced injury in normal and hindlimb-suspended mouse soleus and EDL muscles. J. Appl. Physiol. 77: 1421-1430.

181. Edgerton, V.R., and Roy, R.R. 1994. Neuromuscular adaptation to actual and simulated weightlessness. Pp. 33-67 in Advances in Space Biology and Medicine, Vol. 4 (S.L. Bonting, ed.). JAI Press, Greenwich, Conn.

182. Riley, D.A., Ellis, S., Giometti, C.S., Hoh, J.F.Y., Ilyina-Kakueva, E.I., Oganov, V., Slocum, G.R., Bain, J.L.W., and Sedlak, F.R. 1992. Muscle sarcomere lesions and thrombosis after spaceflight and suspension unloading. J. Appl. Physiol. 73: S33-S43.

183. Riley, D.A., Thompson, J.L., Krippendorf, B.B., and Slocum, G.R. 1995. Review of spaceflight and hindlimb suspension unloading induced sarcomere damage and repair. Basic Appl. Myology 5: 139-145.

184. Riley, D.A., Ellis, S., Slocum, G.R., Sedlak, F.R., Bain, J.L.W., Krippendorf, B.B., Lehman, C.T., Macias, M.Y., Thompson, J.L., Vijayan, K., and DeBruin, J.A. 1996. In-flight and postflight changes in skeletal muscles of SLS-1 and SLS-2 spaceflown rats. J. Appl. Physiol. 81: 133-144.

185. Rezvani, M., Ornatsky, O.I., Connor, M.K., Eisenberg, H.A. and Hood, D.A. 1996. Dystrophin, vinculin, and aciculin in skeletal muscle subject to chronic use and disuse. Med. Sci. Sports Exercise 28: 79-84.

186. Decker, M.L., Janes, D.M., Barclay, M.M., Harger, L., and Decker, R.S. 1997. Regulation of adult cardiocyte growth: Effects of active and passive mechanical loading. Am. J. Physiol. 272: H2902-H2918.

187. Campbell, K.P. 1995. Three muscular dystrophies: Loss of cytoskeleton-extracellular matrix linkage. Cell 80: 675-679.

188. Matsumura, K., and Campbell, K.P. 1994. Dystrophin-glycoprotein complex: Its role in the molecular pathogenesis of muscular dystrophies. Muscle Nerve 17: 2-15.

189. Petrof, B.J., Sharger, J.B., Stedman, H.H., Kelly, A.M., and Sweeney, H.L. 1993. Dystrophin protects the sarcolemma from stresses developed during muscle contraction. Proc. Natl. Acad. Sci. U.S.A. 90: 3710-3714.

190. Stone, M.H. 1990. Muscle conditioning and muscle injuries. Med. Sci. Sports Exercise 22: 457-462.

191. Riley, D.A., Ellis, S., Giometti, C.S., Hoh, J.F.Y., Ilyina-Kakueva, E.I., Oganov, V., Slocum, G.R., Bain, J.L.W., and Sedlak, F.R. 1992. Muscle sarcomere lesions and thrombosis after spaceflight and suspension unloading. J. Appl. Physiol. 73: S33-S43.

192. Riley, D.A., Thompson, J.L., Krippendorf, B.B., and Slocum, G.R. 1995. Review of spaceflight and hindlimb suspension unloading induced sarcomere damage and repair. Basic Appl. Myology 5: 139-145.

193. Riley, D.A., Ellis, S., Slocum, G.R., Sedlak, F.R., Bain, J.L.W., Krippendorf, B.B., Lehman, C.T., Macias, M.Y., Thompson, J.L., Vijayan, K., and DeBruin, J.A. 1996. In-flight and postflight changes in skeletal muscles of SLS-1 and SLS-2 spaceflown rats. J. Appl. Physiol. 81: 133-144.

194. Tidball, J.G., and Quan, D.M. 1992. Reduction in myotendinous junction surface area of rats subjected to 4-day spaceflight. J. Appl. Physiol. 73: 59-64.

195. Russell, B., Dix, D.J., Haller, D.L., and Jacobs-El, J. 1992. Repair of injured skeletal muscle: A molecular approach. Med. Sci. Sports Exercise 24: 189-196.

196. Beall, C.J., Sepanski, M.A., and Fyrberg, E.A. 1989. Genetic dissection of Drosophila myofibril formation: Effects of actin and myosin heavy chain null alleles. Genes Dev. 3: 131-140.

197. Drummond, D.R., Peckham, M., Sparrow, J.C., and White, D.C.S. 1990. Alteration in crossbridge kinetics caused by mutations in actin. Nature 348: 440-442.

198. Rezvani, M., Ornatsky, O.I., Connor, M.K., Eisenberg, H.A., and Hood, D.A. 1996. Dystrophin, vinculin, and aciculin in skeletal muscle subject to chronic use and disuse. Med. Sci. Sports Exercise 28: 79-84.

199. Campbell, K.P. 1995. Three muscular dystrophies: Loss of cytoskeleton-extracellular matrix linkage. Cell 80: 675-679.

200. Matsumura, K., and Campbell, K.P. 1994. Dystrophin-glycoprotein complex: Its role in the molecular pathogenesis of muscular dystrophies. Muscle Nerve 17: 2-15.

201. Danowski, B.A., Imanaka-Yoshida, K., Sanger, J.M., and Sanger, J.W. 1992. Costameres are sites of force transmission to the substratum in adult rat cardiomyocytes. J. Cell Biol. 118: 1411-1420.

202. Straub, V., and Campbell, K.P. 1997. Muscular dystrophies and the dystrophin-glycoprotein complex. Curr. Opin. Neurol. 10: 168-175.

203. Baker, L.P., Daggett, D.F., and Peng, H.B. 1994. Concentration of pp125 focal adhesion kinase (FAK) at the myotendinous junction. J. Cell Sci. 107: 1485-1497.

204. Sadoshima, J., and Izumo, S. 1997. The cellular and molecular response of cardiac myocytes to mechanical stress. Annu. Rev. Physiol. 59: 551-571.

205. Vandenbergh, H.H., Haftafaludy, S., Karlisch, P., and Shansky, J. 1989. Skeletal muscle growth is stimulated by intermittent stretch-relaxation in tissue culture. Am. J. Physiol. 256: C674-C682.

206. Campbell, K.P. 1995. Three muscular dystrophies: Loss of cytoskeleton-extracellular matrix linkage. Cell 80: 675-679.

207. Petrof, B.J., Sharger, J.B., Stedman, H.H., Kelly, A.M., and Sweeney, H.L. 1993. Dystrophin protects the sarcolemma from stresses developed during muscle contraction. Proc. Natl. Acad. Sci. U.S.A. 90: 3710-3714.

208. Straub, V., and Campbell, K.P. 1997. Muscular dystrophies and the dystrophin-glycoprotein complex. Curr. Opin. Neurol. 10: 168-175.

209. Straub, V., and Campbell, K.P. 1997. Muscular dystrophies and the dystrophin-glycoprotein complex. Curr. Opin. Neurol. 10: 168-175.

210. Petrof, B.J., Sharger, J.B., Stedman, H.H., Kelly, A.M., and Sweeney, H.L. 1993. Dystrophin protects the sarcolemma from stresses developed during muscle contraction. Proc. Natl. Acad. Sci. U.S.A. 90: 3710-3714.

211. Warren, G.L., Hayes, D.A., Lowe, D.A., Williams, J.H., and Armstrong, R.B. 1994. Eccentric contraction-induced injury in normal and hindlimb-suspended mouse soleus and EDL muscles. J. Appl. Physiol. 77: 1421-1430.

212. Warren, G.L., Hayes, D.A., Lowe, D.A., Williams, J.H., and Armstrong, R.B. 1994. Eccentric contraction-induced injury in normal and hindlimb-suspended mouse soleus and EDL muscles. J. Appl. Physiol. 77: 1421-1430.

213. Riley, D.A., Ilyina-Kakueva, E.I., Ellis, S., Bain, J.L.W., Slocum, G.R., and Sedlak, F.R. 1990. Skeletal muscle fiber, nerve, and blood vessel breakdown in space-flown rats. FASEB J. 4: 84-91.

214. Riley, D.A., Thompson, J.L., Krippendorf, B.B., and Slocum, G.R. 1995. Review of spaceflight and hindlimb suspension unloading induced sarcomere damage and repair. Basic Appl. Myology 5: 139-145.

215. Riley, D.A., Ellis, S., Slocum, G.R., Sedlak, F.R., Bain, J.L.W., Krippendorf, B.B., Lehman, C.T., Macias, M.Y., Thompson, J.L., Vijayan, K., and DeBruin, J.A. 1996. In-flight and postflight changes in skeletal muscles of SLS-1 and SLS-2 spaceflown rats. J. Appl. Physiol. 81: 133-144.

216. Krippendorf, B.B., and Riley, D.A. 1994. Temporal changes in sarcomere lesions of rat adductor longus muscles during hindlimb reloading. Anat. Rec. 238: 304-310.

217. Stauber, W.T., Fritz, V.K., Burkovskaya, T.E., and Ilyina-Kakueva, E.I. 1993. Effect of injury on mast cells of rat gastrocnemius muscle with respect to gravitational exposure. Exp. Mol. Pathol. 59: 87-94.

218. Stauber, W.T., Fritz, V.K., Burkovskaya, T.E., and Ilyina-Kakueva, E.I. 1993. Effect of injury on mast cells of rat gastrocnemius muscle with respect to gravitational exposure. Exp. Mol. Pathol. 59: 87-94.

219. Taipale, J., and Keski-Oja, J. 1997. Growth factors in the extracellular matrix. FASEB J. 11: 51-59.

220. Berk, M., Desai, S.Y., Heyman, H.C., and Colmenares, C. 1997. Mice lacking the ski proto-oncogene have defects in neurulation, craniofacial patterning, and skeletal muscle development. Genes Dev. 11: 2029-2039.

221. Capetanaki, Y., Milner, D.J., and Weitzer, G. 1997. Desmin in muscle formation and maintenance: Knockouts and consequences. Cell Struct. Funct. 22: 103-116.

222. Straub, V. and Campbell, K.P. 1997. Muscular dystrophies and the dystrophin-glycoprotein complex. Curr. Opin. Neurol. 10: 168-175.

223. Riley, D.A., Ellis, S., Slocum, G.R., Sedlak, F.R., Bain, J.L.W., Krippendorf, B.B., Lehman, C.T., Macias, M.Y., Thompson, J.L., Vijayan, K., and DeBruin, J.A. 1996. In-flight and postflight changes in skeletal muscles of SLS-1 and SLS-2 spaceflown rats. J. Appl. Physiol. 81: 133-144.

224. Riley, D.A., Ellis, S., Slocum, G.R., Sedlak, F.R., Bain, J.L.W., Krippendorf, B.B., Lehman, C.T., Macias, M.Y., Thompson, J.L., Vijayan, K., and DeBruin, J.A. 1996. In-flight and postflight changes in skeletal muscles of SLS-1 and SLS-2 spaceflown rats. J. Appl. Physiol. 81: 133-144.

225. Riley, D.A., Thompson, J.L., Krippendorf, B.B., and Slocum, G.R. 1995. Review of spaceflight and hindlimb suspension unloading induced sarcomere damage and repair. Basic Appl. Myology 5: 139-145.

226. Straub, V., and Campbell, K.P. 1997. Muscular dystrophies and the dystrophin-glycoprotein complex. Curr. Opin. Neurol. 10: 168-175.

227. Straub, V., and Campbell, K.P. 1997. Muscular dystrophies and the dystrophin-glycoprotein complex. Curr. Opin. Neurol. 10: 168-175.

8

Cardiovascular and Pulmonary Systems

INTRODUCTION

The cardiovascular system, which includes the heart and all the body's blood vessels, is responsible for delivering oxygen to all tissues of the body and also provides the transport system for metabolic waste products cleared by other systems, including the kidneys, lungs, and some components of the gastrointestinal tract. Because of variability in demand associated with exercise and changes in the gravity vector associated with various body positions in humans and other bipeds, the cardiovascular system has evolved with a range of responsiveness not found in other systems. Complex monitoring and control functions are required to meet normal physiological challenges. The cardiovascular system itself includes a four-chambered muscular pump (two atria and two ventricles) with both internal and external regulatory mechanisms, a vascular tree with arteries that transport oxygenated blood under relatively high pressures to the body at approximately 80 to 90 mm Hg pressure, and veins that return deoxygenated blood to the heart at lower pressures of 5 to 15 mm Hg. The capacity of the venous system is large, and at least 70 percent of the blood volume in humans is found in the veins. Virtually all components of the cardiovascular system have both internal and external control mechanisms, and a widely distributed system of stretch receptors (small nerve fibers in vessel walls that respond to stretch) that provide information about blood pressure and volume, as well as about heart function. The cardiovascular and pulmonary systems are linked to other systems controlling plasma volume and red blood cell mass through afferent autonomic signaling, and also through neurohormonal substances released in response to chamber and vessel distension, blood flow, and oxygen content at other sites in the body.

The pulmonary system includes the trachea (windpipe), the lungs, the muscles and bones of the rib cage, and the diaphragm. The lungs, like the heart and blood vessels in the neck and chest, are richly supplied with small nerve fibers that respond to stretch and function as low-pressure sensors, providing information about central blood volume. The highly expansive vascular tree is also the capacitance vessel between the right and left ventricles of the heart; it accommodates transient differences in right and left ventricular output. The lungs have a unique gravity dependence in several ways, including blood flow distribution, alveolar gas exchange, and inhaled particulate deposition and clearance. The perfusion pressure of deoxygenated blood entering the pulmonary artery from the right ventricle is relatively low, typically 10 to 20 mm Hg. In an individual standing upright, this pressure may be insufficient to overcome hydrostatic gradients, and so very little flow reaches the upper regions of the lungs, but a relatively large portion of pulmonary blood flow perfuses the dependent portions of the lungs. Air in the alveoli flows somewhat preferentially into the middle and upper regions of the lungs. These regional differences create a mismatch of air ventilation (V) and blood perfusion (Q) and are the basis for the system of classification of lung zones.[1] This classification system does not address differences between central and peripheral ventilation. These ventilation/perfusion (V/Q) mismatches account for the observation that clinical conditions associated with excessive fluid in the lungs (e.g., congestive heart failure) tend to affect the bases or dependent portions of the lungs, whereas airborne infections, such as tuberculosis, occur mainly in the upper regions of the lungs where the oxygen content is relatively high. The lungs have both intrinsic and extrinsic mechanisms to match ventilation and perfusion, but the main mechanism for accommodating changes in gravity and demand is redundancy, which provides a substantial excess capacity for gas exchange.

The lungs also have an elaborate system for clearing inhaled particulate materials existing as aerosols. (An aerosol is any system of solid or liquid particles sufficiently small to maintain stability as a suspension in air.[2,3]) The particles have to be too large to diffuse but must be sufficiently small to remain suspended (0.01 to 10 μm). Dust, smoke, chemicals, and inhaled bacteria all represent common threats to the lungs, both on Earth and, perhaps even more so, in microgravity. Three factors determine the location and extent of deposition of inhaled particles in the lungs: (1) the anatomy of the upper and lower respiratory tract, (2) patterns of inhalation and air flow, and (3) characteristics of the inhaled particles, including size, density, electric charge, and the tendency to absorb water. The effective size of a particle reflects its diameter d and density r, and is described by its aerodynamic diameter d_a,

$$d_a = d\sqrt{\rho}.$$

Particles being deposited by gravitational sedimentation reach a terminal velocity V_t where gravitational acceleration g is balanced by air resistance and is described by the equation

$$V_t = (\rho - \sigma)gd^2/\gamma \quad,$$

where σ and γ are the density and viscosity of air, and ρ and d are the density and diameter of the particle. Particles larger than 5 to 10 μm are filtered by small hairs in the nares. Sharp bends in the nasal passages, trachea, and bronchi also cause these larger particles to impact the airway walls, where they adhere and are cleared by the mucocilary sweeping action of the bronchial lining cells. Inertial impaction of particulates leads to highly localized deposition at airway bifurcations and accounts for the observation that most smoking-related lung cancers occur at such locations. Flow velocity in the lower airways falls progressively as the airway cross-sectional area rises. Particles of 1 to 5 μm are deposited in the terminal bronchioles and alveolar regions mainly by sedimentation, with Brownian diffusion transporting particles smaller than 1 μm.

CARDIOVASCULAR PHYSIOLOGY IN MICROGRAVITY

Given the gravity dependence of the cardiovascular and pulmonary systems, it is not surprising that humans exposed to altered gravity show significant cardiovascular and pulmonary changes. These were observed on the very earliest spaceflights. On entering microgravity, there is an immediate headward shift of fluids in the body. This is easily seen by an impressive swelling in the face and neck that occurs within minutes of reaching orbit. Simultaneously, there is a loss of volume in the lower extremities.[4,5] Studies on the nature and time course of this headward fluid shift have produced some surprises. As expected, the rapid initial loss of leg volume is from the deep venous system of the legs. This is quickly followed by loss of lower-extremity interstitial fluids. Ten percent or more of leg volume may be lost within the first 24 hours of spaceflight. Systemic plasma volume[6] also decreases rapidly, but overall, cardiopulmonary blood volume increases.[7] This is reflected in measured increases in pulmonary capillary blood volume and increased cardiac chamber volumes.[8]

Central venous pressure (CVP) is the pressure of blood as it enters the right atrium. The CVP is the heart's filling pressure and is an important determinant of cardiac output, the amount of blood pumped per minute by the heart. In a supine human on Earth, CVP is normally 4 to 7 mm Hg, but in microgravity, it is paradoxically low, typically around 0 to 2 mm Hg.[9-11] The presence of extensive swelling of the face and neck as well as increased cardiac chamber dimensions would usually indicate a high CVP, but the opposite was found in microgravity. This means that in microgravity there is a change in the relationship between cardiac filling pressure and volume. This, by definition, means that cardiac compliance or "stretchiness" has increased. It is unlikely that the intrinsic properties of the heart muscle itself change with only a few seconds of exposure to microgravity. Therefore, some change in the transmyocardial pressure must be present. This is presumably due to some combination of the following: a loss of hydrostatic pressure gradient within the heart itself, a decreased pressure within the thorax, and less pressure being exerted by the lungs on the heart in microgravity.

This increased heart volume (preload) increases cardiac output, to levels at or somewhat above what is found on Earth when subjects are measured while resting in the supine position.[12-14] Since cardiac output is the amount of blood pumped per beat times the number of beats per minute (stroke volume × heart rate = cardiac output), the basis for the increased cardiac output can be determined. In microgravity, the increased cardiac output is completely accounted for by an increased stroke volume, as pulse rate is usually the same or somewhat slower than what was seen on Earth when measured with the subject in a standing position. The cardiac output gradually declines during flight, and by 7 to 10 days in orbit, values approach, but do not quite reach, outputs seen with humans in the upright position in 1 *g*. There are various causes for the decline in cardiac output over time, including progressive decreases in plasma volume and red blood cell mass, altered autonomic cardiovascular control, and continued changes in body fluid distribution.[15,16] It is of interest that the resting hemodynamic state achieved by humans in microgravity closely approximates that seen on Earth when compared with measurements taken of humans in the upright position. The human cardiovascular system is adapted to the upright, biped position and not the supine or quadruped position.

Blood pressure and its regulation are also altered by exposure to microgravity. The larger stroke volume in microgravity elevates systolic blood pressure.[17-19] Decreased systemic vascular resistance reflects a compensatory vasodilation and leads to a widened pulse pressure and decreased diastolic pressure. Arterial blood pressure decreases over the first few days of spaceflight and approaches the values seen in humans who are in the upright position on Earth. Heart rate responsiveness to experimentally induced changes in carotid arterial pressure decreases in microgravity, demonstrating altered function of the carotid (high pressure) baroreflex (pressure sensing) system.[20,21] The magnitude and direction of baroreflex changes are similar to what is produced on Earth with 7 to 14 days of head-down bed rest.[22] Aerobic exercise capacity, assessed by bicycle ergometry, is well maintained in spaceflight, as is the relationship between systemic oxygen uptake and cardiac output.[23] Orthostatic tolerance (the ability to maintain blood pressure and cerebral perfusion during gravitational stress) deteriorates significantly in orbit, but this is not a problem as long as the astronauts are not exposed to gravitational stress.[24]   Overall, cardiovascular adaptation to microgravity is rapid and highly efficient, with no evidence of any functional impairment during spaceflight.[25-29]

Irregular heart rhythms, consisting primarily of isolated premature atrial or ventricular contractions, have been reported in crew members on U.S. and Soviet flights, especially during extravehicular activity (EVA). Evidence from the Apollo program and sporadic monitoring of U.S. shuttle flights suggest a very low incidence of spaceflight heart rhythm disturbances and little or no clinical or operational significance.[30,31] Separate reports of operationally significant heart rhythm disturbances in 2 Mir cosmonauts have been thoroughly reviewed, and in neither case was there a significant medical problem. The heart rhythm irregularities recorded so far in crew members of the U.S. and Russian space programs would be considered normal in healthy individuals. Increased plasma catecholamines (adrenaline-like compounds) associated with psychological stress; heavy upper-body isometric exercise required to function in partial-pressure, soft-shell space suits during space walks and other EVA; or low serum potassium, elevated serum calcium, thermal loads, and nonspecific cardiac changes have been suggested as etiological factors for the heart rate irregularity, but there is no evidence that these factors have caused cardiac rhythm problems in spaceflight.[32,33] Some of the apparent increase in heart rhythm disturbances during EVA may be due to ascertainment bias, as EVA is more often electrocardiographically monitored than most other activities of spaceflight.

PULMONARY PHYSIOLOGY IN MICROGRAVITY

Alterations in pulmonary air flow and volume characteristics and gas exchange in microgravity have also been well characterized and, like the cardiovascular changes, seem to be adaptive (see Table 8.1). The increased lung blood flow in microgravity is not associated with any adverse effects. Specifically, there is no increase in alveolar fluid, as had been postulated by some investigators.[34] There is a significant (15 percent) decrease in the average volume per breath (tidal volume), but an increase in breathing frequency and decreases in both alveolar (15 percent) and physiologic (18 percent) dead space.[35] These changes, along with increased lung blood flow and more uniform flow distribution, actually improved overall lung function in microgravity. The decreased functional reserve capacity may be due to increased pulmonary blood volume and the upward displacement of abdominal contents associated with the absence of gravity.[36] The decreased alveolar dead space is probably due to the alveoli achieving a more uniform size without the compression that normally occurs in the lung bases as a result of crowding from the increased blood volume.[37,38] Lung air-flow rates decreased initially in flight but returned to normal by 7 to 9 days in flight. It is possible that mild airway swelling was present initially, but the studies that were performed did not detect it.[39,40] What were thought to be gravity-dependent lung V/Q inequalities should have disappeared in microgravity, but their persistence suggested that they were structural in nature.[41,42] Gas exchange, as measured by carbon monoxide diffusing capacity (DL_{co}), improved significantly in microgravity. It was postulated that the absence of gravity gradients in pulmonary blood flow distribution provided a much greater effective capillary surface area and increased alveolar-capillary gas exchange. Analysis of the DL_{co} results showed that increases were due mainly to increased membrane diffusion surface area, while increased pulmonary capillary blood volume had only a minor contribution.[43] End-tidal pCO_2 reflects the concentration of carbon dioxide at the end of expiration and is dependent on alveolar CO_2 concentrations. It was normal when cabin CO_2 levels were normal, but rose 1 to 3 mm Hg when there was less efficient cabin CO_2 removal on a subsequent flight. Increased CO_2 levels within a spacecraft create a gradient that

TABLE 8.1 Changes in Pulmonary Function Observed and the Number of Subjects Studied During Spacelab Missions SLS-1 and D-2

Physiological Response to Microgravity	Number of Subjects	Changes in Microgravity (In-flight vs. Preflight Standing Measurements)
Pulmonary blood flow		
Total pulmonary blood flow (cardiac output)	4	18% increase
Cardiac stroke volume	4	4% increase
Diffusing capacity (carbon monoxide)	4	28% increase
Pulmonary capillary blood volume	4	28% increase
Diffusing capacity of alveolar membrane	4	27% increase
Pulmonary blood flow distribution	7	More uniform but some inequality remained
Pulmonary ventilation		
Respiration frequency	8	9% increase
Tidal volume	8	15% decrease
Alveolar ventilation	8	Unchanged
Total ventilation	8	Small decrease
Ventilatory distribution	7	More uniform but some inequality remained
Maximal peak expiratory flow rate	7	Decreased by $\leq 12.5\%$ early in flight but then returned to normal
Pulmonary gas exchange		
O_2 uptake	8	Unchanged
CO_2 output	8	Unchanged
End-tidal pO_2	8	Unchanged
End-tidal pCO_2	8	Small increase when CO_2 concentration in spacecraft increased
Lung volumes		
Functional residual capacity	4	15% decrease
Residual volume	4	18% decrease
Closing volume	7	Unchanged as measured by argon bolus

NOTE: Pulmonary blood flow in normal subjects is the same as cardiac output, the amount of blood pumped by the heart per minute. Stroke volume is the volume of blood pumped per beat. The ability of carbon monoxide to diffuse into the blood is a standard clinical test of the integrity of the alveolar membrane and its surrounding capillary blood supply. Tidal volume is the air breathed in a single breath. Alveolar ventilation is the volume of air moving into and out of the alveoli of the lung per minute. The data indicate that more alveoli are expanded and ventilated in space than on Earth. Peak expiratory flow rate is the maximal flow of air that can be forcefully exhaled. Oxygen uptake is the oxygen consumed by the subject per minute, while CO_2 output is the carbon dioxide produced per minute. End-tidal gas partial pressures are the respective partial pressures at the end of an expired breath and reflect the concentration of those gases in the alveoli. Functional residual capacity is the volume of gas in the lung that can still be exhaled at the end of a normal breath. Residual volume is the remaining gas in the lung after the subject has made a maximal effort to exhale. Closing volume refers to the volume in the lung where the alveoli close in significant numbers.

SOURCE: West, J.B., Elliott A.R., Guy, H.J.B., and Prisk, G.K. 1997. Pulmonary function in space. J. Am. Med. Assoc. 277: 1957-1961. References for various experiments are included in the text of this report and in West's review article.

decreases CO_2 clearance from the body and can have profound effects on astronauts' health and function.[44] Further studies should be performed to determine acceptable spacecraft CO_2 levels.

Pulmonary gas mixing and alveolar ventilation were assessed by examining the concentrations of exhaled helium (He) and sulfur hexafluride (SF_6).[45] These gases, like all respiratory gases, have similar convective mixing properties, but their molecular weights are 4 and 146, respectively, and so the diffusivity of He is approximately 6 times that of SF_6. Convective mixing of inspired gases occurs mainly in the centrally located alveoli, while peripheral acinar ventilation is much more dependent on diffusion. When measured in both standing and supine subjects on Earth, mixing inhomogeneities were much greater for SF_6, whereas in sustained microgravity, the differences disappeared. During breath-holding in microgravity, SF_6 actually exhibited better mixing properties than did He. These results were quite surprising and contradicted a widely accepted component of pulmonary physiology. The investigators were not able to explain their findings fully but postulated that changes in acinar architecture produced by the absence of gravity may have altered ventilation, especially in the peripheral regions. It is also possible that alterations in gas mixing caused by the normal cyclical compression of the lungs by the heart may also have played a role. Similar changes were not seen when measurements were made during short periods of microgravity (20 to 25 seconds) produced by parabolic flights.[46] All of the observed changes in pulmonary function returned to the preflight baseline almost immediately after reentry to 1 *g*, whereas cardiovascular readaptation to 1 *g* appears to be slower and is associated with physiologically and operationally significant impairment. Aerosol deposition has been studied only during parabolic flight, but measurements taken then show unexpectedly high deposition of small particles compared to predictions. This suggests that the absence of sedimentation allows particles to travel more deeply into the alveolar region of the lung where they deposit.[47] Although the absence of sedimentation still reduces overall deposition, the potential health effects of deeper particle penetration are unknown.

POSTFLIGHT CARDIOVASCULAR PHYSIOLOGY

Orthostatic intolerance, an impaired ability to maintain adequate blood pressure while in an upright position, has been found in almost all astronauts returning from spaceflights of even a few hours' duration. Overall, about two-thirds of the astronauts tested early postflight had hemodynamically significant orthostatic intolerance.[48] Virtually no testing has been performed on shuttle pilots and flight engineers during reentry, but levels of postflight impairment suggest that orthostatic intolerance remains a major, unresolved, clinical and operational problem. The extent of orthostatic intolerance postflight is variable and depends on the duration of flight, interindividual differences in cardiovascular function among the astronauts, and the time and method of postflight testing. Careful hemodynamic measurements of astronauts immediately postflight show that heart rate and stroke volume are relatively well maintained initially, but a failure to increase total peripheral resistance adequately leads to a fall in blood pressure, insufficient brain blood flow, and an inability to stand quietly for more than 5 to 10 minutes postflight.

Decreased red blood cell mass and plasma volume (see the section "Hematology" in Chapter 9) is a major factor but does not fully explain the extent of cardiovascular deconditioning seen postflight. Excessive pooling of blood in the lower extremities, thought likely because of decreased mass and tone in postural muscles, does not seem to be present.[49-51] Carotid baroreflex function is decreased in flight and remains so for some days after return to Earth, but the extent to which it accounts for the observed orthostatic intolerance is not clear. Inadequate vasoconstriction may be due to baroreflex dysfunction, but in-flight experiments thus far have addressed only the heart-rate component of the baroreflex.[52,53]

Other abnormalities of autonomic nervous system regulation may also be present. Many of the autonomic function experiments flown on the recent Neurolab Life Sciences mission were designed to address these and other aspects of neurohumoral control of circulation.

Postflight orthostatic intolerance is also associated with a major decrease in exercise capacity. Maximal heart rates are unchanged compared with preflight test results.[54,55] Blood pressure at the start of exercise is either unchanged or slightly elevated because of increased vasoconstriction. The major hemodynamic defect is an inadequate stroke volume, which is decreased by one-third or more from preflight levels and leads to proportionate decreases in cardiac output and skeletal muscle oxygen delivery. Undoubtedly, some component of skeletal muscle atrophy and decreased neuromuscular coordination contribute to the decreased aerobic capacity, but these are probably minor factors compared with decreased blood volume and autonomic nervous system dysregulation. Operational concerns about postflight orthostatic intolerance and aerobic deconditioning have increased since astronauts began using the heavy, bulky, partial-pressure launch and entry suit (LES) now required for all post-Challenger flights.

Recovery to preflight levels of orthostatic tolerance occurs within a day or so following flights of less than 1 month's duration, but longer recovery is associated with longer-duration flights.[56] Recovery of aerobic capacity is also relatively rapid but takes about a week or so after landing. The time course of cardiovascular recovery consistent with return of autonomic control and intravascular volume is too fast for recovery of skeletal muscle atrophy. The recovery course is also slower than was believed from clinical observations only, pointing out the need for careful, objective measurements of physiological function and capacity for accurate assessment of functional status and capability. Serial, long-term data on recovery following long-duration flights are lacking. Magnetic resonance imaging has documented a 7 to 10 percent decrease in heart muscle mass following spaceflights as short as 10 days.[57-59] The significance of this finding and its recovery time course are unknown.

IN-FLIGHT COUNTERMEASURES

A variety of cardiovascular countermeasures have been proposed and/or instituted to counteract the changes associated with spaceflight.[60-68] Most of these countermeasures have evolved over time and are frequently based on clinical intuition rather than prospectively collected data. The Russians use thigh constriction cuffs (*braselet*) to decrease cephalad fluid shifts at various phases of their long-duration spaceflights (Oleg Yu Atkov, M.D., Cosmonaut-USSR, personal communication). Data on the effectiveness of these cuffs are lacking, but many of the cosmonauts report significant relief from the head congestion and facial edema otherwise associated with microgravity. An aggressive in-flight exercise program seems to be only partially effective in maintaining postflight aerobic capacity, and its effects on orthostatic tolerance are largely unknown. On the Russian space station Mir, cosmonauts and astronauts currently exercise for almost 2 hours daily during flight and use saline loading and anti-g garments to minimize orthostatic intolerance postflight. On the U.S. shuttle, astronauts tend to perform some aerobic exercise, although there is no formal requirement for exercise during flight, and both exercise and fluid-loading regimens differ according to crew preference, so that actual practice is highly variable.

Analysis of the physiological responses to large amounts of intravenous normal saline solution administered to bed-rest subjects or to astronauts in flight showed that the adapted blood volumes were maintained at their reduced bed-rest or 0-g levels and that the saline was cleared from the circulation of these reduced-blood-volume subjects as rapidly as in 1 g.[69,70] Anecdotal reports suggest that much of the fluid load taken by shuttle astronauts just prior to reentry is also rapidly cleared by urination. The

Russian program also employs lower body negative pressure (LBNP) with salt loading in the last weeks of flight to improve orthostatic tolerance. LBNP is produced when a rigid chamber encloses the lower half of the astronaut's body and 10 to 50 mm Hg of suction is applied to shift blood footward, producing fluid shifts similar to what occurs with the standing position in 1 g. There is no evidence that salt loading itself prior to reentry increases intravascular volume, with or without LBNP stimulation. Early shuttle era data suggested that salt loading improved orthostatic tolerance postflight, but measurements were often made several hours after landing, and not all subjects were tested.[71] The important question is whether the countermeasure improves blood pressure and orthostatic tolerance during reentry and in the first few minutes after landing, when an emergency egress might be required. Virtually no data are available from these time frames. A recent study measured multiple hemodynamic variables including cardiac outputs within 4 hours of landing and was unable to document a difference in orthostatic tolerance between those who did or did not fluid load prior to reentry to 1 g.[72] A brief bout of maximal exercise performed just before reentry has been suggested as an intervention to promote fluid and salt retention, but this hypothesis has not been adequately tested.[73] The mineralocorticoid fludrocortisone acetate (*Florinef*) showed promise for promoting fluid retention and elevating blood pressure in some bed-rest studies.[74] However, when it was administered to a few astronauts in flight to produce salt and water retention, the results were disappointing. Further studies have not been conducted.

Liquid-filled cooling garments used during EVA decrease thermally induced increases in skin blood flow and could be used during reentry to maintain cardiac output and blood pressure.[75,76] Modified anti-g suits, similar to those used in high-performance military aircraft, have been employed to decrease the lower extremity and abdominal venous pooling that occurs during reentry hypergravity, and their use appears to produce the desired hemodynamic improvement. In addition to decreasing venous pooling, these suits raise systemic vascular resistance and blood pressure significantly when used at standard operating pressures.[77,78] The Russians wrap the lower body tightly with inelastic strapping (*karkas*) to achieve the same effect as the anti-g suit.

The current exercise countermeasures are broadly applied and do not specifically address muscle atrophy, bone demineralization, aerobic deconditioning, or orthostatic intolerance. It is likely that more efficient, better-focused regimens could be devised if each problem were initially addressed separately, with attention paid to underlying physiological mechanisms. The current Russian exercise countermeasure regimen is enormously expensive in terms of crew time and the amount of additional food and water that must be supplied to support the metabolic requirements associated with daily, prolonged, high-intensity exercise. Also, major differences related to age and gender exist among crew members with respect to bone and muscle metabolism, as well as orthostatic tolerance and aerobic capacity. These differences would seem to require astronaut-specific countermeasure "prescriptions." In addition, current Russian and NASA flight rules require rigid adherence to inadequately tested antiorthostatic countermeasures. This confounds testing of any new interventions and impedes assessment of both current and future countermeasures.

FUTURE DIRECTIONS

Over the last few decades, a number of space physiology reviews and reports have identified important research questions and recommended studies to address them. The cardiopulmonary data discussed above, obtained during a series of Skylab, Salyut, Mir, and Spacelab Life Sciences missions, provide a fairly comprehensive, systems-level view of the nature and time course of cardiovascular and pulmonary changes that occur when humans are exposed to microgravity for periods ranging from a few days to 13 months. Many of the more observational research questions have been answered at least partially, and future cardiopulmonary investigations should focus more on mechanisms. An improved understanding of underlying physiological mechanisms should also make it possible to address important operational questions more effectively. Postflight orthostatic hypotension remains one of the most important operational problems. Given the prospect of lunar or martian long-duration flights, orthostatic hypotension, aerobic deconditioning, and pulmonary particulate behavior in low- or 0-g environments must be understood and addressed. Important steps have already been taken, but current scientific knowledge and medical expertise must be increased to provide the level of security that will be required for such missions to proceed.

Cardiopulmonary Equipment

Much of the hardware needed for microgravity research with humans has already been developed for laboratory use, but flight or even portable units are often not available. Even less flight equipment and fewer facilities exist for animal research. Several techniques should be developed or obtained for space physiology research in both animals and humans. These are described briefly:

• Automated recording devices should be used extensively to capture physiological data with minimal additional astronaut involvement. Examples include exercise equipment that records astronaut identification, date, time, and workload; a simple body-mass measuring device; a dietary log system that does not require a manual logbook entry each time food or drink is consumed; and an automated urine measurement system that records void volumes, but also makes and records simple measures of electrolytes, creatinine levels, and so on. Such devices would greatly enhance the quality and quantity of physiological data obtained and would decrease the crew time required for such measurements.

• Accurate measurements should be made of cardiac output for both humans and animals. A human-rated, noninvasive foreign gas system would be appropriate for humans, but animal systems will require chronically implanted, stable, low-risk technologies, such as electromagnetic or Doppler flow probes.

• Systems for accurate measurement of heart rate and both cuff and beat-to-beat blood pressure should be readily available during all phases of flight. Such systems should be easy to apply and use and, where possible, should include data storage and suitable interfaces for down-link capability. Electrocardiograms and beat-to-beat blood pressure could be analyzed for first-order hemodynamics, variability, spectral content, power, and so on and correlated with other indices of cardiopulmonary function.

• Respiratory gas measurements are critical for most studies of pulmonary function. Equipment such as lightweight, stable mass spectrometers and ultrasonic flow meters, although complex, should be obtained and modified for flight.

• A scintigraphic imaging system should be developed for flight and should include a variety of imaging energy levels to support graphical and metabolic imaging of at least blood, bone, and muscle. Such a device would be multidisciplinary in its usage and would support metabolic, cardiopulmonary, and musculoskeletal discipline experiments. Ultrasound imaging systems should be available on the International Space Station. Magnetic resonance imaging (MRI) and computed tomography (CT) systems would be highly desirable for spaceflight and should be considered if continued advances in technology reduce their size and power consumption sufficiently.

• Exercise equipment should be multifunctional to permit exercise in multiple modalities, as well as testing and recording of astronaut data. Linkage with physiological monitoring equipment is also important.

• A microgravity-qualified system for aerosol generation, sizing, and counting will be needed to perform studies involving aerosols in microgravity.

• An in-suit bubble-detection system will be needed to assess the effectiveness of various regimens to counteract decompression sickness.

Research

Cardiovascular Research Recommendations

• *Investigate the specific mechanisms underlying the inadequate total peripheral resistance during orthostatic stress observed postflight, and determine the effective countermeasures. Both human and instrumented animal models will be required.*

• *Determine the appropriate method for referencing intrathoracic vascular pressures to systemic pressures, given the observed changes in alveolar and thoracic volume and compliance in microgravity, and in this setting determine whether chest compression will produce sufficient intrathoracic pressure to achieve adequate cardiopulmonary resuscitation.*

• *Determine the nature, magnitude, and time course of cardiovascular adjustments to microgravity, including assessment of hemodynamic, neurohumoral, morphological, histological, and molecular changes in the cardiovascular system.*

• *Determine whether microgravity-induced changes in local perfusion cause changes in vascular or vasomotor control (vascular proliferation or atrophy, secretion of endothelially derived vasoactive substances and microcirculatory autoregulatory mechanisms, and so on) or organ function (pulmonary gas exchange, renal clearance mechanisms, blood-brain barrier and cerebral pressure, etc.).*

• *Identify and validate countermeasures that might be effective in combating long-duration spaceflight-associated cardiovascular changes, and determine the mechanisms by which changes in fluid distribution and metabolism affect countermeasures designed to protect against cardiovascular and musculoskeletal deconditioning (e.g., LBNP, fluid rehydration, centrifugation, exercise, drugs, etc.).*

• *Determine the relationship between the cardiovascular adjustments to spaceflight and those occurring in Earth-based models.*

• *Determine the relationship of cardiovascular adjustments to microgravity-induced changes in other systems, especially the neurovestibular, autonomic nervous, hematopoietic (blood forming), and renal-endocrine systems.*

• *Determine whether there are changes in myocardial function associated with microgravity (contractile proteins, excitation-contraction coupling mechanisms, cardiac energetics, atrophy, etc.).*

• *Determine whether there are additional uses of microgravity for the study of terrestrial cardiovascular processes.*

• *Determine whether weightlessness affects cardiovascular pathophysiological mechanisms otherwise commonly found in adults in a 1-g environment (e.g., development of hypertrophy, dilation, regurgitant lesions, endothelial repair, and tissue healing in cardiovascular and other tissues).*

Pulmonary Research Recommendations

• *Determine how microgravity alters lung deposition of aerosol and whether this constitutes a health hazard.*
 —*Does the absence of the sedimentation mechanism cause deeper penetration of particles into the lung?*

—*What is the aerosol concentration, particle size profile, and bacterial contamination in the current spacecraft environments and atmospheres?*

—*Does microgravity alter nonventilatory responses to aerosolized antigens (immune responses, phagocytosis, etc.)?*

• *Determine and characterize long-duration, microgravity-induced topographical changes in pulmonary structure and function at rest and during exercise.*

—*What happens to the gravity-determined topographical differences of blood flow, ventilation, alveolar size, intrapleural pressures, and mechanical stresses in microgravity?*

—*What changes occur in rib cage, diaphragm, and abdominal wall configurations during microgravity?*

—*Are pulmonary function changes associated with long-duration exposure to microgravity different from those seen in short (<1 month) flights?*

—*Does long-duration exposure to microgravity affect pulmonary aging or disease processes?*

• *Determine whether lunar or martian dust particles are associated with mechanical, allergic, or biochemical pulmonary toxicity.*

• *Extend current denitrogenation protocols to address extended EVA work schedules that will occur in the International Space Station era.*

—*Do current denitrogenation/decompression schedules lead to microvascular gas emboli in the lung?*

—*What changes in pulmonary function occur during and after EVA as a result of the combined influences of exercise, oxygen breathing, and pulmonary microvascular gas emboli?*

—*What resuscitation procedures should be used in the event of loss of cabin or EVA suit pressure, and would counterpressure and/or pressure breathing prolong consciousness?*

• *Determine whether respiratory muscle structure and function deteriorate in weightlessness and, if so, whether the changes decrease maximal oxygen uptake.*

• *Determine and characterize intrathoracic lymph flow, and changes in blood pressure and volume associated with microgravity.*

—*If fluid leakage into the alveoli does occur, is this increased by exercise, and is the gas exchange defect exaggerated?*

REFERENCES

1. West, J. 1968. Regional differences in the lung. Postgrad. Med. J. 44: 120-122.
2. Clarke, S.W., and Pavia, D. 1994. Defense mechanisms and immunology: Deposition and clearance. Pp. 313-325 in Textbook of Respiratory Medicine, 2nd ed., Vol. 1 (J.F. Murray and J.A. Nadel, eds.). W.B. Saunders, Philadelphia.
3. Taulbee, D., and Yu, C. 1975. A theory of aerosol deposition in the human respiratory tract. J. Appl. Physiol. 38: 77-85.
4. Thorton, W.E., Moore, T.P., and Pool, S.L. 1987. Fluid shifts in weightlessness. Aviat. Space Environ. Med. 58(9, Suppl.): A86-A90.
5. Moore, T.P., and Thornton, W.E. 1987. Space shuttle in-flight and postflight fluid shifts measured by leg volume changes. Aviat. Space Environ. Med. 58: A91-A96.
6. Leach, C.S., Alfrey, C.P., Suki, W.N., Leonard, J.I., Rambaut, P.C., Inners, L.D., Smith, S.M., Lane, H.W., and Krauhs, J.M. 1996. Regulation of body fluid compartments during short-term spaceflight. J. Appl. Physiol. 81: 105-116.
7. Prisk, G.K., Guy, H.J., Elliott, A.R., Deutschman, R.A., and West, J.B. 1993. Pulmonary diffusing capacity, capillary blood volume, and cardiac output during sustained microgravity. J. Appl. Physiol. 75: 15-26.

8. Buckey, J.C., Jr., Gaffney, F.A., Lane, L.D, Levine, B.D., Watenpaugh, D.E., Wright, S.J., Yancy, C.W., Jr., Meyer, D.M., and Blomqvist, C.G. 1996. Central venous pressure in space. J. Appl. Physiol. 81: 19-25.

9. Kirsch, K.A., Rocker, L., Gauer, O.H., et al. 1984. Venous pressure in man during weightlessness. Science 225: 218-219.

10. Buckey, J.C., Jr., Gaffney, F.A., Lane, L.D., Levine, B.D., Watenpaugh, D.E., and Blomqvist, C.G. 1993. Central venous pressure in space. N. Engl. J. Med. 329: 1853-1854.

11. Foldager, N., Andersen, T.A.E., Jessen, F.B., Ellegaard, P., Stadeager, C., Videbæk, R., and Norsk, P. 1994. Central venous pressure during weightlessness in humans. Pp. 695-696 in Scientific Results of the German Spacelab Mission D-2 (P.R. Sahm, M.H. Keller, and B. Schiewe, eds.). Wissenschaftliche Projektführung D-2, Köln, Germany.

12. Buckey, J.C., Jr., Gaffney, F.A., Lane, L.D, Levine, B.D., Watenpaugh, D.E., Wright, S.J., Yancy, C.W., Jr., Meyer, D.M., and Blomqvist, C.G. 1996. Central venous pressure in space. J. Appl. Physiol. 81: 19-25.

13. Shykoff, B.E., Farhi, L.E., Olszowka, A.J., Pendergast, D.R., Rokitka, M.A., Eisenhardt, C.G., and Morin, R.A. 1996. Cardiovascular response to submaximal exercise in sustained microgravity. J. Appl. Physiol. 81: 26-32.

14. Prisk, G.K., Guy, H.J.B., Elliott, A.R., Deutschman III, R.A., and West, J.B. 1993. Pulmonary diffusing capacity, capillary blood volume, and cardiac output during sustained microgravity. J. Appl. Physiol. 75: 15-26.

15. Leach, C.S., Alfrey, C.P., Suki, W.N., Leonard, J.I., Rambaut, P.C., Inners, D., Smith, S.M., Lane, H.W., and Krauhs, J.M. 1996. Regulation of body fluid compartments during short-term space flight. J. Appl. Physiol. 81: 105-116.

16. Alfrey, C.P., Udden, M.M., Leach-Huntoon, C.S., Driscoll, T., and Pickett, M.H. 1996. Control of red blood cell mass in space flight. J. Appl. Physiol. 81: 98-104.

17. Buckey, J.C., Jr., Gaffney, F.A., Lane, L.D, Levine, B.D., Watenpaugh, D.E., Wright, S.J., Yancy, C.W., Jr., Meyer, D.M., and Blomqvist, C.G. 1996. Central venous pressure in space. J. Appl. Physiol. 81: 19-25.

18. Shykoff, B.E., Farhi, L.E., Olszowka, A.J., Pendergast, D.R., Rokitka, M.A., Eisenhardt, C.G., and Morin, R.A. 1996. Cardiovascular response to submaximal exercise in sustained microgravity. J. Appl. Physiol. 81: 26-32.

19. Prisk, G.K., Guy, H.J.B., Elliott, A.R., Deutschman III, R.A., and West, J.B. 1993. Pulmonary diffusing capacity, capillary blood volume, and cardiac output during sustained microgravity. J. Appl. Physiol. 75: 15-26.

20. Fritsch, J.M., Charles, J.B., Bennett, B.S., Jones, M.M., and Eckberg, D.L. 1992. Short-duration spaceflight impairs human carotid baroreceptor-cardiac reflex responses. J. Appl. Physiol. 73: 664-671.

21. Fritsch-Yelle, J.M., Charles, J.B., Jones, M.M., Beightol, L.A., and Eckberg, D.L. 1994. Spaceflight alters autonomic regulation of arterial pressure in humans. J. Appl. Physiol. 77: 1776-1783.

22. Eckberg, D.L., and Fritsch, J.M. 1992. Influence of ten-day head-down bedrest on human carotid baroreceptor-cardiac reflex function. Acta Physiol. Scand. 604 (Suppl.): 69-76.

23. Shykoff, B.E., Farhi, L.E., Olszowka, A.J., Pendergast, D.R., Rokitka, M.A., Eisenhardt, C.G., and Morin, R.A. 1996. Cardiovascular response to submaximal exercise in sustained microgravity. J. Appl. Physiol. 81: 26-32.

24. Bungo, M.W., Charles, J.B., and Johnson, P.C., Jr. 1985. Cardiovascular deconditioning during spaceflight and the use of saline as a countermeasure to orthostatic intolerance. Aviat. Space Environ. Med. 56: 985-990.

25. Johnston, R., and Dietlen, F., eds. 1977. Biomedical Results from Skylab. NASA-SP-377. National Aeronautics and Space Adminstration, Washington, D.C.

26. Hinghoffer-Szalkay, H. 1996. Physiology of cardiovascular, respiratory, interstitial, endocrine, immune, and muscular systems. Pp 107-153 in Biological and Medical Research in Space (D. Moore, P. Bie, and H. Oser., eds.). Springer, Berlin.

27. Blomqvist, C., and Stone, H. 1983. Cardiovascular adjustments to gravitational stress. Pp. 1025-1063 in Handbook of Physiology, Vol. III (J. Shepard and F. Abboud, eds.). American Physiological Society, Bethesda, Md.

28. Watenpaughm, D.E., and Hargens, A.R. 1995. The cardiovascular system in microgravity. Pp. 631-734 in Handbook of Physiology, Vol. 1 (M.J. Fregly and C.M. Blatteis, eds.). Oxford University Press, New York.

29. Berry, C.A. 1976. Medical legacy of Skylab as of May 9, 1974: The manned Skylab missions. Aviat. Space Environ. Med. 47: 418-424.

30. Berry, C.A. 1974. Medical legacy of Apollo. Aerosp. Med. 45: 1046-1057.

31. Nicogossian, A.E., and Parker, J.F., Jr. 1982. The cardiovascular system. P. 180 in Space Physiology and Medicine. NASA SP-447. National Aeronautics and Space Administration, Washington, D.C.

32. Waligora, J., Sauer, R., and Bredt, J. 1989. Spacecraft life support systems. Pp. 104-120 in Space Physiology and Medicine (A. Nicogossian, C. Huntoon, and S. Pool, eds.). Lea and Febiger, Philadelphia.

33. Helmke, C. 1990. Advances in Soviet extravehicular activity (EVA) suit technology. Air Force Foreign Technology Division Bulletin FTD-2660P-127/38-90.

34. Prisk, G.K., Guy, H.J.B., Elliott, A.R., Deutschman III, R.A., and West, J.B. 1993. Pulmonary diffusing capacity, capillary blood volume, and cardiac output during sustained microgravity. J. Appl. Physiol. 75: 15-26.

35. Prisk, G.K., Guy, H.J.B., Elliott, A.R., Paiva, M., and West, J.B. 1995. Ventilatory inhomogeneity determined from multiple-breath washouts during sustained microgravity on Spacelab SLS-1. J. Appl. Physiol. 78: 597-607.

36. Elliott, A.R., Prisk, G.K., Guy, H.J.B., and West, J.B. 1994. Lung volumes during sustained microgravity on Spacelab SLS-1. J. Appl. Physiol. 77: 2005-2014.

37. Guy, H.J.B., Prisk, G.K., Elliott, A.R., Deutschman III, R.A., and West, J.B. 1994. Inhomogeneity of pulmonary ventilation during sustained microgravity as determined by single-breath washouts. J. Appl. Physiol. 76: 1719-1729.

38. Elliott, A.R., Prisk, G.K., Guy, H.J.B., and West, J.B. 1994. Lung volumes during sustained microgravity on Spacelab SLS-1. J. Appl. Physiol. 77: 2005-2014.

39. Guy, H.J.B., Prisk, G.K., Elliott, A.R., and West, J.B. 1991. Maximum expiratory flow-volume curves during short periods of microgravity. J. Appl. Physiol. 70: 2587-2596.

40. Elliott, A.R., Prisk, G.K., Guy, H.J.B., Kosonen, J.M., and West, J.B. 1996. Forced expirations and maximum expiratory flow-volume curves during sustained microgravity on SLS-1. J. Appl. Physiol. 81: 33-43.

41. Prisk, G.K., Elliott, A. R., Guy, H.J.B., Kosonen, J.M., and West, J.B. 1995. Pulmonary gas exchange and its determinants during sustained microgravity on Spacelabs SLS-1 and SLS-2. J. Appl. Physiol. 79: 1290-1298.

42. Lauzon, A.M., Elliott, A.R., Paiva, M., West, J.B., and Prisk, G.K. 1998. Cardiogenic oscillation phase relationships during single-breath tests performed in microgravity. J. Appl. Physiol. 84: In Press.

43. Prisk, G.K., Guy, H.J.B., Elliott, A.R., Deutschman III, R.A., and West, J.B. 1993. Pulmonary diffusing capacity, capillary blood volume, and cardiac output during sustained microgravity. J. Appl. Physiol. 75: 15-26.

44. Prisk, G.K., Elliott, A.R., Guy, H.J.B., Kosonen, J.M., and West, J.B. 1995. Pulmonary gas exchange and its determinants during sustained microgravity on Spacelabs SLS-1 and SLS-2. J. Appl. Physiol. 79: 1290-1298.

45. Prisk, G.K., Lauzon, A.M., Verbanck, S., Elliot, A.R., Guy, H.J., Paiva, M., and West, J.B. 1996. Anomalous behavior of helium and sulfur hexafluoride during single-breath tests in sustained microgravity. J. Appl. Physiol. 80: 1126-1132.

46. Lauzon, A.M., Prisk, G.K., Elliott, A.R., Verbanck, S., Paiva, M., and West, J.B. 1997. Paradoxical helium and sulfur hexafluoride single-breath washouts in short-term vs. sustained microgravity. J. Appl. Physiol. 82: 859-65.

47. Darquenne, C., Paiva, M., West, J.B., and Prisk, G.K. 1997. Effect of microgravity and hypergravity on deposition of 0.5- to 3-mm-diameter aerosol in the human lung. J. Appl. Physiol. 83: 966-974.

48. Buckey, J.C., Jr., Lane, L.D., Levine, B.D., Watenpaugh, D.E., Wright, S.J., Moore, W.E., Gaffney, F.A., and Blomqvist, C.G. 1996. Orthostatic intolerance after spaceflight. J. Appl. Physiol. 81: 7-18.

49. Buckey, J.C., Jr., Lane, L.D., Levine, B.D., Watenpaugh, D.E., Wright, S.J., Moore, W.E., Gaffney, F.A., and Blomqvist, C.G. 1996. Orthostatic intolerance after spaceflight. J. Appl. Physiol. 81: 7-18.

50. Moore, T.P., and Thornton, W.E. 1987. Space shuttle inflight and postflight fluid shifts measured by leg volume changes. Aviat. Space Environ. Med. 58: A91-A96.

51. Thornton, W.E., Moore, T.P., and Pool, S.L. 1987. Fluid shifts in weightlessness. Aviat. Space Environ. Med. 58: A86-A90.

52. Fritsch-Yelle, J.M., Charles, J.B., Jones, M.M., Beightol, L.A., and Eckberg, D.L. 1994. Spaceflight alters autonomic regulation of arterial pressure in humans. J. Appl. Physiol. 77: 1776-1783.

53. Eckberg, D.L., and Fritsch, J.M. 1992. Influence of ten-day head-down bedrest on human carotid baroreceptor-cardiac reflex function. Acta Physiol. Scand. 604 (Suppl.): 69-76.

54. Levine, B.D., Lane, L.D., Watenpaugh, D.E., Gaffney, F.A., Buckey, J.C., and Blomqvist, C.G. 1996. Maximal exercise performance after adaptation to microgravity. J. Appl. Physiol. 81: 686-694.

55. Shykoff, B.E., Farhi, L.E., Olszowka, A.J., Pendergast, D.R., Rokitka, M.A., Eisenhardt, C.G., and Morin, R.A. 1996. Cardiovascular response to submaximal exercise in sustained microgravity. J. Appl. Physiol. 81: 26-32.

56. Watenpaugh, D.E., and Hargens, A.R. 1995. The cardiovascular system in microgravity. Pp. 631-674 in Handbook of Physiology, Vol. 1 (M.J. Fregly and C.M. Blatteis, eds.). Oxford University Press, New York.

57. Buckey, J.C., Jr., Lane, L.D., Levine, B.D., Watenpaugh, D.E., Wright, S.J., Moore, W.E., Gaffney, F.A., and Blomqvist, C.G. 1996. Orthostatic intolerance after spaceflight. J. Appl. Physiol. 81: 7-18.

58. Charles, J.B., and Jones, M.M. 1992. Cardiovascular orthostatic function of space shuttle astronauts during and after return from orbit. In 43rd International Astronautics Federation Congress, Paris. Reprint 92-0262. International Astronautics Federation/International Astronautical Association, Paris.

59. Blomqvist C.G., Lane L.D., Wright S.J., Meny G.M, Levine B.D., Buckey J.C., Peashock R.M., Weatherall P., Stray-Gundersen J., Gaffney F.A., Watenpaugh D.E., Arbeille Ph. R.M., and Baisch, F. 1995. Cardiovascular regulation at microgravity. Pp. 688-690 in Proceedings of the Nordeney Symposium on Scientific Results of the German D-2 Spacelab Mission. (P.R. Sahm, M.H. Keller, and B. Schieve, eds.). WPF and Deutsche Agentur für Raumfahrtangelegenheiten (DARA), Bonn & Deutsche Forschungsanstalt für Luft- und Raumfahrt, Köln, Germany.

60. Mikhailov, V.M., Pometov, Y.D., and Andretsov, V.A. 1984. [LBNP training of crew members of the Saliut-6 orbital station]. Kosm. Biol. Aviakosmicheskaya Med. 18: 29-33.

61. Greenleaf, J.E., Bulbulian, R., Bernauer, E.M., Haskell, W.L., and Moore, T. 1989. Exercise-training protocols for astronauts in microgravity. J. Appl. Physiol. 67: 2191-204.

62. Arbeille, P., Gauquelin, G., Pottier, J.M., Pourcelot, L., Guell, A., and Gharib, C. 1992. Results of a 4-week head-down tilt with and without LBNP countermeasure: II. Cardiac and peripheral hemodynamics—comparison with a 25-day spaceflight. Aviat. Space Environ. Med. 63: 9-13.

63. Fortney, S.M. 1991. Development of lower body negative pressure as a countermeasure for orthostatic intolerance. J. Clin. Pharmacol. 31: 888-892.

64. Charles, J.B., and Lathers, C.M. 1994. Summary of lower body negative pressure experiments during space flight. J. Clin. Pharmacol. 34: 571-583.

65. Cavanagh, P.R., Davis, B.L., and Miller, T.A. 1992. A biomechanical perspective on exercise countermeasures for long term spaceflight. Aviat. Space Environ. Med. 63: 482-485.

66. Convertino, V.A. 1996. Exercise as a countermeasure for physiological adaptation to prolonged spaceflight. Med. Sci. Sports Exercise 28: 999-1014.

67. Hargens, A.R. 1994. Recent bed rest results and countermeasure development at NASA. Acta Physiol. Scand. 616 (Suppl.): 103-114.

68. Bungo, M.W., Charles, J.B., and Johnson, P.C., Jr. 1985. Cardiovascular deconditioning during spaceflight and the use of saline as a countermeasure to orthostatic intolerance. Aviat. Space Environ. Med. 56: 985-990.

69. Gaffney, F.A., Buckey, J.C., Lane, L.D., Hillebrecht, A., Schulz, H., Meyer, M., Baisch, F., Beck, L., Heer, M., and Maass, H. 1992. The effects of a 10-day period of head-down tilt on the cardiovascular responses to intravenous saline loading. Acta Physiol. Scand. 604 (Suppl.): 121-30.

70. Blomqvist C.G., Lane L.D., Wright S.J., Meny G.M, Levine B.D., Buckey J.C., Peashock R.M., Weatherall P., Stray-Gundersen J., Gaffney F.A., Watenpaugh D.E., Arbeille Ph. R.M., and Baisch, F. 1995. Cardiovascular regulation at microgravity. Pp. 688-690 in Proceedings of the Nordeney Symposium on Scientific Results of the German D-2 Spacelab Mission (P.R. Sahm, M.H. Keller, and B. Schieve, eds.). WPF and Deutsche Agentur für Raumfahrtangelegenheiten (DARA), Bonn & Deutsche Forschungsanstalt für Luft- und Raumfahrt (DLR), Köln.

71. Bungo, M.W., Charles, J.B., and Johnson, P.C. Cardiovascular deconditioning during space flight and the use of saline as a countermeasure to orthostatic intolerance. Aviat. Space Environ. Med. 56: 985-990.

72. Buckey, J.C., Jr., Lane, L.D., Levine, B.D., Watenpaugh, D.E., Wright, S.J., Moore, W.E., Gaffney, F.A., and Blomqvist, C.G. 1996. Orthostatic intolerance after spaceflight. J. Appl. Physiol. 81: 7-18.

73. Convertino, V.A., Polet, J.L., Engelke, K.A., Hoffler, G.W., and Lane, L.D., and Blomqvist, C.G. 1997. Evidence of increased beta-adrenoreceptor responsiveness induced by 14 days of simulated microgravity in humans. Am. J. Physiol. 273: R93-R99.

74. Vernikos, J., Dallman, M.F., Van Loon, G., and Keil, L.C. 1991. Drug effects on orthostatic intolerance induced by bedrest. J. Clin. Pharmacol. 31: 974-984.

75. Carson, M.A., Rouen, M.N., Lutz, C.C., McBarron II, J.W. 1975. Extravehicular mobility unit. Pp. 545-569 in Biomedical Results of APOLLO (R.S. Johnston, R.F. Dietlein, and C.A. Berry, eds.). NASA SP368. National Aeronautics and Space Administration, Washington, D.C.

76. Raven, P., Saito, M., Gaffney, F., Schutte, J., and Blomqvist, C. 1980. Interactions between surface cooling and LBNP-induced central hypovolemia. Aviat. Space Environ. Med. 51: 497-503.

77. Gaffney, F.A., Thal, E.R., Taylor, W.F., Bastian, B.C., Weigelt, J.A., Atkins, J.M., and Blomqvist, C.G. 1981. Hemodynamic effects of medical anti-shock trousers (MAST garment). J. Trauma 21: 931-937.

78. Pawelczyk, J., Query, R., Zuckerman, J., McMinn, S., and Raven, P. 1994. Sympathetic activation with abdominal compression: A mechanism for the action medical anti-shock trousers. FASEB J. 8: A604.

9

Endocrinology

INTRODUCTION

The endocrine, nervous, and immune systems regulate the human response to spaceflight and the readjustment processes that follow landing. There is a multidirectional flow of information among these three systems, which together are responsible for maintaining homeostasis and reestablishing homeostasis if it is perturbed.[1-6] Quite apart from the intrinsic scientific interest of the underlying mechanisms, a basic knowledge and understanding of the effects of spaceflight on the endocrine and neuroendocrine systems are essential to the rational development of countermeasures.

Aspects of endocrine influence on specific organ systems are addressed in several chapters of this report. This chapter stresses integrative homeostatic functions of the endocrine and neuroendocrine systems. The principal spaceflight responses with a significant endocrine contribution are fluid shifts, perturbation of the circadian rhythms, loss of red blood cell mass, possible changes in the immune system, loss of bone and muscle, and maintenance of energy balance. The last three are chronic responses, whereas fluid shifts occur only after entry into a microgravity environment and again after return to Earth. Whether perturbations of circadian rhythms and loss of red blood cell mass are short- or long-term problems is not known.

In the space station era, systems physiology research will shift from the investigation of the acute responses to spaceflight to the long-term effects.[7] The former are associated with large, immediate changes in certain hormones. Thus, the change in blood volume, red cell mass, and the associated fluid shifts are acute responses to the novel environment, with appropriately large and immediate changes in hormone levels such as ADH (antidiuretic hormone, or vasopressin), aldosterone, and norepinephrine. What is needed is a focus on the problems of long-term spaceflight, specifically calcium loss from bone, muscle atrophy, energy balance, and the possible perturbation of circadian rhythms.

Many biological investigations involve endocrine measurements. Such measurements are often not the primary focus of the study but are necessary to identify the perturbation mechanism that is the actual subject of investigation. Studies of bone calcium losses are an example. In contrast with other systems such as fluid and electrolyte balance and circadian rhythms, and for some muscle experiments, endocrine measurements are the primary focus of the study. Thus many of the chapters in this report refer to making endocrine measurements.

In the past, postflight measurements have been used to extrapolate back to the in-flight situation. Postflight hormone measurements are not an adequate substitute for in-flight measurements because they measure only the postflight situation. The half-life of most hormones in the plasma is too short—usually on the order of minutes or less—to have any bearing on any situation except the one that exists at the time the sample was collected.

CURRENT STATUS OF RESEARCH

Effects of Spaceflight on Humans

The human body seems to have little difficulty in restoring and maintaining homeostasis in those systems that must respond rapidly—specifically, the resetting of the cardiovascular, vestibular, and hematopoietic systems, together with the regulation of fluid and electrolyte balance. The problems that arise are in systems that acclimate slowly to a new situation, particularly muscle, bone, and energy balance, and possibly the circadian timing system. Failure to maintain bone and muscle mass and regulate energy metabolism at 1 *g* are major health care problems.

Our current knowledge of the effects of spaceflight on humans is based largely on the results obtained on Skylab years ago and on the more recent SLS-1 and SLS-2 (Space Life Sciences 1 and 2) shuttle missions. The experiments that were conducted and samples collected on these missions match those expected from high-quality ground studies. New information of comparably high quality is expected when the results become available from the 1996 Life and Microgravity Sciences and the 1998 Neurolab shuttle missions. Neurolab was the last of a series of space shuttle Spacelab missions with a complement of experiments in the life sciences.

Entry into Earth orbit elicits a metabolic stress response.[8,9] The metabolic response of the human body to a physiological stress consists of an orderly and tightly controlled series of reactions. These responses include activation of the hypothalamic-pituitary-adrenal (HPA) axis, increases in the whole-body protein turnover rate and acute phase protein synthesis, increased gluconeogenesis, substrate cycling, proinflammatory cytokine activity, basal energy expenditure, and loss of body protein.[10-15] Collectively, these reactions limit the extent of the injury or stress, protect the rest of the organism against any further stresses by mobilizing host defense mechanisms, and initiate various processes aimed at restoring the organism's homeostatic balance. The process has been termed the metabolic stress or hypermetabolic response and is regulated by the combined effects of the central nervous system and neuroendocrine and immune systems.[16-19]

Hypothalamic-Pituitary-Adrenal Axis

Most of these responses result from activation of the hypothalamic-pituitary-adrenal (HPA) axis.[20-23] Secretion of the anterior pituitary hormones is regulated by the release of neurohormones from the hypothalamus, by both direct and indirect feedback from target tissue hormones, and by the immune system via cytokines.[24-26]

The secretion of the anterior pituitary hormones is regulated by the release of neurohormones from the hypothalamus and by the direct/indirect feedback from target tissue hormones and by the immune system. Of particular significance for the response to stress is the enhanced release of adrenocorticotropic hormone (ACTH) from the pituitary, which stimulates glucocorticoid release by the adrenal cortex. An increase in cortisol production is a good marker for a metabolic stress response.

During the first day or two of spaceflight, there is increased cortisol secretion, protein turnover, acute phase protein synthesis, and activation of the proinflammatory cytokine IL-6.[27] Protein breakdown is increased more than synthesis, and so the net effect is a loss of body protein.[28,29] The rise in protein turnover reflects the increased synthesis of proteins involved in host defenses. Thus, fibrinogen synthesis is increased even though there was no actual injury.[30] All of these measurements indicate that entry into orbit is associated with a metabolic stress response, which is over after about 1 week in space.[31]

There is currently little interpretable in-flight data on the endocrine system beyond the initial response period (less than 1 week), and obtaining reliable data has been difficult for the following reasons: (1) Observations from spaceflight experiments may have three components: mission specific, microgravity specific, and genetic. Mission-specific components include low dietary intakes, varying levels of activity, mission-specific elements of stress, and failure to allow the subjects to adapt their circadian rhythms to the mission. Attention should be paid to separating these effects and minimizing any interference from mission-specific components. (2) Conflicting experiments. Careful preflight planning can minimize the occurrence of this potential difficulty. (3) The endocrine changes associated with the chronic effects of spaceflight (bone calcium loss, muscle atrophy, and possible shifts in energy balance) are likely to be small and therefore difficult to detect except under very carefully controlled conditions. Nevertheless, the cumulative effects of small changes in endocrine response can result in the characteristic bone loss and muscle atrophy that are found.

The absence of reliable endocrine measurements can preclude drawing conclusions about the in-flight mechanism. Collection of data of the quality needed to detect small changes has proven problematic for individual experiments because of the time and resources needed to collect a quality set of controlled endocrine measurements for an individual experiment. Without these data, an understanding of the human response and adaptation to spaceflight will be impossible. Integrating experiments encompassing the best of several different investigator groups into one program offer the best hope of obtaining quality data using limited in-flight resources.

Even though the details of the mechanism are not known, spaceflight clearly induces multiple changes within the endocrine system. These can be characterized as changes in the interrelationship among hormones and alterations in the sensitivity of responding systems. Obtaining a baseline data set will be more complex than just obtaining a series of controlled blood and urine samples, because many hormones are released in a pulsatile manner; the frequency of the pulses, the quantity per pulse, and the total released over a 24-hour period are likely to be important factors in defining the endocrine responses to spaceflight.

Defining what should constitute the baseline set of hormone measurements and when in flight the measurements should be made should be left to a future group of experts in this field. Making these measurements properly once, so that the question does not have to be revisited, should be given very high priority. Measurements should be made both early and late in the mission. Serial blood sampling will be required, and the measurement period would probably extend over 2 to 5 days. Full use should be made of novel technologies now available for making precise measurements on small blood samples.

Alterations in endocrine activity can be expressed at the level of the concentration of the circulating hormone, receptor activity, and the subsequent signal transduction. To date, only hormone concentrations have been measured during spaceflight. Although it is reasonable to assume that any spaceflight-induced alteration in the circulating level of a hormone has some physiological significance, the absence of an observable change may not always imply that the effect of the hormone in question is unaltered. There is always the real possibility that the response may be altered at the receptor or transduction stages. These issues require invasive studies and as a result can only be addressed in ground models.

MODELS

Most ground-based animal studies to date have been done on rats and monkeys. In contrast to the human situation where some in-flight data is available, there is little useful in-flight animal endocrine data available to compare against the data derived from ground-based models. There is data from some of the Russian Bion missions, but most of it (and earlier Soviet data) is difficult to interpret because there was no in-flight monitoring of the test animals. The animals are not recovered until several hours postflight, and the landing is rough and may cause alterations and even injury or damage. Postflight data primarily reflect the stresses involved in landing and recovery and cannot be extrapolated to the in-flight situation.[32]

Assessing the relevance of rodent studies for humans is difficult. The situation with humans may be different for a number of reasons:[33,34] (1) Humans do not vegetate in space,[35] they make a conscientious effort to maintain health, and they exercise and are aware of the need to eat. (2) Space vehicles are designed to minimize changes from the normal lifestyle for astronauts, whereas the animals are often housed in confining conditions and their movement is often restricted. The environment may be quite stressful for reasons not related to spaceflight. Space life support systems are necessarily noisy, and the noise may disturb the rats. (3) Some flight studies use immature rats, because their smaller size allows fitting more rats into the cages. As a result, the body weight and muscle changes found include a "rate-of-growth" component. In contrast, human flight crew members are mature adults. (4) Psychological factors may be important, because emotions can influence the hormonal response (see Chapter 12). With these caveats, the rodent may be a better subject for endocrine studies designed to characterize the basic effects of microgravity, simply because more can be done with rodents than with humans. Any extrapolation of the results to humans must take the above factors into account.

Although flight experiments should be the primary focus, there is a continuing need for ground-based studies with the appropriate models. Currently available models range from whole body (human bed rest, rodent hindlimb unloading, centrifugation) to tissue culture systems.[36] Ground-based studies are not subject to the numerous restrictions and variations of spaceflight investigations and therefore provide the opportunity to elucidate mechanisms at the cellular and molecular levels. However, care must be taken to ensure that the model studied is appropriate.

Recommendations

The following recommendations are listed in order of priority.

1. Studies should obtain a baseline in-flight human hormone profile early and late in flight. As a control, the measurement set should include preflight measurements on the same individual over an extended period of time.

2. Studies should also continue to evaluate the relevance of ground-based models to spaceflight.

3. Researchers should continue to construct the human component of the life sciences database so that the issue of human variability can be addressed by modern statistical techniques.

ENERGY METABOLISM AND BALANCE

Energy expenditure for comparable ground-based activity is reduced in space for both humans and monkeys.[37-39] The principal hormone regulating energy metabolism is thyroxine (T_3). Urine gives a time-averaged value, and in-flight urine analyses on the SLS-1 and SLS-2 crew showed a consistent decrease of about 40 percent in the urinary excretion of T_3. Astronauts are anabolic during the postflight period; as expected for an anabolic state, there has been a consistent finding of increased plasma T_3 concentrations.[40-43] Likewise, most but not all thyroid-stimulating-hormone (TSH) measurements were increased postflight.[44-47]

The complicating role of nutrition tends to confound the interpretation of human experiments involving endocrine studies. Most hormones that are of interest are sensitive to nutritional intake and status. There may be a problem in maintaining energy balance during spaceflight in missions where there is a high level of physical activity.

Two in-flight energy balance studies have been done,[48-50] the first on Skylab using a combination of dietary intake and body composition measurements,[51] and the second on the Life and Microgravity Sciences (LMS) shuttle mission using the doubly labeled water method.[52] In both cases, astronauts were in negative energy balance. On the long-term Skylab mission, the deficit was about 3 kcal kg^{-1} day^{-1}, with the deficit being greatest during the first month and tapering off toward the end of the mission (Table 9.1).[53] On the 16-day LMS mission, the average energy deficit for the four payload crew members was ~10 kcal kg^{-1} day^{-1}.[54] The deficit on the LMS mission was due to a shortfall in energy intake (24 kcal kg^{-1} day^{-1}) rather than an increase in energy expenditure (34 kcal kg^{-1} day^{-1}).[55]

The negative energy balance and consequent greater N (protein) losses found in astronauts on the Skylab and LMS missions are a mission-specific response rather than a general response to spaceflight. Both missions had a heavy exercise component, whereas SLS-1 and SLS-2 did not. This raises some questions: (1) Is it possible to maintain energy balance with a substantial exercise regimen? On the ground, some exercise studies have reported a cachectic effect from intensive exercise.[56-58] (2) What contribution does negative energy balance make to muscle protein loss? As on the ground, if energy

TABLE 9.1   Comparison of Energy Intake, Expenditure, and N Balance for First 12 Days of Spaceflight

Vehicle	Energy intake (kcal kg^{-1} day^{-1})	N balance (mg N × kg^{-1} day^{-1})	Energy balance (kcal kg^{-1} day^{-1})
Skylab[a-c]	37 ± 1 (9)	−19 ± 6 (9)	−3
Shuttle, SLS-1, SLS-2[d]	30 ± 2 (11)	16 ± 3 (11)	~0
Shuttle, LMS[e]	24 ± 2 (4)		−10

[a]The energy balance value for Skylab is averaged over the duration of the Skylab missions (28 to 84 days).

[b]Rambaut, P.C., Leach, C.S., and Leonard, J.I. 1977. Observations in energy balance in man during spaceflight. Am. J. Physiol. 233:R208-R212.

[c]Leonard, J.I., Leach, C.S., and Rambaut, P.C. 1983. Quantitation of tissue loss during prolonged spaceflight. Am. J. Clin. Nutr. 38: 667-679.

[d]Stein, T.P., Leskiw, M.J., and Schluter, M.D. 1996. Diet and nitrogen metabolism during spaceflight on the shuttle. J. Appl. Physiol. 81: 82-92.

[e]Stein, T.P., Schluter, M.D., Leskiw, M.J., Gretebeck, R.J., Lane, H.W., and Hoyt, R.W. 1998. 1 year report of the LMS Shuttle mission. NASA, Washington, D.C.

intake is inadequate, exercise will probably exacerbate the loss of body protein during spaceflight.[59-61] (3) Does a negative energy balance also affect bone homeostasis? (4) Is there a dietary problem? Dietary intake on the recent LMS mission in 1996 (24 kcal kg^{-1} day^{-1}) was remarkably low, although exercise levels were much greater on this mission than on SLS-1 and SLS-2. The poor dietary intake raises questions as to whether enough time was allocated for eating on this busy mission, whether there were "taste" problems with diet, or whether something more serious occurs when there is a heavy exercise schedule. Construction and maintenance of the International Space Station will require many hours of extravehicular activity (EVA) where energy costs are high (200+ kcal/hr).[62]

On missions with a heavy exercise component, there is apparently a problem in maintaining energy balance. Clearly, an inadequate energy intake will lead to a starvation response, and the hormonal associations with that response may well mask any microgravity-induced changes. Attention needs to be paid to diet; for the data to have any validity, they must be collected under conditions where subjects are in or at least near to energy balance.

Beyond its importance for interpretation of scientific data, it is essential for crew health that astronauts be in approximate energy balance. Ensuring that this is the case for all astronauts should be given a very high priority, since prolonged periods of negative energy balance create serious health concerns. Apart from the loss of muscle and decreased physical performance,[63,64] there is progressively increasing susceptibility to infection.[65-67] Decreased immunocompetence during spaceflight has been reported.[68-70] Wound healing is also compromised, which may be a problem if injury ever occurs during spaceflight.[71-73]

Recommendations

• *Ensure adequate dietary input during spaceflight. Energy intake must meet needs, and physiological measurements must be made on subjects that are in approximate energy balance so that measurements are not confounded by an undernutrition response. Furthermore, a chronic negative energy balance is detrimental to overall health. The relationship between the amount of exercise and the protein and energy balance in flight needs to be investigated.*

• *Of lesser priority is the need to address the question of the quality and acceptability of in-flight diets.*

REPRODUCTION

The success of reproduction is highly dependent upon a diverse set of hormonal changes. A single aberration in either males or females can result in the inability to reproduce. The ultimate sign of acclimation to a new environment is the ability to reproduce the species. Alterations of the reproductive hormones during spaceflight are poorly understood. As yet, there is no information about the effects of exposure to spaceflight on the endocrine profile of females. In males, a reduction in testosterone levels has been reported during flight and postflight in both rats[74-76] and humans.[77] The reduction in testosterone in rats during spaceflight is not associated with changes in spermatogenesis in relatively short flights,[78] and the ability of humans to reproduce following spaceflight does not appear to be an issue. However, the ability of mammals to reproduce during spaceflight has yet to be evaluated. The rearing of multiple generations in space and a clear understanding of the changes in hormones important in maturation and reproduction will be essential to the acclimation to spaceflight and the success of colonization of remote planets. (For additional discussion of developmental biology, see Chapter 3.)

FLUID AND ELECTROLYTE BALANCE

Upon entry into orbit, the body loses water, and water also shifts from the lower to the upper body. The cephalic fluid shifts result in facial edema,[79,80] decreased leg volume,[81] and decreased plasma volume.[82] It is believed that these changes are effected to reduce a postulated increase in intrathoracic blood volume by the Gauer-Henry response,[83] (the inhibition of ADH secretion by an increase in arterial pressure). Leach's SLS-1 and SLS-2 experiment showed that if the Gauer-Henry mechanism is operative, it is over within the first day in orbit.[84]

Most of the initial reduction in body mass is from a loss of water. Electrolytes are lost, and there are some changes in blood hormone concentrations and hormones excreted in the urine. Changes in fluid-electrolyte metabolism and kidney function are secondary to the hypohydration status, which stimulates the hormonal systems responsible for fluid-electrolyte homeostasis. The principal hormone changes are summarized below. All of the in-flight fluid and electrolyte data are from human studies.

Hormonal Systems and Changes

Anti-Diuretic Hormone

On the ground, a reduction in plasma volume and an increase in blood osmolality are the main physiologic stimulants of antidiuretic hormone (ADH) secretion. The osmolality increases that occur with spaceflight do not seem to be large enough to cause consistent increases in ADH secretion during flight.[85] Thus, overall urinary ADH decreased during both the long-term Skylab[86] and Salyut-7 missions.[87] On a later Salyut flight, a transient increase of <100 percent was found.[88] Smith suggested that this increase may result from the combination of exercise and the high ambient temperature and CO_2 levels aboard.[89] During the first day on the shuttle, ADH levels increased in both plasma and urine.[90,91] On SLS-1 and SLS-2, ADH returned to baseline by the end of the first day in space and remained unchanged for the rest of the mission.[92]

Aldosterone

The renin-angiotensin-aldosterone system is also involved in the regulation of fluid and electrolyte balance. Most studies of aldosterone have been done on urine, and consistent increases in aldosterone levels have been reported.[93,94] On Skylab, urinary aldosterone levels were elevated for the entire flight.[95] Similar findings were made on Mir during the 25-day Mir-Aragatz flight[96] and at the end of the 8-month Salyut-7/Soyuz-T mission.[97] Overall plasma aldosterone was elevated during the first month on Skylab and then returned to baseline for the remaining 2 months in space.[98,99] The results from shuttle experiments are conflicting, with increases in aldosterone levels being reported on some missions[100] but (except for the first day in orbit) not on SLS-1 and SLS-2.[101]

The postflight data on plasma aldosterone levels are conflicting. Plasma aldosterone levels were increased after Skylab,[102] shuttle missions,[103-105] a Russian short-term mission,[106] and some of the long-duration Russian flights.[107,108] But after other long-term Soviet Salyut missions, plasma aldosterone levels were lower than preflight.[109] In contrast, urinary aldosterone has consistently been elevated in the immediate postflight period.[110-114] This increase has been ascribed to the need to retain sodium postflight.[115] The lack of significant changes during the SLS-1 and SLS-2 missions may have been because crew members were encouraged to maintain fluid intake. This result suggests that the aldosterone changes are intake related and are no longer of any great interest.

Angiotensin

Angiotensin concentrations are usually assessed indirectly by measuring renin activity. Plasma renin activity of Skylab crew members decreased for the first 2 days in space and then increased.[116,117] During the first month on Skylab, plasma renin levels were elevated most of the time.[118] Similarly, on SLS-1 and SLS-2, plasma renin levels were elevated on some but not all days.[119] With increasing in-flight time on Skylab, the fluctuations stabilized after the first month, with angiotensin 1 levels being consistently elevated.[120] Levels were also elevated during the eighth month of the Soviet Salyut-7 mission.[121] As with ADH, the fluctuations are within the normal range and appear to have little physiological significance.

Atrial Natrurietic Factor

Atrial natrurietic factor (ANF) has been measured in both urine and plasma. Urinary ANF was reduced at most sampling times during the first 8 days of two early shuttle flights, although there were "spikes" where the in-flight values were above the mean.[122] The decrease occurs as early as 5 hours in flight, probably as a result of the decrease in central venous pressure.[123] On the more recent SLS-1 and SLS-2 missions, where sample collection was more controlled, a constant decrease in ANF was found with a trend increasing toward the preflight value by the end of the mission.[124]

Catecholamines

On Earth, catecholamines have widespread effects on systems ranging from blood pressure to substrate metabolism. Most studies found no consistent change in epinephrine with spaceflight.[125,126] A Russian report from the long-duration Salyut-7 mission found increases in catecholamine levels toward the end of the mission, but these did not appear to be physiologically significant.[127] Norepinephrine is decreased, although the significance of this is not clear.[128-131] It is likely that much more will be known about the neuroendocrine system as a result of the 1998 Neurolab mission.

Summary Comments

Hundreds of humans have flown in space with no apparently serious consequences from the fluid shifts. The observed variations in hormonal response can be attributed to differences between missions and other mission-related factors such as dietary salt content and water consumption.[132] The SLS-1 and SLS-2 data provided the best values to date, and they showed little change in endocrine status. It is not likely that more studies will help us learn much more about the endocrine control of fluid balance. From an endocrine standpoint, the fluid shifts are unlikely to be of much importance for long-term missions, since there have been no reports of clinical consequences from the readjustment of fluid balance. Further study of the renal-endocrine relationships in microgravity is unlikely to contribute significantly to the study of disturbances in fluid and electrolyte equilibrium on Earth.

Models

Bed rest, particularly with 6° head-down tilt, has been the model of choice for investigating fluid shifts. Because the model is so convenient and reproduces the symptoms of spaceflight reasonably well, there has been little interest in other models.

Recommendation

Although not of high priority, further studies on the endocrine changes associated with the fluid shifts in microgravity may be a useful adjunct to studies of the fluid shifts that occur on return to Earth.

HEMATOLOGY

In-flight Observations

Spaceflight anemia characterized by decreased red blood cell (RBC) mass has been noted since the early days of human spaceflight.[133,134] Typically, there is a 1 percent daily loss of RBC mass, resulting in as much as a 10 percent decrease in total blood volume. Erythropoietin, the hormone responsible for stimulating RBC production, decreases in flight.[135] Along with the decreased reticulocytes[136] and the absence of any evidence of hemolysis in flight, this suggests that the observed anemia results from decreased RBC production rather than increased RBC destruction. Comprehensive experiments by Alfrey et al. conducted aboard SLS-1 and SLS-2 provided an understanding of the mechanisms underlying spaceflight anemia.[137,138]

These studies also provided new information on the role of erythropoietin in both the production and maturation of RBCs. Preflight and in-flight isotopic labeling studies were performed on six astronauts. In these studies, plasma volume (PV) decreased approximately 17 percent on the first day of flight, and although it increased slightly by 8 to 14 days in orbit, it remained significantly below preflight levels. Plasma volumes returned to normal by about 1 week after flight. Both RBC count and hemoglobin increased in flight, as would be expected with the marked decrease in plasma volume. Serum erythropoietin showed substantial variability, but a statistically and physiologically significant decrease was detected early in flight, and a significant rise occurred postflight. Serum iron levels in spaceflight and on Earth did not differ, but in-flight incorporation of ^{59}Fe was 66 percent of preflight levels when measured after 22 hours in orbit. Although RBC production decreased in flight, the survival rate of labeled RBCs was not changed by spaceflight.[139,140]

Spaceflight anemia appears to be a self-limiting and appropriate response to fluid shifts associated with microgravity. Space motion sickness induces decreases in fluid intake. This, along with increased regional perfusion with increased vascular permeability, causes shifts of fluid from plasma out of the intravascular compartment. The relative increases in RBC mass and hemoglobin concentration cause decreased erythropoietin production, which persists until RBC mass normalizes to a level appropriate for the persistently reduced in-flight plasma volume. Over time, the plasma volume and hemoglobin concentration are normal; however, when plasma volume increases postflight, an anemia is seen, because the reduced RBC mass is now distributed in a "normal" plasma volume. Increased erythropoietin levels quickly return RBC mass to normal within 1 to 2 weeks of return to gravity.

Perhaps the most surprising result from this study was the extent to which RBC production decreased with the decrease in erythropoietin. It had been generally thought that erythropoietin regulates the number of blast-forming erythroid units, which determine the number of proerythroblasts and therefore the number of RBCs produced.[141] Koury and Bondurant noted that there appears to be a substantial excess of blast-forming erythroid units and that erythropoietin is required for their survival.[142] Thus, in their model, erythropoietin actually regulates apoptosis in the bone marrow. The data from SLS-1 and SLS-2 are entirely consistent with this hypothesis and provide some of the best experimental support available for understanding the role that erythropoietin plays in RBC formation.

The Koury and Bondurant study[143] was important for the space program. It was one of the first instances in which a systematic scientific approach to a spaceflight-induced change not only resulted in an explanation of why the observed change occurred, but also it added to our knowledge of how red blood cell volume is controlled. It was the first study to show that the control of the red blood cell mass resides in part outside the bone marrow.

Models

The rat has been used as a general model for studying erythropoiesis in both ground-based and in-flight studies.[144] However, from the limited data available to date (from SLS-1 and SLS-2), it appears that there may be important differences between rat and human erythropoiesis. In both cases, spaceflight affects the bone marrow progenitor cells; but in rats, there appears to be no effect on peripheral erythroid parameters. The reason for this discrepancy is not understood.

Recommendations

• *Further hematology research is not of the highest priority. As a result of the SLS-1 and SLS-2 studies, the hematological changes must be regarded as one of the best-understood physiological changes that occur during spaceflight. Where the opportunity is available, it would be useful to refine the details of these processes, particularly that of sequestration in bone.*

• *Investigations of how erythropoiesis is regulated by hyper- and hypogravity should be continued with human and rodent models.*

ENDOCRINE ASPECTS OF MUSCLE LOSS

A decrease in muscle mass has been a consistent finding in humans and animals after short- and long-term missions. As expected, the decrease is found principally in muscles with antigravity functions in the back and the legs. It is likely that spaceflight induces muscle loss through remodeling, with the adaptation and the retention of functions needed for the new status. If there is an energy deficit, the remodeling response will also include a simple starvation-type muscle loss component. The anatomical, physiological, and biochemical changes in muscle are discussed in detail in Chapter 7. This section focuses on the nutritional and endocrine involvement.

As yet there are no fully effective measures for preventing the muscle atrophy associated with spaceflight. Exercise provides some protection, but whether some sort of exercise program alone will suffice or if other approaches need to be considered is not known. It is clear that current exercise procedures are not wholly effective, for the problem persists. Further development of in-flight measures requires that the nature of the atrophy be defined at a biochemical level in humans, and this requires a detailed knowledge of the endocrine changes.

As described in detail in Chapter 7, much is now known about the molecular and cellular aspects of responses of rodent muscles to unloading. Muscle atrophy and remodeling involve changes in the synthesis and breakdown of muscle proteins. These two processes are regulated primarily by hormones; however, at 1 *g* there are many pathways by which changes in human muscle protein content can be carried out. Pathways that have been proposed as relevant to the spaceflight situation include a bed rest/atrophy response,[145,146] an energy deficit,[147-149] a metabolic stress response,[150,151] altered activity of the HPA axis,[152-154] tension-induced remodeling,[155,156] and altered neuroendocrine signals.[157] Each of these proceeds via different mechanisms, and there are different ground-based models for each. For atrophy, there is bed rest and hindlimb suspension; for an energy deficit, starvation; for altered HPA axis function, hindlimb suspension with manipulation of the HPA axis hormones; for remodeling, applying tension to muscle cells in tissue culture and possibly also hindlimb suspension; and for the neuroendocrine hypotheses, denervation.

Good progress is being made in tracing the mechanisms of the various models to the molecular level in ground-based studies (Chapter 7). However, the relationships between in-flight data and data obtained from ground-based models, and between data on animal systems (particularly in-flight rodent experiments) and on humans in space, are not known.

The principal hormones and hormone systems that have been implicated to date in humans are insulin, hormones of the HPA axis, and prostaglandins, as well as the hormones that regulate energy metabolism (T_3) previously discussed.

Hormones Involved

Insulin

Insulin is an important factor in the regulation of muscle protein synthesis and breakdown, and insulin activity and resistance are therefore likely to be closely associated with muscle protein loss.[158] Many ground-based studies have shown a correlation between insulin levels, insulin resistance, and decreased nitrogen balance.[159-162]

Insulin data are difficult to interpret in the absence of parallel dietary data. The Skylab data are conflicting. After an initial drop in insulin levels, there was a trend toward an increase, followed by a very sharp drop and then a spike between the third and fourth weeks in flight.[163] For the remaining two months of the mission, plasma insulin levels were less than preflight levels. A study on a single subject suggested that glucose tolerance might be impaired during spaceflight, implying the development of insulin resistance.[164] On SLS-1 and SLS-2, insulin secretion increased with time in orbit, even though dietary intake was significantly less than preflight.[165] Other in-flight studies have demonstrated both insulin resistance together with variable increased insulin activity and increased insulin secretion during spaceflight. There is a good correlation between the increased insulin secretion and the protein loss.[166] Progressively increasing insulin secretion and decreased nitrogen retention are consistent with the development of insulin resistance and with a role for the insulin resistance in the etiology of the poorer nitrogen balance found toward the end of the mission.[167]

Postflight measurements by the Skylab investigators and Soviet scientists showed that postflight plasma insulin levels were increased, an increase that persisted for as long as 2 weeks after landing.[168,169] This increase was probably associated with an increase in dietary intake and so was part of an anabolic response.

Adrenocorticotropic Hormone

The Skylab investigators found that plasma adrenocorticotropic hormone (ACTH) levels were depressed early and late in flight, while growth hormone (GH) levels were elevated early in flight and then declined toward the end of the mission relative to preflight baseline.[170,171] On the early shuttle missions, there was an initial increase followed by a return to the preflight value of plasma ACTH.[172] The one subject studied by Gauquelin on MIR showed an increase in ACTH level.[173] Under the more carefully controlled conditions of the SLS-1 and SLS-2 missions, ACTH levels were unchanged during the 2 weeks of the mission and increased on landing which is in agreement with prior findings.[174]

Cortisol

Elevated cortisol levels in flight have been a frequent but not invariable finding across multiple missions.[175-179]   On the ground, the acute infusion of cortisol leads to protein breakdown,[180] but whether cortisol alone will lead to protein breakdown over time is not known. A number of observations argue against cortisol having a primary role in the protein loss: (1) The increased cortisol secretion occurs in the absence of any increase in ACTH.[181-184] A characteristic feature of a metabolic stress response is an increase in cortisol secretion, which is preceded by and associated with increased secretion of ACTH.[185]   The earlier findings of a lack of a correlation between ACTH and cortisol were confirmed on the recent SLS-1 and SLS-2 studies.[186] (2) Increased cortisol is found in some[187-188] but not all[189-191] bed-rest studies. (3) An increase in cortisol production is a systemic response and so cannot by itself account for the specificity of the muscle and bone losses that are found. An alternate possibility is that the increased cortisol response reflects the subject's emotional state rather than metabolic state.[192-195]   (4) The protein loss associated with hypercortisolemia is a systemic response, and the muscle loss is primarily localized to the antigravity muscles. (5) Skeletal muscle consists of two fiber types, red slow-twitch and white fast-twitch fibers. The cortisol-activated proteolysis primarily affects white fast-twitch fibers,[196] whereas with spaceflight it is the slow-twitch muscle fibers that show the greatest degradation in rats (see Chapter 7).

Growth Hormone

The available in-flight human data, principally from Skylab and SLS-1 and SLS-2, shows little evidence for any gross changes in growth hormone (GH), even though rodent studies suggest that spaceflight may have a unique effect on the HPA axes. Sawchenko and Hymer and their colleagues investigated the effect of spaceflight on the secretory activity of the rat anterior pituitary, specifically for GH and prolactin, in agreement with data from tail suspension.[197-199] They found postflight evidence for decreased secretion of GH and prolactin with both spaceflight and tail suspension.

GH and prolactin secretion are reduced in pituitary cells obtained from rats that have flown, and these reductions persist for as long as 2 weeks postflight.[200,201]   With the bioassay, both GH and prolactin secretion were consistently lower across three missions (Spacelab-3 and the Russian Cosmos 1887 and 2044), whereas the results from the GH immunoassay were inconsistent.[202,203] The finding of apparent depression of secretory function in anterior pituitary has been reproduced. It was hypothesized that exposure to microgravity (or unweighting) altered the secretory activity of a subpopulation of GH cells, perhaps mediated through changes in the intracellular repackaging of the hormone molecules.[204,205] The interpretation and relevance of these observations to the role of the pituitary in the response to microgravity and spaceflight remain unclear.

Much interest has focused on GH because there is some evidence for in-flight changes in pituitary function (see above). GH treatment increases body and muscle weights in ground-based studies in hypophysectomized rats.[206]   Exercise protocols have been only partially successful in attenuating muscle loss in hindlimb-suspended rats.[207,208] But the combination of exercise and supplemental GH or IGF-1[209] was more effective in attenuating muscle protein loss than either parameter alone.[210]

It is apparent that GH has major effects on muscle mass and energy metabolism in the rodent hypogravity model of hindlimb unloading. Given the observed cellular changes in space-flown cultured pituitary cells, this might be a useful cellular model for investigating how cells detect changes in gravity levels. The rodent HPA findings appear to be different from the human flight data for reasons that are not clear. A possible explanation is that if there are changes, they may be too small to be detected above the background noise. Also possible is that these changes may be less marked in humans because countermeasures may mitigate the negative effects of microgravity. The inference is that there are limits to the utility of the rodent hindlimb suspension model.

The HPA axis plays a central role in regulating homeostasis. Although it seems highly likely that it is involved in modulating the responses of the principal systems affected by spaceflight, there is currently little information about the HPA axis's role. Elucidation of the role of the HPA axis in the human response to spaceflight is central to any understanding of these changes. This may be a formidable problem, because the response to chronic changes may be small and changes specific to spaceflight rather than activity levels or diet may be very difficult to detect given the small numbers of subjects. Inflight collection of a series of carefully controlled blood samples offers the best hope for obtaining the necessary human data.

Summary Comments

The 1987 Goldberg report pointed out that "the identification of the cellular and molecular mechanisms that link active muscle contraction and changes in contractile protein synthesis remains a fundamental question of muscle biology."[211]   These changes are regulated in part by various hormones, including insulin, the somatomedins, cytokines, prostaglandins; but their precise role in hypertrophy and atrophy still needs to be determined and there is no in-flight information. There is also a need to determine the coupling mechanism between muscle and bone. Although the two systems are necessarily integrated, the exact coupling mechanism needs to be determined. This is as true now as it was in 1987.

Models

Bed rest continues to be the principal human model for studying the endocrine and metabolic changes associated with muscle (and bone) loss. Other human models include the limb-suspension model developed by Tesch,[212] and the dry immersion model.[213,214] There are several animal and cell culture models as well. The most frequently used animal model has been that of the hindlimb-suspended rodent (see Chapter 7).

Recommendation

Most of the high-priority recommendations for muscle are given in Chapter 7 (muscle). However, interpretation of in-flight changes in muscle, as well as bone, requires measurement of the associated hormone changes. Therefore, the determination of the human baseline hormone profile is of great importance.

BONE

The decalcification of bone has been the single most invariant finding of the effects of spaceflight on humans. It is also potentially the most serious. Although there have been many descriptive studies of the changes that occur in bone calcium content during spaceflight, there have been no successful investigations of the in-flight mechanism(s) involved. The unanswered questions are, (1) Is the mechanism for the in-flight calcium loss the same as that for osteoporosis on Earth? (2) Is there sexual dimorphism? (3) Is the process self-limiting or continuous? The endocrine aspects of these issues are discussed in detail in Chapter 6.

CIRCADIAN RHYTHMS

Circadian rhythms serve to coordinate the physiology and behavior of an animal so that they are in synchrony with environmental needs. There are daily, monthly, and even annual rhythms, with out-of-phase circadian rhythms leading to chronic fatigue and impaired performance on the ground.[215,216] There is some evidence that spaceflight affects circadian rhythms.[217-221] This is potentially of serious concern because it could lead to avoidable "human" errors (see Chapter 12).

The monkey has proven to be a useful model for studying circadian rhythms, but the future of this program is currently uncertain. Fuller et al. studied the effects of the spaceflight environment on circadian rhythms of male rhesus monkeys flown on the Russian Bion missions.[222] During flight, the animals were maintained on a 24-hour (light [16-hr]–dark [8-hr]) cycle. This enabled the animals to maintain normal heart rate cycling and motor activity, even though actual heart rate was decreased, presumably because of the lower rate of energy expenditure.[223,224] In contrast, several studies have now documented delays in the phasing of body temperature rhythms in monkeys and rats.[225-227] This discrepancy probably reflects the presence of more than one pacemaker in the body.[228] Perturbed circadian rhythms are not unique to monkeys; the free-running activity rhythm of beetles was decreased,[229] as was spore formation in the fungus *Neurospora crassa*.[230]

Circadian rhythms are important and should be assigned a high priority. Quite apart from performance and psychosocial effects (which are discussed in Chapter 12), there may be other unrecognized effects of out-of-phase circadian rhythms—for example, the perturbation of growth-hormone secretion that occurs when sleep patterns are disrupted. Although there are several ways of determining circadian rhythms, including temperature monitoring, plasma metabolite levels, and the diurnal variation of the secretion of various hormones, there are few useful human data from flight experiments. The spaceflight-induced changes in human circadian rhythms need to be determined and countermeasures developed if warranted. There is good evidence from ground-based studies that countermeasures using a combination of light therapy and melatonin treatment are effective in resetting the human biological clock.[231]

Recommendations

The study of circadian rhythms is a high-priority area for research. In order of priority, the recommendations are as follows:

1. The effect of spaceflight on human circadian rhythms needs to be determined.
2. If significant degradation of performance is found, and it can be attributed to the disturbed circadian rhythm, the use of countermeasures (including the use of a combination of light and melatonin) should be explored.

GENDER

An important issue about which little is known is whether there are gender differences in the human response to spaceflight. The differences in body composition, particularly the degree of musculature and bone mass, could affect the response of women astronauts to spaceflight. Stress and strenuous exercise can induce amenorrhea, which (like menopause) has been shown to cause bone loss. Furthermore, on Earth the incidence of osteoporosis is much higher in women because women have less bone mass to start with. Women may therefore be at higher risk following microgravity-induced bone loss.

There is reason to suspect that differences will be found in systems other than the obvious ones of bone and muscle. In a ground-based study,[232] Vernikos and colleagues showed that male and female subjects showed marked differences in endocrine responses to 7 days of 6° head-down bed rest. All of these changes are regulated by various aspects of the endocrine system. Therefore, as part of its program to include women on missions, NASA should monitor the data to investigate whether there are significant gender differences in the human response to spaceflight. Should evidence of differences become apparent, a high priority must be given to investigating their significance.

Recommendation

NASA should continue to examine data from in-flight and ground-based model experiments, for gender differences in the response to microgravity.

REFERENCES

1. Bellomo, R. 1992. The cytokine network in the critically ill. Anaesth. Intensive Care 20: 288-302.
2. Besedovsky, H.O., and DelRey, A. 1982. Immune-neuroendocrine circuits: Integrative role of cytokines. Front. Neuroendocrinol. 13: 61-94.
3. Nistico, G., and De Sarro, G. 1991. Is interleukin-2 a neuromodulator in the brain? Trends Neurosci. 14: 146-150.
4. Madden, K.S., and Felten, D. 1995. Experimental basis for neural immune reaction. Physiol. Rev. 75: 77-106.
5. Hughes-Fulford, M. 1993. Review of the biological effects of weightlessness on the human endocrine system. Receptor 3(3): 145-154.
6. Berg, H.E., and Tesch, P.A. 1996. Changes in muscle function in response to 10 days of lower limb unloading in humans. Acta Physiol. Scand. 157: 63-70.
7. Space Science Board, National Research Council. 1987. A Strategy for Space Biology and Medical Sciences for the 1980s and 1990s. National Academy Press, Washington, D.C.
8. Stein, T.P., Leskiw, M.J., and Schluter, M.D. 1993. The effect of spaceflight on human protein metabolism. Am. J. Physiol. 264: E824-E828.
9. Stein, T.P., Leskiw, M.J., and Schluter, M.D. 1996. Diet and nitrogen metabolism during spaceflight on the shuttle. J. Appl. Physiol. 81: 82-92.
10. Cerra, F.B. 1987. Hypermetabolism, organ failure and metabolic support. Surgery 101: 1-14.
11. Lowry, S.F. 1992. Modulating the metabolic response to infection. Proc. Nutr. Soc. 51: 267-277.
12. Kern, K.A., and Norton, J.A. 1988. Cancer cachexia. J. Parenter. Enteral. Nutr. 12: 286-298.
13. Weissman, C. 1990. The metabolic response to stress: An overview and an update. Anesthesiology 73: 308-327.
14. Stein, T.P., Leskiw, M.J., Oram-Smith, J.C., Wallace, H.W., and Blakemore, W.S. 1977. Changes in protein synthesis after trauma: Importance of nutrition. Am. J. Physiol. 233: 348-355.
15. Gore, D.G., Jahoor, F., Wolfe, R.R., and Herndon, D.N. 1993. Acute response of human muscle protein to catabolic hormones. Ann. Surg. 218: 679-684.
16. Lowry, S.F. 1992. Modulating the metabolic response to infection. Proc. Nutr. Soc. 51: 267-277.
17. Kern, K.A., and Norton, J.A. 1988. Cancer cachexia. J. Parenter. Enteral Nutr. 12: 286-298.
18. Naito, Y., Tamai, S., Shingu, K., Shindo, K., Matsui, T., Segawa, H., Nakai, Y., and Mori, K. 1992. Responses of plasma adrenocorticotrophic hormone, cortisol and cytokines during and after upper abdominal surgery. Anesthesiology 77: 426-431.
19. Besedovsky, H.O., and DelRey, A. 1982. Immune-neuroendocrine circuits: Integrative role of cytokines. Front. Neuroendocrinol. 13: 61-94.
20. Besedovsky, H.O., and DelRey, A. 1982. Immune-neuroendocrine circuits: Integrative role of cytokines. Front. Neuroendocrinol. 13: 61-94.
21. Naito, Y., Tamai, S., Shingu, K., Shindo, K., Matsui, T., Segawa, H., Nakai, Y., and Mori, K. 1992. Responses of plasma adrenocorticotrophic hormone, cortisol and cytokines during and after upper abdominal surgery. Anesthesiology 77: 426-431.
22. Weissman, C. 1990. The metabolic response to stress: An overview and an update. Anesthesiology 73: 308-327.

23. Turnbull A.V., and Rivier, C. 1995. Regulation of the HPA axis by cytokines. Brain Behav. Immunol. 9: 253-275.

24. Besedovsky, H.O., and DelRey, A. 1982. Immune-neuroendocrine circuits: Integrative role of cytokines. Front. Neuroendocrinol. 13: 61-94.

25. Dunn, A.J. 1993. Infection as a stressor: A cytokine-mediated activation of the hypothalamo-pituitary-adrenal axis? Ciba Found. Symp. 172: 226-239.

26. Spangelo, B.L., Judd, A.M., MacCleod, R.M., Goodman, D.W., and Isakson, P.C. 1990. Endotoxin-induced release of interleukin-6 from rat medial basal hypothalami. Endocrinology 127: 1779-1785.

27. Stein, T.P., and Schluter, M.D. 1994. Excretion of IL6 by astronauts during spaceflight. Am. J. Physiol. 266: E448-E454.

28. Gore, D.G., Jahoor, F., Wolfe, R.R., and Herndon, D.N. 1993. Acute response of human muscle protein to catabolic hormones. Ann. Surg. 218: 679-684.

29. Stein, T.P., Leskiw, M.J., Oram-Smith, J.C., Wallace, H.W., and Blakemore, W.S. 1977. Changes in protein synthesis after trauma: Importance of nutrition. Am. J. Physiol. 233: 348-355.

30. Stein, T.P., Leskiw, M.J., and Schluter, M.D. 1996. Diet and nitrogen metabolism during spaceflight on the shuttle. J. Appl. Physiol. 81: 82-92.

31. Stein, T.P., Leskiw, M.J., and Schluter, M.D. 1996. Diet and nitrogen metabolism during spaceflight on the shuttle. J. Appl. Physiol. 81: 82-92.

32. Riley, D.A., Ellis, S., Slocum, G.R., Sedlak, F.R., Bain, J.L., Krippendorf, B.B., Lehman, C.T., Macias, M.Y., Thompson, J.L., Vijayan, K., and De Bruin, J.A. 1996. In-flight and postflight changes in skeletal muscles of SLS-1 and SLS-2 spaceflown rats. J. Appl. Physiol. 81: 133-144.

33. Dietlein, L.F. 1977. Skylab: A beginning. Pp. 408-418 in Biomedical Results from Skylab (R.S. Johnston and L.F. Dietlein, eds.). NASA SP-377. National Aeronautics and Space Administration, Washington, D.C.

34. Kinney, J.M., and Elwyn, D.H. 1983. Protein metabolism and injury. Ann. Rev. Nutr. 3: 433-466.

35. Riley, D.A., Ellis, S., Slocum, G.R., Sedlak, F.R., Bain, J.L., Krippendorf, B.B., Lehman, C.T., Macias, M.Y., Thompson, J.L., Vijayan, K., and De Bruin, J.A. 1996. In-flight and postflight changes in skeletal muscles of SLS-1 and SLS-2 spaceflown rats. J. Appl. Physiol. 81: 133-144.

36. Tipton, C.M. 1996. Animal models and their importance to human physiological responses in microgravity. Med. Sci. Sports Exercise. 28: S94-S100.

37. Stein, T.P., Dotsenko, M.A., Korolkov, V.I., Griffin, D.W., and Fuller, C.A. 1996. Measurement of energy expenditure in rhesus monkeys during spaceflight using doubly labeled water ($^2H_2{}^{18}O$). J. Appl. Physiol. 81: 201-207.

38. Stein, T.P., Leskiw, M.J., and Schluter, M.D. 1996. Diet and nitrogen metabolism during spaceflight on the shuttle. J. Appl. Physiol. 81: 82-92.

39. Stein, T.P., Schluter, M.D., Leskiw, M.J., Gretebeck, R.J., Lane, H.W., and Hoyt, R.W. 1998. 1 year report of the LMS Shuttle mission. National Aeronautics and Space Administration, Washington, D.C.

40. Huntoon, C.L., Cintron, N.M., and Whitson, P.A. 1994. Endocrine and biochemical functions. Pp. 334-351 in Space Medicine and Physiology, 3rd ed. (A.E. Nicogossian, C.L. Huntoon, and S.L. Pool, eds.). Lea and Febiger, Philadelphia.

41. Kalita, N.F., and Tigranian, R.A. 1986. Endocrine status of cosmonauts following long-term space missions. Space Biol. Aerosp. Med. 20(4): 84-86.

42. Popova, I.A., Vetrova, E.G., and Ruatamyan, L.A. 1991. Evaluation of energy metabolism in cosmonauts. Physiologist 34: S98-S99.

43. Grigoriev, A.I., Noskov, V.B., Atkov, O.Y., Afonin, B.V., Sukjanov, Y.V., Lebedev, V.I., and Boiko, T.A. 1991. Fluid-elecrolyte homeostasis and hormonal regulation in a 237 day space flight. Space Biol. Aerosp. Med. 25: 15-18.

44. Huntoon, C.L., Cintron, N.M., and Whitson, P.A. 1994. Endocrine and biochemical functions. Pp. 334-351 in Space Medicine and Physiology, 3rd ed. (A.E. Nicogossian, C.L. Huntoon, and S.L. Pool, eds.). Lea and Febiger, Philadelphia.

45. Kalita, N.F., and Tigranian, R.A. 1986. Endocrine status of cosmonauts following long-term space missions. Space Biol. Aerosp. Med. 20(4): 84-86.

46. Popova, I.A., Vetrova, E.G., and Ruatamyan, L.A. 1991. Evaluation of energy metabolism in cosmonauts. Physiologist 34: S98-S99.

47. Gauquelin, G., Maillet A., Allevard, A.M., Vorobiev D., Grigoriev, A.I., and Gharib, C. 1990. Volume regulating hormones, fluid and electrolyte modifications during the Aragatz mission (Mir station). Pp. 603-608 in Proceedings of the Fourth European Symposium on Life Sciences Research in Space, Trieste, Italy, May 28-June 1, 1990. ESA-SP-307. European Space Agency, Paris.

48. Rambaut, P.C., Leach, C.S., and Leonard, J.I. 1977. Observations on energy balance in man during space flight. Am. J. Physiol. 233: R208-R212.

49. Leonard, J.I., Leach, C.S., and Rambaut, P.C. 1983. Quantitation of tissue loss during prolonged space flight. Am. J. Clin. Nutr. 38: 667-679.

50. Stein, T.P., Schluter, M.D., Leskiw, M.J., Gretebeck, R.J., Lane, H.W., and Hoyt, R.W. 1998. 1 year report of the LMS Shuttle mission. National Aeronautics and Space Administration, Washington, D.C.

51. Rambaut, P.C., Leach, C.S., and Leonard, J.I. 1977. Observations in energy balance in man during space flight. Am. J.

52. Leonard, J.I., Leach, C.S., and Rambaut, P.C. 1983. Quantitation of tissue loss during prolonged space flight. Am. J. Clin. Nutr. 38: 667-679.

53. Rambaut, P.C., Leach, C.S., and Leonard, J.I. 1977. Observations on energy balance in man during space flight. Am. J. Physiol. 233: R208-R212.

54. Stein, T.P., Schluter, M.D., Leskiw, M.J., Gretebeck, R.J., Lane, H.W., and Hoyt, R.W. 1998. 1 year report of the LMS Shuttle mission. National Aeronautics and Space Administration, Washington, D.C.

55. Stein, T.P., Schluter, M.D., Leskiw, M.J., Gretebeck, R.J., Lane, H.W., and Hoyt, R.W. 1998. 1 year report of the LMS Shuttle mission. National Aeronautics and Space Administration, Washington, D.C.

56. Kissileff, H.R., Pi-Sunyer, F.X., Segal, K., Meltzer, S., and Foelsch, P.A. 1990. Acute effects of exercise on food intake in obese and nonobese women. Am. J. Clin. Nutr. 52: 240-245.

57. King, N.A., Burley, V.J., and Blundell, J.E. 1994. Exercise-induced suppression of appetite: Effects on food intake and implications for energy balance. Eur. J. Clin. Nutr. 48: 715-724.

58. King, N.A., Lluch, A., Stubbs, R.J., and Blundell, J.E. 1997. High dose exercise does not increase hunger or energy intake in free living males. Eur. J. Clin. Nutr. 51: 478-83.

59. Friedl, K.E., Moore, R.J., Martinez-Lopez, L.J., Vogel, J.A., Askew, E.W., Marchitelli, L.J., Hoyt, R.W., and Gordon, G.C. 1994. Lower limit of body fat in healthy active men. J. Appl. Physiol. 77: 933-940.

60. Iyengar, A., and Rao, B.S. 1979. Effect of varying energy and protein intake on nitrogen balance in adults engaged in heavy manual labor. Br. J. Nutr. 41: 19-25.

61. Ku, Z., and Thomason, D.B. 1994. Soleus muscle nascent polypeptide chain elongation slows protein synthesis rate during non-weight bearing activity. Am. J. Physiol. 267: C115-C126.

62. Powell, M.R., Horrigan, D.R., Waligora, J.M., and Norfleet, W.T. 1994. Extravehicular activities. Pp 128-140 in Space Medicine and Physiology, 3rd ed. (A.E. Nicogossian, C.L. Huntoon, and S.L. Pool eds.). Lea and Febiger, Philadelphia.

63. Berg, H.E., Dudley, G.A., Haggmark, M.E., Ohlsen, K., and Tesch, P.A. 1992. Effects of lower limb unloading on skeletal muscle mass and function in humans. J. Appl. Physiol. 70: 1882-1885.

64. Convertino, V.A. 1990. Physiological adaptations to weightlessness: Effects on exercise and work performance. Exercise Sport Sci. Rev. 18: 119-166.

65. Askenazi, J.A., Weissman, C., Rosenbaum, S.A., Elwyn, D.H., and Kinney, J.M. 1982. Nutrition and the respiratory system. Crit. Care Med. 10: 163-187.

66. Chandra, R.K. 1990. McCollum award lecture. Nutrition and immunity: Lessons from the past and new insights into the future. Am. J. Clin. Nutr. 53: 1087-1101.

67. Keusch, G.T., and Farthing, M.J.G. 1986. Nutrition and infection. Ann. Rev. Nutr. 6: 131-154.

68. Hughes-Fulford, M. 1993. Review of the biological effects of weightlessness on the human endocrine system. Receptor 3(3): 145-154.

69. Gmunder, F.K., Konstantinova, I., Cogoli, A., Lesnyak, A., Bogulov, W., and Grachov, A.W. 1994. Cellular immunity in cosmonauts during long duration spaceflight on board the orbital MIR station. Aviat. Space Environ. Med. 65: 419-423.

70. Hughes-Fulford, M.H. 1991. Altered cell function in microgravity. Exp. Gerontol. 26: 247-256.

71. Kinney, J.M., and Elwyn, D.H. 1983. Protein metabolism and injury. Ann. Rev. Nutr. 3: 433-466.

72. Stein, T.P., and Gaprindachvili, T. 1994. Spaceflight and protein metabolism, with special reference to man. Am. J. Clin. Nutr. 80: 806S-819S.

73. Young, V.R., and Marchini, J.S. 1990. Mechanisms and nutritional significance of metabolic responses to altered intakes of protein and amino acids with reference to nutritional adaptation in humans. Am. J. Clin. Nutr. 51: 270-289.

74. Plakhuta-Plakutina, G.I. 1977. State of spermatogenesis in rats flown aboard the biosatellite Cosmos-690. Aviat. Space Environ. Med. 48: 12-15.

75. Serova, L.V. 1989. Effect of weightlessness on the reproductive system of mammals. Kosm. Biol. Aviakosmicheskaya Med. 23: 11-16.

76. Deaver, D.R., Amann, R.P., Hammerstedt, R.H., Ball, R., Veeramachaneni, D.N., and Musacchia, X.J. 1992. Effects of caudal elevation on testicular function in rats. Separation of effects on spermatogenesis and steroidogenesis. J. Androl. 13: 224-231.

77. Tigranjan, R.A., Haase, H., Kalita, N.F., Ivanov, V.M., Pavlova, E.A., Afonin, B.V., Voronin, L.I., and Jarsumbeck, B. 1982. Results of endocrinolgic studies of the 3rd international crew of the scientific orbital station complex; Soyuz 29–Salyut 6–Soyuz 31 (joint space flight enterprise of the USSR-GDR). 2. Hormones and biologically active substances in blood. Endokrinologie 80: 37-41.

78. Deaver D.R., Amann, R.P., Hammerstedt, R.H., Ball, R., Veeramachaneni, D.N., and Musacchia, X.J. 1992. Effects of caudal elevation on testicular function in rats: Separation of effects on spermatogenesis and steroidogenesis. J. Androl. 13: 224-31.

79. Thornton, W.E., and Rummel, J.A. 1977. Muscular deconditioning and its prevention in space flight. Pp. 191-197 in Biomedical Results from Skylab (R.S. Johnston and L.F. Dietlein, eds.). NASA SP-377. National Aeronautics and Space Administration, Washington, D.C.

80. Thornton, W.E., Moore, T.P., and Pool, S.L. 1987. Fluid shifts in weightlessness. Aviat. Space Environ. Med. 58: A86-90.

81. Moore, T.P., and Thornton, W.E. 1987. Space shuttle inflight and post flight fluid shifts measured by leg volume changes. Aviat. Space Environ. Med. 58: A91-A96.

82. Leach, C.S., Alfrey, C.P., Suki, W.N., Leonard, J.I., Rambaut, P.C., Inners, L.D., Smith, S.M., Lane, H.W., and Krauhs, J.M. 1996. Regulation of body fluid compartments during short-term spaceflight. J. Appl. Physiol. 81(1): 105-16.

83. Gauer, O.H., and Henry, J.P. 1963. Circulatory basis of fluid volume control. Physiol. Rev. 43: 423-481.

84. Moore, T.P., and Thornton, W.E. 1987. Space shuttle inflight and post flight fluid shifts measured by leg volume changes. Aviat. Space Environ. Med. 58: A91-A96.

85. Smith, S.M., Krauhs J.M., and Leach C.S. 1996. Regulation of body fluid volume and electrolyte concentrations in spaceflight. Pp. 123-165 in Advances in Space Biology and Medicine, Vol. 6 (S.J. Bonting. ed.). JAI Press, Greenwich, Conn.

86. Leach, C.S., and Rambaut, P.C. 1977. Biochemical responses of the Skylab crewmen: An overview. Pp. 204-216 in Biomedical Results from Skylab (R.S. Johnston and L.F. Dietlein, eds.). NASA SP-377. National Aeronautics and Space Administration, Washington, D.C.

87. Vorobyev, Y.I., Gazenko, O.G., Shulzhenko, Y.B., Grigoryev, A.I., Barer, A.S., Yegorov, A.D., and Skiba, I.A. 1986. Preliminary results of medical investigations during 5-month spaceflight aboard Salyut 7 Soyuz-T orbital complex. Space Biol. Aerosp. Med. 20: 27-34.

88. Grigoriev, A.I., Noskov, V.B., Atkov, O.Y., Afonin, B.V., Sukjanov, Y.V., Lebedev, V.I., and Boiko, T.A. 1991. Fluid-electrolyte homeostasis and hormonal regulation in a 237 day space flight. Space Biol. Aerosp. Med. 25: 15-18.

89. Smith, S.M., Krauhs J.M., and Leach C.S. 1996. Regulation of body fluid volume and electrolyte concentrations in spaceflight. Pp. 123-165 in Advances in Space Biology and Medicine, Vol. 6 (S.J. Bonting, ed.). JAI Press, Greenwich, Conn.

90. Huntoon, C.L., Cintron, N.M., and Whitson, P.A. 1994. Endocrine and biochemical functions. Pp. 334-351 in Space Medicine and Physiology, 3rd ed. (A.E. Nicogossian, C.L. Huntoon, and S.L. Pool, eds.). Lea and Febiger, Philadelphia.

91. Moore, T.P., and Thornton, W.E. 1987. Space shuttle inflight and post flight fluid shifts measured by leg volume changes. Aviat. Space Environ. Med. 58: A91-A96.

92. Moore, T.P., and Thornton, W.E. 1987. Space shuttle inflight and post flight fluid shifts measured by leg volume changes. Aviat. Space Environ. Med. 58: A91-A96.

93. Leach, C.S., and Rambaut, P.C. 1977. Biochemical responses of the Skylab crewmen: An overview. Pp. 204-216 in Biomedical Results from Skylab (R.S. Johnston and L.F. Dietlein, eds.). NASA SP-377. National Aeronautics and Space Administration, Washington, D.C.

94. Grigoriev, A.I., Noskov, V.B., Atkov, O.Y., Afonin, B.V., Sukjanov, Y.V., Lebedev, V.I., and Boiko, T.A. 1991. Fluid-electrolyte homeostasis and hormonal regulation in a 237 day space flight. Space Biol. Aerosp. Med. 25: 15-18.

95. Huntoon, C.L., Cintron, N.M., and Whitson, P.A. 1994. Endocrine and biochemical functions. Pp. 334-351 in Space Medicine and Physiology, 3rd ed. (A.E. Nicogossian, C.L. Huntoon, and S.L. Pool, eds.). Lea and Febiger, Philadelphia.

96. Gauquelin, G., Maillet A., Allevard, A.M., Vorobiev D., Grigoriev, A.I., and Gharib, C. 1990. Volume regulating hormones, fluid and electrolyte modifications during the Aragatz mission (Mir station). Pp. 603-608 in Proceedings of the Fourth European Symposium on Life Sciences Research in Space, Trieste, Italy, May 28-June 1, 1990. ESA-SP-307. European Space Agency, Paris.

97. Vorobyev, Y.I., Gazenko, O.G., Shulzhenko, Y.B., Grigoryev, A.I., Barer, A.S., Yegorov, A.D., and Skiba, I.A. 1986. Preliminary results of medical investigations during 5-month spaceflight aboard Salyut 7 Soyuz-T orbital complex. Space Biol. Aerosp. Med. 20: 27-34.

98. Leach, C.S., and Rambaut, P.C. 1977. Biochemical responses of the Skylab crewmen: An overview. Pp. 204-216 in Biomedical Results from Skylab (R.S. Johnston and L.F. Dietlein, eds.). NASA SP-377. National Aeronautics and Space Administration, Washington, D.C.

99. Huntoon, C.L., Cintron, N.M., and Whitson, P.A. 1994. Endocrine and biochemical functions. Pp. 334-351 in Space Medicine and Physiology, 3rd ed. (A.E. Nicogossian, C.L. Huntoon, and S.L. Pool, eds.). Lea and Febiger, Philadelphia.

100. Grigoriev, A.I., Noskov, V.B., Atkov, O.Y., Afonin, B.V., Sukjanov, Y.V., Lebedev, V.I., and Boiko, T.A. 1991. Fluid-electrolyte homeostasis and hormonal regulation in a 237 day space flight. Space Biol. Aerosp. Med. 25: 15-18.

101. Moore, T.P., and Thornton, W.E. 1987. Space shuttle inflight and post flight fluid shifts measured by leg volume changes. Aviat. Space Environ. Med. 58: A91-A96.

102. Huntoon, C.L., Cintron, N.M., and Whitson, P.A. 1994. Endocrine and biochemical functions. Pp. 334-351 in Space Medicine and Physiology, 3rd ed. (A.E. Nicogossian, C.L. Huntoon, and S.L. Pool, eds.). Lea and Febiger, Philadelphia.

103. Leach, C.S. 1983. Medical results from STS 1-4.: Analysis of body fluids. Aviat. Space Environ. Med. 54: 550-554.

104. Grigoriev, A.I., Noskov, V.B., Atkov, O.Y., Afonin, B.V., Sukjanov, Y.V., Lebedev, V.I., and Boiko, T.A. 1991. Fluid-electrolyte homeostasis and hormonal regulation in a 237 day space flight. Space Biol. Aerosp. Med. 25: 15-18.

105. Moore, T.P., and Thornton, W.E. 1987. Space shuttle inflight and post flight fluid shifts measured by leg volume changes. Aviat. Space Environ. Med. 58: A91-A96.

106. Leach, C.S., and Rambaut, P.C. 1977. Biochemical responses of the Skylab crewmen: An overview. Pp. 204-216 in: Biomedical Results from Skylab (R.S. Johnston and L.F. Dietlein, eds.). NASA SP-377. National Aeronautics and Space Administration, Washington, D.C.

107. Vorobyev, Y.I., Gazenko, O.G., Shulzhenko, Y.B., Grigoryev, A.I., Barer, A.S., Yegorov, A.D., and Skiba, I.A. 1986. Preliminary results of medical investigations during 5-month spaceflight aboard Salyut 7 Soyuz-T orbital complex. Space Biol. Aerosp. Med. 20: 27-34.

108. Grigoriev, A.I., and Egorov, A.D. 1992. General mechanisms of the effect of weightlessness on the human body. Pp. 43-83 in Advances in Space Biology and Medicine, Vol. 2 (S.J. Bonting. ed.). JAI Press, Greenwich, Conn.

109. Kalita, N.F., and Tigranian, R.A. 1986. Endocrine status of cosmonauts following long-term space missions. Space Biol. Aerosp. Med. 20(4): 84-86.

110. Leach, C.S., and Rambaut, P.C. 1977. Biochemical responses of the Skylab crewmen: An overview. Pp. 204-216 in Biomedical Results from Skylab (R.S. Johnston and L.F. Dietlein, eds.). NASA SP-377. NASA, Washington, D.C.

111. Huntoon, C.L., Cintron, N.M., and Whitson, P.A. 1994. Endocrine and biochemical functions. Pp. 334-351 in Space Medicine and Physiology, 3rd ed. (A.E. Nicogossian, C.L. Huntoon, and S.L. Pool, eds.). Lea and Febiger, Philadelphia.

112. Moore, T.P., and Thornton, W.E. 1987. Space shuttle inflight and post flight fluid shifts measured by leg volume changes. Aviat. Space Environ. Med. 58: A91-A96.

113. Grigoriev, A.I., Popova, I.A., and Ushakov, A.S. 1987. Metabolic and hormonal status of crewmembers in short-term spaceflights. Aviat. Space Environ. Med. 58: A121-A125.

114. Grigoriev, I.A., Noskov, V.B., Atkov, O.Y., Afonin, B.V., Sukjanov, Y.V., Lebedev, V.I., and Boiko, T.A. 1991. Fluid-elecrolyte homeostasis and hormonal regulation in a 237 day space flight. Space Biol. Aerosp. Med. 25: 15-18.

115. Grigoriev, A.I., Noskov, V.B., Atkov, O.Y., Afonin, B.V., Sukjanov, Y.V., Lebedev, V.I., and Boiko, T.A. 1991. Fluid-elecrolyte homeostasis and hormonal regulation in a 237 day space flight. Space Biol. Aerosp. Med. 25: 15-18.

116. Leach, C.S., and Rambaut, P.C. 1977. Biochemical responses of the Skylab crewmen: An overview. Pp. 204-216 in Biomedical Results from Skylab (R.S. Johnston and L.F. Dietlein, eds.). NASA SP-377. National Aeronautics and Space Administration, Washington, D.C.

117. Huntoon, C.L., Cintron, N.M., and Whitson, P.A. 1994. Endocrine and biochemical functions. Pp. 334-351 in Space Medicine and Physiology, 3rd ed. (A.E. Nicogossian, C.L. Huntoon, and S.L. Pool, eds.). Lea and Febiger, Philadelphia.

118. Huntoon, C.L., Cintron, N.M., and Whitson, P.A. 1994. Endocrine and biochemical functions. Pp. 334-351 in Space Medicine and Physiology, 3rd ed. (A.E. Nicogossian, C.L. Huntoon, and S.L. Pool, eds.). Lea and Febiger, Philadelphia.

119. Leach, C.S., Alfrey, C.P., Suki, W.N., Leonard, J.I., Rambaut, P.C., Inners, L.D., Smith, S.M., Lane, H.W., and Krauhs, J.M. 1996. Regulation of body fluid compartments during short-term spaceflight. J. Appl. Physiol. 81(1): 105-116.

120. Huntoon, C.L., Cintron, N.M., and Whitson, P.A. 1994. Endocrine and biochemical functions. Pp. 334-351 in Space Medicine and Physiology, 3rd ed. (A.E. Nicogossian, C.L. Huntoon, and S.L. Pool, eds.). Lea and Febiger, Philadelphia.

121. Grigoriev, A.I., and Egorov, A.D. 1992. General mechanisms of the effect of weightlessness on the human body. Pp. 43-83 in Advances in Space Biology and Medicine, Vol. 2 (S.J. Bonting, ed.). JAI Press, Greenwich, Conn.

122. Cintron, N.M., Lane, H.W., and Leach, C.S. 1990. Metabolic consequences of fluid shifts induced by microgravity. Physiologist 33: S16-S19.

123. Kirsch, K.A., Haenel, F., and Rocker, L. 1986. Venous pressure in microgravity. Naturwissenschaften 73: 447-449.

124. Moore, T.P., and Thornton, W.E. 1987. Space shuttle inflight and post flight fluid shifts measured by leg volume changes. Aviat. Space Environ. Med. 58: A91-A96.

125. Huntoon, C.L., Cintron, N.M., and Whitson, P.A. 1994. Endocrine and biochemical functions. Pp. 334-351 in Space Medicine and Physiology, 3rd ed. (A.E. Nicogossian, C.L. Huntoon, and S.L. Pool, eds.). Lea and Febiger, Philadelphia.

126. Leach, C.S., and Rambaut, P.C. 1977. Biochemical responses of the Skylab crewmen: An overview. Pp. 204-216 in Biomedical Results from Skylab (R.S. Johnston and L.F. Dietlein, eds.). NASA SP-377. National Aeronautics and Space Administration, Washington, D.C.

127. Davydova N.A., Kvetnanski, R., and Ushakov, A.S. 1989. Sympathetic-adrenal responses of cosmonauts after long term space flights on Salyut-7. Kosm. Biol. Aviakosmicheskaya Med. 23: 14-20.

128. Smith, S.M., Krauhs, J.M., and Leach, C.S. 1996. Regulation of body fluid volume and electrolyte concentrations in spaceflight. Pp. 123-165 in Advances in Space Biology and Medicine, Vol. 6 (S.J. Bonting, ed.). JAI Press, Greenwich, Conn.

129. Leach, C.S., and Rambaut, P.C. 1977. Biochemical responses of the Skylab crewmen: An overview. Pp. 204-216 in Biomedical Results from Skylab (R.S. Johnston and L.F. Dietlein, eds.). NASA SP-377. National Aeronautics and Space Administration, Washington, D.C.

130. Leach, C.S., Alfrey, C.P., Suki, W.N., Leonard, J.I., Rambaut, P.C., Inners, L.D., Smith, S.M., Lane, H.W., and Krauhs, J.M. 1996. Regulation of body fluid compartments during short-term spaceflight. J. Appl. Physiol. 81(1): 105-116.

131. Robertson D., Convertino, V.A., and Vernokos, J. 1994. The sympathetic nervous system and the physiologic consequences of space flight. Med. Sci. 308: 126-132.

132. Navidi, M., Wolinsky, I., Fung, P., and Arnaud, S.B. 1995. Effect of excess dietary salt on calcium metabolism and bone mineral in a spaceflight rat model. J. Appl. Physiol. 78: 70-75.

133. Johnson, P.C., Driscoll, T.B., and Fisher, C.L. 1977. Blood volume changes. Pp. 235-241 in Biomedical Results from Skylab (R.S. Johnston and L.F. Dietlein, eds.). NASA-SP-377. NASA, Washington, D.C.

134. Leach, C.S., and Johnson, P.C. 1984. Influence of spaceflight on erythrokinetics in man. Science 225: 216-218.

135. Udden, M.M., Driscoll, T.B., Pickett, M.H., Leach-Huntoon, C.S., and Alfrey, C.P. 1995. Decreased production of red blood cells in human subjects exposed to microgravity. J. Lab. Clin. Med. 125: 442-449.

136. Johnson, P.C., Driscoll, T.B., and Fisher, C.L. 1977. Blood volume changes. Pp. 235-241 in Biomedical Results from Skylab (R.S. Johnston and L.F. Dietlein, eds.). NASA SP-377. National Aeronautics and Space Administration, Washington, D.C.

137. Udden, M.M., Driscoll, T.B., Pickett, M.H., Leach-Huntoon, C.S., and Alfrey, C.P. 1995. Decreased production of red blood cells in human subjects exposed to microgravity. J. Lab. Clin. Med. 125: 442-449.

138. Alfrey, C.P., Udden, M.M., Leach-Huntoon, C.S., Driscoll, T., and Pickett, M.H. 1996. Control of red blood cell mass in spaceflight. J. App. Physiol. 81: 98-104.

139. Leach, C.S., and Johnson, P.C. 1984. Influence of spaceflight on erythrokinetics in man. Science 225: 216-218.

140. Alfrey, C.P., Udden, M.M., Leach-Huntoon, C.S., Driscoll, T., and Pickett, M.H. 1996. Control of red blood cell mass in spaceflight. J. App. Physiol. 81: 98-104.

141. Erslev, V. 1990. Erythropoietin. Leuk. Res. 14: 683-688.

142. Koury, M.J., and Bondurant, M.C. 1990. Erythropoietin retards DNA breakdown and prevents programmed cell death in erythroid progenitor cells. Science 248: 378-381.

143. Koury, M.J., and Bondurant, M.C. 1990. Erythropoietin retards DNA breakdown and prevents programmed cell death in erythroid progenitor cells. Science 248: 378-381.

144. Allebban, Z., Gibson, L.A., Lange, R.D., Jago, T.L., Strickland, K.M., Johnson, D.L., and Ichiki, A.T. 1996. Effects of spaceflight on rat erythroid parameters. J. Appl. Physiol. 81: 123-132.

145. Nicogossian, A.E. 1994. Microgravity simulations and analogs. Pp. 363-374 in Space Medicine and Physiology, 3rd ed. (A.E. Nicogossian, C.L. Huntoon, and S.L. Pool, eds.). Lea and Febiger, Philadelphia.

146. Whedon, G.D. 1982. Changes in weightlessness in calcium metabolism and in the musculoskeletal system. Physiologist 25: S41-S44.

147. Rambaut, P.C., Leach, C.S., and Leonard, J.I. 1977. Observations in energy balance in man during space flight. Am. J. Physiol. 233: R208-R212.

148. Rambaut, P.C., Smith, M.C., Leach, C.S., Whedon, G.D., and Reid, J. 1977. Nutrition and responses to zero gravity. Fed. Proc. 36: 1678-1682.

149. Leonard, J.I., Leach, C.S., and Rambaut, P.C. 1983. Quantitation of tissue loss during prolonged space flight. Am. J. Clin. Nutr. 38: 667-679.

150. Leach, C.S., Altschuler, S.I., and Cintron, N.M. 1982. The endocrine and metabolic response to space flight. Med. Sci. Sports Exercise 15: 432-440.

151. Stein, T.P., Leskiw, M.J., and Schluter, M.D. 1993. The effect of spaceflight on human protein metabolism. Am. J. Physiol. 264: E824-E828.

152. Sawchenko, P.E., Arias, C., Kransnov, I., Grindeland, R.E., and Vale, W. 1992. Effects of spaceflight on hypothalamic peptide systems controlling pituitary growth hormone dynamics. J. Appl. Physiol. 73: 158S-165S.

153. Hymer, W.C., Grindeland, R.E., Salada, T., Nye, P., Grossman, E.J., and Lane, P.K. 1996. Experimental modification of rat pituitary growth hormone cell function during and after spaceflight. J. Appl. Physiol. 80: 955-970.

154. Hymer, W.C., Salada, T., Avery, L., and Grindeland, R.E. 1996. Experimental modification of rat pituitary prolactin cell function during and after spaceflight. J. Appl. Physiol. 80: 971-980.

155. Vandenbergh, H.H., Haftafaludy, S., Sohar, I., and Shansky, J. 1990. Stretch induced prostaglandins and protein turnover in cultured skeletal muscle. Am. J. Physiol. 259: C232-C240.

156. Palmer, R.M., Reeds, P.J., Atkinson, T., and Smith, R.H. 1983. The influence of changes in tension on protein synthesis and prostaglandin release in isolated rabbit muscles. Biochem. J. 214: 1011-1014.

157. Edgerton, V.R., and Roy, R.R. 1997. Response of skeletal muscle to spaceflight. Pp. 105-120 in Fundamentals of Space Life Sciences, Vol. 1 (S. Churchill, ed.). Krieger Publishing Company, Malabar, Fla.

158. Fulks, R.M., Li, J.B., and Goldberg, A.L. 1975. Effects of insulin, glucose and amino acids on protein turnover in rat diaphragm. J. Biol. Chem. 250: 290-298.

159. Shangraw, R.E., Stuart, C.A., Prince, M.J., Peters, E.J., and Wolfe, R.R. 1988. Insulin responsiveness of protein metabolism in vivo following bedrest in humans. Am. J. Physiol. 255(4, Pt.1): E548-E558.

160. Bonen, A., Elde, G.C.B., and Tan, M.H. 1986. Hind limb suspension increases insulin binding and glucose metabolism. J. Appl. Physiol. 65: 1833-1839.

161. Tischler, M.E., Satarug, S., Eisenfeld, S.H., Henriksen, E.J., and Rosenberg, S.B. 1990. Insulin effects in denervated and non-weight bearing rat soleus muscle. Muscle Nerve 13: 593-600.

162. Thomason, D.B., and Booth, F.W. 1990. Atrophy of soleus muscle by hind limb unweighting. J. Appl. Physiol. 68: 1-12.

163. Leach, C.S., and Rambaut, P.C. 1977. Biochemical responses of the Skylab crewmen: An overview. Pp. 204-216 in Biomedical Results from Skylab (R.S. Johnston and L.F. Dietlein, eds.). NASA SP-377. National Aeronautics and Space Administration, Washington, D.C.

164. Alexandrov, A., Gharib, C., Grigoriev, A.I., Guell, A., Kovarinov, Y., Ruinva, L., and Smirnov, K.V. 1985. Tests d'hyperglycemie provoque par voie orale chez l'homme au cours d'un vol spatial de 150 jours (Salyut 7-Soyuz T9). C.R. Soc. Biol. 179: 192-195.

165. Stein, T.P., Schluter, M.D., and Boden, G. 1994. Development of insulin resistance during spaceflight. Aviat. Space Environ. Med. 65: 1091-1096.

166. Stein, T.P., Schluter, M.D., and Boden, G. 1994. Development of insulin resistance during spaceflight. Aviat. Space Environ. Med. 65: 1091-1096.

167. Stein, T.P., and Schluter, M.D. 1994. Excretion of IL6 by astronauts during spaceflight. Am. J. Physiol. 266 (Endocrinol. Metab.): E448-E452.

168. Grigoriev, A.I., Popova, I.A., and Ushakov, A.S. 1987. Metabolic and hormonal status of crewmembers in short-term spaceflights. Aviat. Space Environ. Med. 58 (9, Suppl.): A121-A125.

169. Leach, C.S., and Rambaut, P.C. 1977. Biochemical responses of the Skylab crewmen: An overview. Pp. 204-216 in Biomedical Results from Skylab (R.S. Johnston and L.F. Dietlein, eds.). NASA SP-377. National Aeronautics and Space Administration, Washington, D.C.

170. Leach, C.S., Altschuler, S.I., and Cintron, N.M. 1982. The endocrine and metabolic response to space flight. Med. Sci. Sports Exercise 15: 432-440.

171. Leach, C.S., and Rambaut, P.C. 1977. Biochemical responses of the Skylab crewmen: An overview. Pp. 204-216 in Biomedical Results from Skylab (R.S. Johnston and L.F. Dietlein, eds.). NASA SP-377. NASA, Washington, D.C.

172. Huntoon, C.L., Cintron, N.M., and Whitson, P.A. 1994. Endocrine and biochemical functions. Pp. 334-351 in Space Medicine and Physiology, 3rd ed. (A.E. Nicogossian, C.L. Huntoon, and S.L. Pool eds.). Lea and Febiger, Philadelphia.

173. Gauquelin, G., Maillet A., Allevard, A.M., Vorobiev D., Grigoriev, A.I., and Gharib, C. 1990. Volume regulating hormones, fluid and electrolyte modifications during the Aragatz mission (Mir station). Pp. 603-608 in Proceedings of the Fourth European Symposium on Life Sciences Research in Space, Trieste, Italy, May 28-June 1, 1990. ESA-SP-307. European Space Agency, Paris.

174. Leach, C.S., Alfrey, C.P., Suki, W.N., Leonard, J.I., Rambaut, P.C., Inners, L.D., Smith, S.M., Lane, H.W., and Krauhs, J.M. 1996. Regulation of body fluid compartments during short-term spaceflight. J. Appl. Physiol. 81(1): 105-116.

175. Kalita, N.F., and Tigranian, R.A. 1986. Endocrine status of cosmonauts following long-term space missions. Space Biol. Aerosp. Med. 20(4): 84-86.

176. Leach, C.S. 1983. Medical results from STS 1-4: Analysis of body fluids. Aviat. Space Environ. Med. 54: 550-554.

177. Huntoon, C.L., Cintron, N.M., and Whitson, P.A. 1994. Endocrine and biochemical functions. Pp. 334-351 in Space Medicine and Physiology, 3rd ed. (A.E. Nicogossian, C.L. Huntoon, and S.L. Pool, eds.). Lea and Febiger, Philadelphia.

178. Grigoriev, A.I., Noskov, V.B., Atkov, O.Y., Afonin, B.V., Sukjanov, Y.V., Lebedev, V.I., and Boiko, T.A. 1991. Fluid-electrolyte homeostasis and hormonal regulation in a 237 day space flight. Space Biol. Aerosp. Med. 25: 15-18.

179. Stein, T.P., and Schluter, M.D. 1994. Excretion of IL6 by astronauts during spaceflight. Am. J. Physiol. 266 (Endocrinol. Metab.): E448-E452.

180. Darmaun, D., Matthews, D.E., and Bier, D.M. 1988. Physiological hypercortisolemia increases proteolysis, glutamine, and alanine production. Am. J. Physiol. 255: E366-73.

181. Leach, C.S., Alfrey, C.P., Suki, W.N., Leonard, J.I., Rambaut, P.C., Inners, L.D., Smith, S.M., Lane, H.W., and Krauhs, J.M. 1996. Regulation of body fluid compartments during short-term spaceflight. J. Appl. Physiol. 81(1): 105-16.

182. Huntoon, C.L., Cintron, N.M., and Whitson, P.A. 1994. Endocrine and biochemical functions. Pp. 334-351 in Space Medicine and Physiology, 3rd ed. (A.E. Nicogossian, C.L. Huntoon, and S.L. Pool, eds.). Lea and Febiger, Philadelphia.

183. Tipton, C.M., Greenleaf, J.E., and Jackson, C.G.R. 1996. Neuroendocrine and immune system responses with space-flights. Med. Sci. Sports Exercise 28: 988-998.

184. Grigoriev, A.I., and Egorov, A.D. 1992. General mechanisms of the effect of weightlessness on the human body. Pp. 43-83 in Advances in Space Biology and Medicine, Vol. 2 (S.E. Bonting, ed.). JAI Press, Greenwich, Conn.

185. Lowry, S.F. 1992. Modulating the metabolic response to infection. Proc. Nutr. Soc. 51: 267-277.

186. Leach, C.S., Alfrey, C.P., Suki, W.N., Leonard, J.I., Rambaut, P.C., Inners, L.D., Smith, S.M., Lane, H.W., and Krauhs, J.M. 1996. Regulation of body fluid compartments during short-term spaceflight. J. Appl. Physiol. 81(1): 105-116.

187. Gmunder, F.K., Baisch, F., Bechler, B., Cogoli, A., Cogoli, M., Joller, P.W., Maass, H., Muller, J., and Ziegler, W.H. 1992. Effect of head-down tilt bedrest (10 days) on lymphocyte reactivity. Acta Physiol. Scand. 604 (Suppl.): 131-141.

188. Vernikos, J., Dallman, M.F., Keil, L.C., O'Hara, D., and Convertino, V.A. 1993. Gender differences in endocrine responses to posture and 7 days of 6 degrees head-down bed rest. Am. J. Physiol. 265(1, Pt. 1): E153-E161.

189. Vernikos, J. 1986. Metabolic and endocrine changes. Pp. 99-121 in Inactivity: Physiological Effects (H. Sandler and J. Vernikos, eds.). Academic Press, Orlando, Fla.

190. Ferrando, A.A., Lane, H.W., Stuart, C.A., Davis-Street, J., and Wolfe, R.R. 1996. Prolonged bed rest decreases skeletal muscle and whole body protein synthesis. Am. J. Physiol. (Endocrinol. Metab.) 270: E627-E633.

191. Tipton, C.M., Greenleaf, J.E., and Jackson, C.G.R. 1996. Neuroendocrine and immune system responses with space-flights. Med. Sci. Sports Exercise 28: 988-998.

192. Vernikos, J. 1986. Metabolic and endocrine changes. Pp. 99-121 in Inactivity: Physiological Effects (H. Sandler and J. Vernikos, eds.). Academic Press, Orlando, Fla.

193. Tiao, G., Lieberman, M., Fischer, J.E., and Hasselgren, P.O. 1997. Intracellular regulation of protein degradation during sepsis is different in fast and slow twitch muscles. Am. J. Physiol. (Reg. Integ. Comp. Physiol.) 41: R846-R849.

194. Tipton, C.M., Greenleaf, J.E., and Jackson, C.G.R. 1996. Neuroendocrine and immune system responses with space-flights. Med. Sci. Sports Exercise 28: 988-998.

195. Berg, H.E., and Tesch, P.A. 1996. Changes in muscle function in response to 10 days of lower limb unloading in humans. Acta Physiol. Scand. 157: 63-70.

196. Tiao, G., Lieberman, M., Fischer, J.E., and Hasselgren, P.O. 1997. Intracellular regulation of protein degradation during sepsis is different in fast and slow twitch muscles. Am. J. Physiol. (Reg. Integ. Comp. Physiol.) 41: R846-R849.

197. Sawchenko, P.E., Arias, C., Kransnov, I., Grindeland, R.E., and Vale, W. 1992. Effects of spaceflight on hypothalamic peptide systems controlling pituitary growth hormone dynamics. J. Appl. Physiol. 73 (Suppl.): 158S-165S.

198. Hymer, W.C., Grindeland, R.E., Salada, T., Nye, P., Grossman, E.J., and Lane, P.K. 1996. Experimental modification of rat pituitary growth hormone cell function during and after spaceflight. J. Appl. Physiol. 80: 955-970.

199. Hymer, W.C., Salada, T., Avery, L., and Grindeland, R.E. 1996. Experimental modification of rat pituitary prolactin cell function during and after spaceflight. J. Appl. Physiol. 80: 971-980.

200. Hymer, W.C., Grindeland, R.E., Salada, T., Nye, P., Grossman, E.J., and Lane, P.K. 1996. Experimental modification of rat pituitary growth hormone cell function during and after spaceflight. J. Appl. Physiol. 80: 955-970.

201. Hymer, W.C., Grindeland, R., Krasnov, I., Victorov, K., Motter, P., Mukherjee, K., Shellenzerger, K., and Vasques, M. 1992. Effects of spaceflight on rat pituitary cell function. J. Appl. Physiol. 73S: 151S-157S.

202. Hymer, W.C., Grindeland, R.E., Salada, T., Nye, P., Grossman, E.J., and Lane, P.K. 1996. Experimental modification of rat pituitary growth hormone cell function during and after spaceflight. J. Appl. Physiol. 80: 955-970.

203. Hymer, W.C., Grindeland, R., Krasnov, I., Victorov, K., Motter, P., Mukherjee, K., Shellenzerger, K., and Vasques, M. 1992. Effects of spaceflight on rat pituitary cell function. J. Appl. Physiol. 73S: 151S-157S.

204. Hymer, W.C., Salada, T., Avery, L., and Grindeland, R.E. 1996. Experimental modification of rat pituitary prolactin cell function during and after spaceflight. J. Appl. Physiol. 80: 971-980.

205. Hymer, W.C., Grindeland, R.E., Salada, T., Nye, P., Grossman, E.J., and Lane, P.K. 1996. Experimental modification of rat pituitary growth hormone cell function during and after spaceflight. J. Appl. Physiol. 80: 955-970.

206. Linderman, J.K., Gosselink, K.L., Booth, F.W., Mukku, V.K., and Grindeland, R.E. 1994. Resistance exercise and growth hormone as countermeasures for skeletal muscle atrophy in hindlimb suspended rats. Am. J. Physiol. 267: R365-R371.

207. Booth, F.W., and Kirby, C.R. 1992. Changes in skeletal muscle gene expression consequent to altered weight bearing. Am. J. Physiol. 262: R329-R332.

208. Kirby, C.R., Ryan, M.J., and Booth, F.W. 1992. Eccentric exercise training as a countermeasure to non-weight-bearing soleus muscle atrophy. J. Appl. Physiol. 73: 1894-1899.

209. Roy, R.R., Tri, C., Grossman, E.J., Talmadge, R.J., Grindeland, R.E., Mukku, V.R., and Edgerton, V.R. 1996. IGF-1, growth hormone, and/or exercise effects on non-weight bearing soleus of hypophysectomized rats. J. Appl. Physiol. 81: 302-311.

210. Linderman, J.K., Gosselink, K.L., Booth, F.W., Mukku, V.K., and Grindeland, R.E. 1994. Resistance exercise and growth hormone as countermeasures for skeletal muscle atrophy in hindlimb suspended rats. Am. J. Physiol. 267: R365-R371.

211. Space Science Board, National Research Council. 1987. A Strategy for Space Biology and Medical Sciences for the 1980s and 1990s. National Academy Press, Washington, D.C., p. 23.

212. Berg, H.E., and Tesch, P.A. 1996. Changes in muscle function in response to 10 days of lower limb unloading in humans. Acta Physiol. Scand. 157: 63-70.

213. Ivakhnov, A. 1987. A model of weightlessness: New research on the human organism's resources during prolonged flight. Translated from Isvestiya (Sept. 26, 1986, p. 6), Pp. 129-139 in document JPRS-USP-87-001. Joint Publications Research Service, Washington, D.C.

214. Shul'zenko, E.B., and Williams, I.F. 1976. Possibility long-term water-immersion using "dry"-immersion method. Kosm. Biol. Aviakosmicheskaya. Med. 19: 82-84.

215. Czeisler, C.A. 1995. The effect of light on the human circadian pacemaker. Ciba Found. Symp. 183: 254-302.

216. Gundel, A., Polyakov, V.V., and Zulley, J. 1997. The alteration of human sleep and circadian rhythms during spaceflight. J. Sleep Res. 6: 1-8.

217. Gundel, A., Nalishiti, E., Reucher, M., Vejvoda, M., and Zulley, J. 1993. Sleep and circadian rhythm during a short space mission. Clin. Invest. 71: 718-724.

218. Gundel, A., Polyakov, V.V., and Zulley, J. 1997. The alteration of human sleep and circadian rhythms during spaceflight. J. Sleep Res. 6: 1-8.

219. Hahn, P.M., Hoshizaki, T., and Adey, W.R. 1971. Circadian rhythms of the *Macaca nemestrina* monkey in Biosatellite III. Aerosp. Med. 42: 295-304.

220. Sulzman, F.M., Ferraro, J.S., Fuller, C.A., Moore-Ede, M.C., Klimovitsky, V., Magedov, V., and Alpatove, A.M. 1992. Thermoregulatory responses of rhesus monkeys during spaceflight. Physiol. Behav. 51: 585-591.

221. Alpatov, A.M., Reitveld, W.J., and Oryntaeva, L.B. 1994. Impact of microgravity and hypergravity on free running circadian rhythm of the desert beetle *Trigonoscelis gigas* Reitt. Biol. Rhythm Res. 25: 168-177.

222. Fuller, C.A., Hoban-Higgins, T.M., Klimovitsky, V.L., Griffin, D.W., and Alpatov, A.M. 1996. Primate circadian rhythms during spaceflight: Results from Cosmos 2044 and 2229. J. Appl. Physiol. 81: 188-193.

223. Young, V.R., and Marchini, J.S. 1990. Mechanisms and nutritional significance of metabolic responses to altered intakes of protein and amino acids with reference to nutritional adaptation in humans. Am. J. Clin. Nutr. 51: 270-289.

224. Moore-Ede, M.C., Sulzman, F.M., and Fuller, C.A. 1982. The Clocks That Time Us: The Circadian Timing System in Mammals. Harvard University Press, Cambridge, Mass.

225. Hahn, P.M., Hoshizaki, T., and Adey, W.R. 1971. Circadian rhythms of the *Macaca nemestrina* monkey in Biosatellite III. Aerosp. Med. 42: 295-304.

226. Fuller, C.A., Hoban-Higgins, T.M., Klimovitsky, V.L., Griffin, D.W., and Alpatov, A.M. 1996. Primate circadian rhythms during spaceflight: Results from Cosmos 2044 and 2229. J. Appl. Physiol. 81: 188-193.

227. Fuller, C.A. 1985. Homeostasis and biological rhythms in the rat. Physiologist 28: S199-S200.

228. Moore-Ede, M.C., Sulzman, F.M., and Fuller, C.A. 1982. The Clocks That Time Us: The Circadian Timing System in Mammals. Harvard University Press, Cambridge, Mass.

229. Alpatov, A.M., Reitveld, W.J., and Oryntaeva, L.B. 1994. Impact of microgravity and hypergravity on free running circadian rhythm of the desert beetle *Trigonoscelis gigas* Reitt. Biol. Rhythm Res. 25: 168-177.

230. Sulzman, F.M., Ellman, D., Fuller, C.A., Moore-Ede, M.C., and Wassner, G. 1984. Neurospora circadian rhythm in space. A reexamination of the endogenous/exogenous question. Science 225:232-234.

231. Czeisler, C.A., Allan, J.S., Strogatz, S.H., Rnda, J.M., Sanchez, R., Rios, C.D., Freitag, W.O., Richardson, G.S., and Kronauer, R.E. 1986. Bright light rests the human circadian pacemaker independent of the timing of the sleep-wake cycle. Science 233: 667-671.

232. Vernikos, J., Dallman, M.F., Keil, L.C., O'Hara, D., and Convertino, V.A. 1993. Gender differences in endocrine responses to posture and 7 days of 6 degrees head-down bed rest. Am. J. Physiol. 265(1, Pt. 1): E153-E161.

10

Immunology

INTRODUCTION

Spaceflight produces marked effects on several parameters of immune responses. Neither the biological or biomedical significance nor the mechanism(s) of induction of these changes have yet been established.

Multiple factors could be involved in the effects of spaceflight on immune responses. In spaceflight as on Earth, the immune system interacts dynamically with other body systems. Factors that must be considered during spaceflight include changes induced by microgravity; changes induced in the neuroendocrine stress hormone system and other stress responses; and changes induced by exposure to radiation, by alterations in nutritional intake, by alterations in levels of 1,25-dihydroxyvitamin D_3 (i.e., changes in calcium), by acceleration and deceleration forces, and by possible alterations in prolactin and growth hormone levels.[1-3] These factors may act independently or combine with effects of exposure to microgravity to produce additive or synergistic effects on the immune system. For example, changes in diet and workload during spaceflight could trigger hormonal responses that in turn affect immune responses. The interaction of the immune response with the hypothalamic-pituitary-adrenal (HPA) axis could also play a major role. The discipline of psychoneuroimmunology has recently been established to facilitate the study of these interactions.[4] It appears that there is major interaction between these two systems, and factors such as stress can have a great impact on immune responses and resistance to infection. These interactions could play a major role as one potential mechanism for the effects of spaceflight on immune responses. Additional and as yet unknown factors could also play some role.

A variety of observations made during and after spaceflights have generated an interest in the effects of spaceflight on immune responses. Many of the early studies reporting possible infections may actually have observed "space motion sickness" and headward fluid shifts that produced symptoms that at that time were indistinguishable from those of the common cold and influenza.[5] Astronauts who were isolated before flight in Apollo missions had decreased problems with upper respiratory tract infections compared to those not isolated.[6,7] During the Apollo 13 mission, one astronaut developed a urinary tract infection from *Pseudomonas aeruginosa*.[8,9] One Mir cosmonaut was removed from the space station because of possible but unconfirmed upper respiratory tract infections (I. Konstantinova, personal communication, 1992).[10] These results are evidence of limited problems with infectious diseases during spaceflight.[11] To date, there has been no evidence of extensive problems with infections during and after spaceflight. However, as plans for very long term missions and colonies in space develop, the potential for problems with infectious diseases increases and should be explored. In addition, the role of microgravity in spaceflight-induced alterations of in vivo immune responses has not yet been established.

SPACEFLIGHT EXPERIMENTS

Animal Studies

Animal studies have focused on cell-mediated immunity and have attempted to evaluate the greatest possible number of immunological parameters affected by spaceflight.[12,13] Studies on antibody responses, mechanistic studies, and studies to determine the biological and biomedical significance of spaceflight-induced alterations in immunological parameters have been few in number and limited in scope, owing primarily to technical difficulties in carrying out experiments and not to lack of interest or potential effects. Most spaceflight studies conducted with animals have involved specimens shared by multiple research groups in different disciplines. Experiments have had to be designed so that they would not interfere with or compromise other experiments. Therefore, studies involving resistance to infections, sensitization to an antigen, and determination of antibody levels have not been carried out.

Access to and experimentation on animals during flight in current spacecraft is also difficult and has been conducted in only a limited fashion. Most studies have been carried out immediately upon return of the animals to Earth. Reentry acceleration and other forces could have affected results obtained after landing.[14,15]

The principal experimental mammalian animal for spaceflight studies has been the rat. Housing suitable for spaceflight has been developed for rats. In addition, because of its larger size, the rat has been most useful for studies requiring sharing of tissues among different experimental groups. Multiple modifications currently under development will be required before housing for spaceflight will be available for use with mice. However, use of the mouse in spaceflight studies has several benefits, including (1) the potential for using more animals in one study, (2) the availability of many immunological reagents that are specific for mice and unavailable for rats, and (3) the ability to use genetically unique strains of mice (including "knockout" and transgenic mice) for study. A limited number of studies have been carried out with rhesus monkeys.[16,17]

Experiments have been carried out using both Russian and U.S. spacecraft. Although housing, in-flight environmental conditions, duration and apogee of flight, and landing conditions have differed, the results have generally been consistent.[18-30] All flights carrying experimental animals have been of relatively short duration, usually from 1 to 2 weeks. Almost all experiments, except where noted, have been carried out immediately after return to Earth (usually within 2 to 4 hours, but sometimes as late as 24 hours after return).

Early studies indicated involution of the thymus after spaceflight.[31] Alterations in thymus and other tissue were later confirmed but shown to possibly be transient in nature.[32,33]

Although early experiments carried out in the Russian Cosmos biosatellite to determine the effects of spaceflight on rat leukocyte blastogenesis showed no effects of spaceflight,[34] later studies found compartmentalization of the effects of spaceflight on blastogenesis. Leukocytes obtained from lymph nodes were affected differently from cells obtained from systemic tissue.[35,36]

The majority of studies have involved acquired immunity (specific immunity to pathogens and tumors requiring previous exposure to the pathogen or tumor). An experiment with rats flown on the Space Shuttle mission SL-3 studied production of cytokines (soluble mediators such as interleukins and interferons that carry messages between cells of the immune response) after landing.[37] Spleens were removed from the rats within a few hours after landing and challenged with mitogen to induce cytokine production. Interferon-γ production was greatly reduced; however, interleukin-3 measurements made of the same culture's supernatant fluids showed no decrease in interleukin-3 production. Later work showed decreases in the production of other cytokines, including interleukin-2, only at certain times after spaceflight.[38,39] Production of other cytokines such as interleukin-3 and interleukin-6 also increased after spaceflight.[40] However, there was compartmentalization of these responses; production of interleukin-6 by spleen cells was unaffected, but production of interleukin-6 by thymus cells increased. Compartmentalization indicates that there is no overall blunting of the immune response after spaceflight, but selective effects instead.[41]

Although studies of the effects of spaceflight on cytokine production have been carried out, there has never been an attempt to establish whether or not spaceflight induces changes in Th1 and Th2 cytokine production profiles of helper T lymphocytes. A Th1 cytokine production profile indicates a prevalence toward development of cell-mediated immune reactions, whereas a Th2 cytokine production profile shows a prevalence toward development of antibody-mediated immunity. Shifts between Th1 and Th2 profiles of cytokine production could indicate spaceflight-induced regulatory changes in immune responses and could be of great significance.

Some studies have been conducted on the effects of spaceflight on innate immunity (nonspecific immunity always present). Initial studies on cells from rats flown on the Russian Cosmos biosatellite showed a decrease in the ability of spleen natural killer cells to kill target tumor cells compared with controls, but later studies showed that this was selective; the ability of natural killer cells from animals flown in space to kill different target tumor cells was unaffected.[42] These measurements were made in cells from animals euthanized immediately after flight. In later experiments, rats were euthanized in flight aboard the Space Shuttle mission SLS-2.[43] This was the first in-flight animal study, and the animals were euthanized 1 day before landing. Spleens were removed and refrigerated, and the assay was carried out after landing. In this case, the ability of spleen cells from the rats euthanized in space to kill both types of tumor target cells was inhibited compared with that of spleen cells from controls euthanized on Earth and maintained at 4°C for the same length of time as the in-flight samples. When spleen cells from animals flown aboard SLS-2 but euthanized immediately upon return to Earth were tested, only the ability to kill 1 type of target cells was affected,[44] repeating the results of previous studies. These important data indicate that in-flight sampling is extremely important, as some immunological changes occurring during flight could reverse very rapidly after a return to Earth from space.

Experiments have also centered on colony stimulating factor responsiveness and leukocyte subset distribution. The response to colony stimulating factors of bone marrow cells from rats flown on several Russian Cosmos biosatellite missions has been examined.[45,46] The response of cells from rats flown in space to both granulocyte-macrophage colony stimulating factor and macrophage colony stimulating factor was greatly reduced after spaceflight, compared with the response of cells from control animals housed in normal caging or in caging designed to simulate conditions in the space capsule.[47,48] Some alterations in leukocyte subset distribution were also noted after spaceflight of rats on the Cosmos capsule and the space shuttle, most notably an increase in the level of CD4+ helper T cells.[49-52]

A control experiment was carried out using centrifugation of rats on Earth in housing conditions similar to those used in the Cosmos biosatellite flight studies to determine if hypergravity could affect immune parameters. Hypergravity had no effect on any of the immune parameters shown to be affected during the Cosmos biosatellite flights.[53]

Results from one flight experiment using rhesus monkeys have been generally consistent with those observed using rats; in rhesus monkeys flown in a Russian Cosmos biosatellite mission, interleukin-1 levels, interleukin-2 receptor levels, and bone marrow cell responses to granulocyte-macrophage colony stimulating factor decreased.[54]

Although this evidence shows that immunological parameters are affected by spaceflight, no evidence to date indicates that these changes are of biological or medical significance. The use of animal models could help to resolve this uncertainty and could influence future considerations as to whether spaceflight-induced changes in immune parameters are an important issue for the health and safety of space travelers. This issue should be given the highest priority. One reason that this issue has not been addressed before is that animals have had to be shared among investigators in multiple disciplines, and so carrying out functional immunological studies could affect the results of other investigators in other disciplines. It is hoped that research in the International Space Station era will allow assignment of specific animals for these important immunological studies. Mice would be the species of choice for the studies because of the plethora of immunological reagents available. Functional immunological studies could also be performed in the rat, the rodent species used most frequently in previous spaceflight studies. If studies using monkeys are carried out again, immunological experiments could be included using this species.

Recommendations

The biological and biomedical significance of spaceflight-induced changes in immune responses should be investigated in both short- and long-term studies, preferably in mice or, if necessary, in rats. The types of animal studies that should be carried out, listed in priority order, include the following:

1. Resistance to infection. Studies of resistance to infection might create some difficulty if carried out entirely in the space station environment, because of the difficulty in bringing potential pathogens into that environment. Therefore, studies of infection should not be carried out on board the space station or the space shuttle but should be performed on animals flown in space immediately upon their return to Earth. This would model the period immediately after return to Earth when crews leave closed environments, which could be the time of crew members' greatest risk from infection. All other animal studies should be carried out both on the space station and upon return to Earth.

Animals exposed to spaceflight conditions or suspension modeling for short- or long-term periods should be infected with viruses (e.g., influenza and herpes viruses) or bacteria (e.g., Salmonella typhimurium) and spaceflight-induced changes in resistance to infection determined. These would test the relevance of spaceflight effects on both antibody and cell-mediated immunity.

2. Acquired immune responses

• *Studies of antigen sensitization and resistance to infections should be carried out to determine if spaceflight affects the ability to mount a new immune response and resistance to infection. Sensitization studies could be carried out in flight using innocuous sensitizing antigens such as keyhole limpet hemocyanin.*

- *Humoral immune responses should be tested directly after immunization and challenge by determination of antibody specificity and type. Th1 and Th2 cytokine profiles should be established, not only in nonimmune animals but also following specific immune challenge.*
- *Cell-mediated immunity should be tested with both delayed-type hypersensitivity responses to contact-sensitivity agents, and generation of cytotoxic T lymphocyte activity.*

3. Interactions with the HPA-axis and other body systems. When measurements are taken for immune responses, simultaneous determination of stress-related mediators (including stress hormones, catecholamines, and neuropeptide Y) should be carried out. Measurements of blood pressure, heart rate, orthostatic intolerance, and other appropriate variables/parameters should also be collated and made available to allow determination of interactions of the immune system with other body systems, including the musculoskeletal system.

Human Studies

Few studies have been carried out to determine the effects of spaceflight on immune responses of humans. Human experimentation has been restricted because of necessary limited access to crew members. However, it has been possible to study the effects of long-term spaceflights of several months to greater than 1 year.[55]

Several alterations in immunological parameters have been observed in crew members after short-term space shuttle flights. These studies were carried out after landing, and so the relative roles played by microgravity, psychological stress, and landing stresses in inducing the alterations cannot be differentiated. The results presented here are a consensus of data from most flight studies, although contradictory data do exist. The alterations reported include decreases in lymphocyte number, decreases in leukocyte blastogenesis, increases in leukocyte number, alterations in the relative percentage of B and T lymphocytes, decreases in monocytes, increases in helper T lymphocytes, decreases in cytotoxic T lymphocytes, and an increase in the ratio of CD4+/CD8+ lymphocytes.[56-59] These alternatives may be associated with certain neuroendocrine system changes, particularly in catecholamines such as epinephrine and norepinephrine.[60] Again, it must be noted that there are individual flights where there have been no changes or changes with shifts in the opposite direction from those reported in the summary above. Conditions, including flight duration, have varied for every flight. This variability makes it difficult to obtain a standard effect of spaceflight on every immunological parameter. Regardless of the individual flight-specific differences in results, it is clear from the data to date that spaceflight can alter immunological parameters.

Changes observed in cells obtained from cosmonauts immediately after Russian short-term flights include decreases in natural killer cell activity and decreases in production of interferon-α/β.[61-63] Interestingly, the decreased interferon-α/β production was from the same mission and the same cosmonauts whose peripheral blood leukocytes were placed in culture and challenged with an interferon inducer in the in vitro space experiment of Tálas et al., in which interferon-α/β production was markedly enhanced.[64] These results reiterate the point that in vitro spaceflight results may not be representative of the in vivo situation, because cells in culture are not in their normal environment involving interactions with other body systems.

When testing was carried out on samples from cosmonauts immediately upon their return to Earth after long-duration spaceflight, the most prominent effects were alterations in natural killer cell activity, alterations in leukocyte blastogenesis, and alterations in interleukin-2 production.[65,66] The same studies also reported increases in the level of serum immunoglobulins, particularly total serum IgA and IgM. The adaptability of the human immune system to long-term spaceflight conditions has yet to be established.

Some recent attempts have been made to address the issue of sampling of spaceflight effects on the immune response of humans in midflight.[67,68] These are significant studies, as they allow for differentiation between spaceflight effects and landing stress effects on immune responses. Additional studies of this nature are needed. These results were originally reported with data obtained from astronauts during relatively short-term space shuttle flight[69] but have been extended to longer-term Mir space station experiments.[70] These experiments involved delayed hypersensitivity skin testing to common recall antigens and did show an effect of spaceflight. There was a marked decrease in the skin-test response to these antigens when the crews were tested during spaceflight.

Recommendations

Immunological measurements and testing of humans should be carried out to look at parameters with potential functional consequences. Additional human immunological studies and development of potential countermeasures should be conducted *only* if the animal infection studies and the initial human functional studies show that spaceflight-induced changes in immune parameters are of biological and biomedical importance. The following studies, listed in order of priority, should be carried out:

1. Acquired immunity
  • *Delayed-type hypersensitivity responses should be tested in flight and postflight to expand this database.*
  • *Influenza vaccine responses should be monitored in spaceflight and after return to Earth to test both T helper cell function and antibody production.*
  • *Antibody responses to latent viral epitopes such as oral herpes virus or Epstein-Barr virus should be determined.*
  • *If the animal infection studies and the above-mentioned human functional immunity studies show biologically and biomedically important results, then both ground-based and in-flight studies of Th1 (cell-mediated) cytokines (e.g., interleukin-2, interferon-γ, interleukin-12) and Th2 (humoral) cytokines (e.g., interleukin-4, interleukin-10) should be conducted. Each subject should be used as his or her own control, to ensure that individual variation is taken into consideration for each experiment and to allow all subjects to be followed longitudinally over time.*

2. Innate immunity. Natural killer cell numbers and activity, as well as neutrophil function (oxygen burst activity) using peripheral blood samples should be measured.

3. Epidemiological studies. To assess potential risk from infection (and in particular, tumors), complete, longitudinal, and comprehensive records should be kept for all space travelers and should be made available for epidemiological studies and risk assessment.

4. Interactions with the HPA-axis and other body systems. In humans exposed to ground-based modeling and to spaceflight, when measurements are taken for immune responses, simultaneous determination of stress-related mediators (including stress hormones, catecholamines, and neuropeptide Y) should be carried out. Measurements of blood pressure, heart rate, orthostatic intolerance, and other appropriate parameters should also be collated and made available to allow determination of interactions of the immune system with other body systems, including the musculoskeletal system.

Cell Culture Studies

In cell culture studies of the immune system, cells are isolated from their normal interactions with other body systems such as the neuroendocrine, cardiovascular, and musculoskeletal systems. Results of these studies are not necessarily representative of events occurring in vivo.

All space-based studies to date of cells in culture have been limited by technical limitations of the equipment available, leading to limitations on the design of experiments (see Chapter 2). Early studies indicated that mitogen-mediated blastogenesis of human peripheral blood leukocytes grown in culture during spaceflights was severely inhibited.[71-74] This was attributed to a possible direct effect of microgravity on lymphocytes. However, other mechanisms could possibly be involved. For example, blastogenesis of lymphocytes requires the assistance of macrophages as accessory cells. Changes in fluid convection due to microgravity could have prevented necessary interactions between lymphocytes and macrophages that could have led to inhibited blastogenesis after spaceflight.[75,76] There have also been some reports of hypergravity enhancing in vitro leukocyte blastogenesis, although the mechanism remains open to debate.[77] Changes in T cell and macrophage function have also been reported.[78-81] Killing of target cells by tumor necrosis factor-α was also inhibited in spaceflight by a mechanism involving protein kinase C; however, no effect on production of superoxide was noted.[82,83] Studies have shown enhanced interferon-δ, interleukin-1, interleukin-2, and tumor necrosis factor-α production from cultures of lymphoid cells flown in space.[84-90] All of these experiments require additional studies to confirm results.

Recommendation

Although cell culture results may not exactly mirror events in vivo, they are potentially useful for exploration of mechanisms. These studies should be carried out after confirmation of the biological and biomedical importance of the effects of spaceflight on the immune systems.

Hypothesis-driven tissue culture studies of immune cells in spaceflight should be carried out only after there is proof of biological significance and with availability of reliable tissue culture hardware, including in-flight centrifuge controls.

GROUND-BASED MODELS OF THE EFFECTS OF SPACEFLIGHT ON IMMUNE RESPONSES

Ground-based models of some of the effects of spaceflight on immune responses have become a major step in the design of spaceflight experiments. Because opportunities to carry out flight experiments are relatively rare and very expensive, modeling is useful in planning for flight experiments. Since it is impossible to faithfully model microgravity on Earth, all models have deficiencies and strengths that must be taken into consideration in evaluating results.

Models that have been used include hindlimb unloading in rodents,[91-93] chronic bed rest in humans,[94-96] and rotation of cells in a clinostat.[97] Hindlimb unloading of rodents has been used to simulate some aspects of the effects of microgravity on immune responses.[98] Two different varieties of the model have been used: suspension by the tail and harness suspension.[99] There are benefits and drawbacks to both varieties of the model; however, studies with both involving immune responses have shown similar results to date.[100-105] Suspension of rats and mice with no load bearing on the hindlimbs and with head-down tilt (usually 15° to 20°) leads to bone and muscle disuse and a fluid shift to the head.

Although this model has many limitations because microgravity cannot be truly modeled on Earth, it is the best model available for rodent studies. One positive aspect of the use of hindlimb unloading with a head-down tilt is that rodents suspended without head-down tilt (i.e., no-head-down tilt rodents) can be used as a control for the stress of the model.

Hindlimb unloading with a head-down tilt in rats has been shown to yield changes in several dynamic immune parameters, but to have little to no effect on more static immune parameters.[106,107] Immune parameters altered after hindlimb unloading with a head-down tilt of rodents have included interferon-α/β and -γ production, the response of bone marrow cells from suspended rats to exogenous colony stimulating factors, macrophage function, and interleukin production.[108-111] There was no correlation between corticosterone levels and alterations in immune responses.[112] The situation with neutrophil function is less clear; in mice, neutrophil function was inhibited after hindlimb unloading with a head-down tilt,[113] whereas in rats, there was no effect on neutrophil function.[114] There are several possible explanations for these differences, including differences in the type of hindlimb unloading used and the response of different species to hindlimb unloading. Immunological parameters such as leukocyte subset distribution were not affected by hindlimb unloading with a head-down tilt.[115-117] The reason for differential effects on immune responses is not clear.

Studies have shown that resistance to infection is altered by hindlimb unloading with a head-down tilt.[118] Female mice normally resistant to infection with the [D] variant of encephalomyocarditis virus became susceptible to infection after 4 days of hindlimb unloading with a head-down tilt.[119] The decreased resistance is correlated with the drop in interferon production seen after hindlimb unloading. In contrast, mice subjected to hindlimb unloading with a head-down tilt demonstrated enhanced immunological memory and resistance to infection with *Listeria monocytogenes*.[120-121] These differences in resistance responses to bacteria and viruses may be related to differences in the effects of hindlimb unloading on the function of lymphocytes (for the virus) and macrophages (for the bacteria).[122-124]

Benefits can also be obtained using human models of spaceflight effects on immune responses. Several models have been used for immunological studies, including isolation, high altitude, and chronic bed rest.[125-131] Chronic bed rest is the most frequently used model and for humans is equivalent to hindlimb unloading with a head-down tilt in rodents. Bed rest eliminates load bearing on limbs and uses head-down tilt to model fluid shifts.[132] Results of the various model studies cited above have demonstrated alterations in interferon and interleukin production. These include increases in interleukin-1 production (which could affect bone), decreases in interleukin-2 production, and changes in leukocyte subset distribution and in neutrophil and macrophage function.

A clinostat is an apparatus that allows for rotation of cells so that the direction of the gravity vector is not constant.[133] Its use for studies on cells important in the immune response, which do not have classical receptors for gravity, has been hotly debated. Nevertheless, rotation of cells in a clinostat has shown decreased T-lymphocyte activation[134] and decreased leukocyte migration through type 1 collagen.[135] Mechanisms that cause these changes in the clinostat may be different from those that could occur during spaceflight; however, the data obtained could prove useful in planning experiments to be carried out in space.

REFERENCES

1. Nicogossian, A.E., Huntoon, C.L., and Pool, S., eds. 1989. Space Physiology and Medicine, 2nd ed. Lea and Febiger, Philadelphia.

2. Churchill, S.E. 1997. Fundamentals of Space Life Sciences, Vols. 1 and 2. Krieger Publishing Co., Malabar, Fla.

3. Sonnenfeld, G., and Taylor, G.R. 1991. Effect of microgravity on the immune system. SAE Technical Paper Series No. 911515. Society of Automotive Engineers, Warrendale, Pa.

4. Ader, R., Cohen, N., and Felten, D. 1995. Psychoneuroimmunology: Interactions between the nervous system and the immune system. Lancet 345: 99-103.
5. Taylor, G.R., Konstantinova, I.V., Sonnenfeld, G., and Jennings, R. 1997. Changes in the immune system during and after spaceflight. Adv. Space Biol. Med. 6: 1-32.
6. Taylor, G.R., Konstantinova, I.V., Sonnenfeld, G., and Jennings, R. 1997. Changes in the immune system during and after spaceflight. Adv. Space Biol. Med. 6: 1-32.
7. Taylor, G.R. 1974. Recovery of medically important microorganisms from Apollo astronauts. Aerosp. Med. 45: 824-828.
8. Taylor, G.R., Konstantinova, I.V., Sonnenfeld, G., and Jennings, R. 1997. Changes in the immune system during and after spaceflight. Adv. Space Biol. Med. 6: 1-32.
9. Taylor, G.R. 1974. Recovery of medically important microorganisms from Apollo astronauts. Aerosp. Med 45: 824-828.
10. Konstantinova, I.V., Rykova, M.P., Lesnyak, A.T., and Antropova, E.A. 1993. Immune changes during long duration missions. J. Leukocyte Biol. 54: 189-201.
11. Taylor, G.R., Konstantinova, I.V., Sonnenfeld, G., and Jennings, R. 1997. Changes in the immune system during and after spaceflight. Adv. Space Biol. Med. 6: 1-32.
12. Taylor, G.R., Konstantinova, I.V., Sonnenfeld, G., and Jennings, R. 1997. Changes in the immune system during and after spaceflight. Adv. Space Biol. Med. 6: 1-32.
13. Konstantinova, I.V., Rykova, M.P., Lesnyak, A.T., and Antropova, E.A. 1993. Immune changes during long duration missions. J. Leukocyte Biol. 54: 189-201.
14. Taylor, G.R., Konstantinova, I.V., Sonnenfeld, G., and Jennings, R. 1997. Changes in the immune system during and after spaceflight. Adv. Space Biol. Med. 6: 1-32.
15. Konstantinova, I.V., Rykova, M.P., Lesnyak, A.T., and Antropova, E.A. 1993. Immune changes during long duration missions. J. Leukocyte Biol. 54: 189-201
16. Taylor, G.R., Konstantinova, I.V., Sonnenfeld, G., and Jennings, R. 1997. Changes in the immune system during and after spaceflight. Adv. Space Biol. Med. 6: 1-32.
17. Konstantinova, I.V., Rykova, M.P., Lesnyak, A.T., and Antropova, E.A. 1993. Immune changes during long duration missions. J. Leukocyte Biol. 54: 189-201
18. Sonnenfeld, G., Mandel, A.D., Konstantinova, I.V., Berry, W.D., Taylor, G.R., Lesnyak, A.D., Fuchs, B.B., and Rakhmilevich, A. 1992. Spaceflight alters immune cell function and distribution. J. Appl. Physiol. 73: 191S-195S.
19. Durnova, G.N., Kaplansky, A.S., and Portugalov, V.V. 1977. Effect of 22-day spaceflight on lymphoid organs of rats. Aviat. Space Environ. Med. 47: 588-591.
20. Congdon, C.C., Allebban, Z., Gibson, L.A., Jago, T.L., Strickland, K.M., Johnson, D.L., Lange, R.D., and Ichiki, A.T. 1996. Lymphatic changes in rats flown on Spacelab Life Sciences-2. J. Appl. Physiol. 81: 172-177.
21. Ichiki, A.T., Gibson, L.A., Jago, T.L., Strickland, K.M., Johnson, D.L., Lange, R.D., and Allebban, Z. 1996. Effects of spaceflight on rat peripheral blood leukocytes and bone marrow progenitor cells. J. Leukocyte Biol. 60: 37-43.
22. Mandel, A.D., and Balish, E. 1977. Effect of spaceflight on cell-mediated immunity. Aviat. Space Environ. Med. 48: 1051-1057.
23. Nash, P., Konstantinova, I.V., Fuchs, B., Rakhmilevich, A., Lesynak, A., and Mastro, A.M. 1992. Effect of spaceflight on lymphocyte proliferation and interleukin-2 production. J. Appl. Physiol. 73: 186S-190S.
24. Nash, P., and Mastro, A.M. 1992. Variable lymphocyte responses in rats after spaceflight. Exp. Cell Res. 202: 125-131.
25. Gould, C.L., Lyte, M., Williams, J.A., Mandel, A.D., and Sonnenfeld, G. 1987. Inhibited interferon-gamma but normal interleukin-3 production from rats flown on the Space Shuttle. Aviat. Space Environ. Med. 58: 983-986.
26. Miller, E.S., Koebel, D.A., and Sonnenfeld, G. 1995. Influence of spaceflight on the production of interleukin-3 and interleukin-6 by rat spleen and thymus cells. J. Appl. Physiol. 78: 810-813.
27. Sonnenfeld, G. and Miller, E.S. 1993. The role of cytokines in immune changes induced by spaceflight. J. Leukocyte Biol. 54: 253-258.
28. Rykova, M., Sonnenfeld, G., Lesnyak, A.D., Taylor, G.R., Meshkov, D., Mandel, A.D., Medvedev, A., Berry, W.D., Fuchs, B.B., and Konstantinova, I.V. 1992. Effect of spaceflight on natural killer cell activity. J. Appl. Physiol. 73: 196S-200S.
29. Lesnyak, A.D., Sonnenfeld, G., Avery, L., Konstantinova, I.V., Rykova, M., Meshkov, D., and Orlova, T. 1996. Effects of SLS-2 spaceflight on immunologic parameters. J. Appl. Physiol. 81: 178-182.
30. Sonnenfeld, G., Mandel, A.D., Konstantinova, I.V., Taylor, G.R., Berry, W.D., Wellhausen, S.R., Lesnyak, A.T., and Fuchs, B.B. 1990. Effects of spaceflight on levels of activity of immune cells. Aviat. Space Environ. Med. 61: 648-653.
31. Durnova, G.N., Kaplansky, A.S., and Portugalov, V.V. 1997. Effect of 22-day spaceflight on lymphoid organs of rats. Aviat. Space Environ. Med. 47: 588-591.

32. Congdon, C.C., Allebban, Z., Gibson, L.A., Kaplansky, A., Strickland, K.M., Jago, T.L., Johnson, D.L., Lange, R.D., and Ichiki, A.T. 1996. Lymphatic changes in rats flown on Spacelab Life Sciences-2. J. Appl. Physiol. 81: 172-177.

33. Ichiki, A.T., Gibson, L.A., Jago, T.L., Strickland, K.M., Johnson, D.L., Lange, R.D., and Allebban, Z. 1996. Effects of spaceflight on rat peripheral blood leukocytes and bone marrow progenitor cells. J. Leukocyte Biol. 60: 37-43.

34. Mandel, A.D., and Balish, E. 1977. Effect of spaceflight on cell-mediated immunity. Aviat. Space Environ. Med. 48: 1051-1057.

35. Nash, P., Konstantinova, I.V., Fuchs, B., Rakhmilevich, A., Lesnyak, A., and Mastro, A.M. 1992. Effect of spaceflight on lymphocyte proliferation and interleukin-2 production. J. Appl. Physiol. 73: 186S-190S.

36. Nash, P., and Mastro, A.M. 1992. Variable lymphocyte responses in rats after spaceflight. Exp. Cell Res. 202: 125-131.

37. Gould, C.L., Lyte, M., Williams, J.A., Mandel, A.D., and Sonnenfeld, G. 1987. Inhibited interferon-gamma but normal interleukin-3 production from rats flown on the Space Shuttle. Aviat. Space Environ. Med. 58: 983-986.

38. Nash, P., Konstantinova, I.V., Fuchs, B., Rakhmilevich, A., Lesnyak, A., and Mastro, A.M. 1992. Effect of spaceflight on lymphocyte proliferation and interleukin-2 production. J. Appl. Physiol. 73: 186S-190S.

39. Nash, P., and Mastro, A.M. 1992. Variable lymphocyte responses in rats after spaceflight. Exp. Cell Res. 202: 125-131.

40. Miller, E.S., Koebel, D.A., and Sonnenfeld, G. 1995. Influence of spaceflight on the production of interleukin-3 and interleukin-6 by rat spleen and thymus cells. J. Appl. Physiol. 78: 810-813.

41. Sonnenfeld, G., and Miller, E.S. 1993. The role of cytokines in immune changes induced by spaceflight. J. Leukocyte Biol. 54: 253-258.

42. Rykova, M., Sonnenfeld, G., Lesnyak, A.D., Taylor, G.R, Meshkov, D., Mandel, A.D., Medvedev, A., Berry, W.D., Fuchs, B.B., and Konstantinova, I.V. 1992. Effect of spaceflight on natural killer cell activity. J. Appl. Physiol. 73: 196S-200S.

43. Lesnyak, A., Sonnenfeld, G., Avery, L., Konstantinova, I., Rykova, M., Meshkov, D., and Orlova, T. 1996. Effect of SLS-2 spaceflight on immunologic parameters. J. Appl. Physiol. 81: 178-182.

44. Lesnyak, A., Sonnenfeld, G., Avery, L., Konstantinova, I., Rykova, M., Meshkov, D., and Orlova, T. 1996. Effect of SLS-2 spaceflight on immunologic parameters. J. Appl. Physiol. 81: 178-182.

45. Sonnenfeld, G., Mandel, A.D., Konstantinova, I.V., Berry, W.D., Taylor, G.R., Lesnyak, A.D., Fuchs, B.B., and Rakhmilevich, A. 1992. Spaceflight alters immune cell function and distribution. J. Appl. Physiol. 73: 191S-195S.

46. Sonnenfeld, G., Mandel, A.D., Konstantinova, I.V., Taylor, G.R., Berry, W.D., Wellhausen, S.R., Lesnyak, A.T., and Fuchs, B.B. 1990. Effects of spaceflight on levels and activity of immune cells. Aviat. Space Environ. Med. 61: 648-653.

47. Sonnenfeld, G., Mandel, A.D., Konstantinova, I.V., Berry, W.D., Taylor, G.R., Lesnyak, A.D., Fuchs, B.B., and Rakhmilevich, A. 1992. Spaceflight alters immune cell function and distribution. J. Appl. Physiol. 73: 191S-195S.

48. Sonnenfeld, G., Mandel, A.D., Konstantinova, I.V., Taylor, G.R., Berry, W.D., Wellhausen, S.R., Lesnyak, A.T., and Fuchs, B.B. 1990. Effects of spaceflight on levels and activity of immune cells. Aviat. Space Environ. Med. 61: 648-653.

49. Sonnenfeld, G., Mandel, A.D., Konstantinova, I.V., Berry, W.D., Taylor, G.R., Lesnyak, A.D., Fuchs, B.B., and Rakhmilevich, A. 1992. Spaceflight alters immune cell function and distribution. J. Appl. Physiol. 73: 191S-195S.

50. Ichiki, A.T., Gibson, L.A., Jago, T.L., Strickland, K.M., Johnson, D.L., Lange, R.D., and Allebban, Z. 1996. Effects of spaceflight on rat peripheral blood leukocytes and bone marrow progenitor cells. J. Leukocyte Biol. 60: 37-43.

51. Sonnenfeld, G., Mandel, A.D., Konstantinova, I.V., Taylor, G.R., Berry, W.D., Wellhausen, S.R., Lesnyak, A.T., and Fuchs, B.B. 1990. Effects of spaceflight on levels and activity of immune cells. Aviat. Space Environ. Med. 61: 648-653.

52. Allebban, Z., Ichiki, A.T., Gibson, L.A., Jones, J.B., Congdon, C.C., and Lange, R.D. 1994. Effects of spaceflight on the number of rat peripheral blood leukocytes and lymphocyte subsets. J. Leukocyte Biol. 55: 209-213.

53. Sonnenfeld, G., Koebel, D.A., and Davis, S. 1995. Effects of hypergravity on immunologic function. Micrograv. Sci. Technol. 7: 323-326.

54. Sonnenfeld, G., Davis, S., Taylor, G.R., Mandel, A.D., Konstantinova, I.V., Lesnyak, A., Fuchs, B.B., Peres, C., Tkackzuk, J., and Schmitt, D.A. 1996. Effect of spaceflight on cytokine production and other immunologic parameters of rhesus monkeys. J. Interferon Cytokine Res. 16: 409-415.

55. Konstantinova, I.V., Rykova, M.P., Lesnyak, A.T., and Antropova, E.A. 1993. Immune changes during long duration missions. J. Leukocyte Biol. 54: 189-201.

56. Taylor, G.R., and Dardano, J.R. 1983. Human cellular responsiveness following spaceflight. Aviat. Space Environ. Med. 54:S55-S59.

57. Taylor, G.R. 1993. Immune changes in short-duration missions. J. Leukocyte Biol. 54: 202-208.

58. Meehan, R.T., Neale, L.S., Kraus, E.T., Stuart, C.A., Smith, M.L., Cintron, N.M., and Sams, C.F. 1992. Alterations in human mononuclear leucocytes following spaceflight. Immunology 76: 491-497.

59. Meehan, R.T., Whitson, P., and Sams, C. 1993. The role of psychoneuroendocrine factors on spaceflight-induced immunological alterations. J. Leukocyte Biol. 54: 236-244.

60. Meehan, R.T., Whitson, P., and Sams, C. 1993. The role of psychoneuroendocrine factors on spaceflight-induced immunological alterations. J. Leukocyte Biol. 54: 236-244.

61. Konstantinova, I.V., Rykova, M.P., Lesnyak, A.T., and Antropova, E.A. 1993. Immune changes during long duration missions. J. Leukocyte Biol. 54: 189-201.

62. Konstantinova, I.V., and Fuchs, B.B. 1991. The Immune System in Space and Other Extreme Conditions. Harwood Academic Publishers, Chur, Switzerland.

63. Tálas, M., Bátkai, L. Stöger, I., Nagy, K., Hiros, L., Konstantinova, I., Rykova, M., Mozogovava, J., Gusev, O., and Kozharinov, V. 1984. Results of the space experiment program "Interferon." Acta Astronautica 11: 379-386.

64. Tálas, M., Bátkai, L. Stöger, I., Nagy, K., Hiros, L., Konstantinova, I., Rykova, M., Mozogovava, J., Gusev, O., and Kozharinov, V. 1984. Results of the space experiment program "Interferon." Acta Astronautica 11: 379-386

65. Konstantinova, I.V., Rykova, M.P., Lesnyak, A.T., and Antropova, E.A. 1993. Immune changes during long duration missions. J. Leukocyte Biol. 54: 189-201.

66. Konstantinova, I.V., and Fuchs, B.B. 1991. The Immune System in Space and Other Extreme Conditions. Harwood Academic Publishers, Chur, Switzerland.

67. Taylor, G.R., and Janney, R.P. 1992. In vivo testing confirms a blunting of the human cell-mediated immune mechanism during spaceflight. J. Leukocyte Biol. 51: 129-132.

68. Gmünder, F.K., Konstantinova, I., Cogoli, A., Lesnyak, A., Bogomolov, W., and Grachov, A.W. 1994. Cellular immunity in cosmonauts during long duration spaceflight on board the orbital MIR station. Aviat. Space Environ. Med. 65: 419-423.

69. Taylor, G.R., and Janney, R.P. 1992. In vivo testing confirms a blunting of the human cell-mediated immune mechanism during spaceflight. J. Leukocyte Biol. 51: 129-132.

70. Gmünder, F.K., Konstantinova, I., Cogoli, A., Lesnyak, A., Bogomolov, W., and Grachov, A.W. 1994. Cellular immunity in cosmonauts during long duration spaceflight on board the orbital MIR station. Aviat. Space Environ. Med. 65: 419-423.

71. Cogoli, A., and Gmünder, F.K. 1991. Gravity effects on single cells: Techniques, findings and theories. Pp. 183-248 in Advances in Space Biology and Medicine, Vol. 1 (S.E. Bonting, ed.). JAI Press, Greenwich, Conn.

72. Cogoli, A., Valuchi-Morf, M., Müller, M., and Breigleb, W. 1980. The effect of hypogravity on human lymphocyte activation. Aviat. Space Environ. Med. 51: 29-34.

73. Cogoli, A., Tschopp, A., and Fuchs-Bislin, P. 1984. Cell sensitivity to gravity. Science 225: 228-230.

74. Cogoli, A. 1993. The effect of hypogravity and hypergravity on cells of the immune system. J. Leukocyte Biol. 54: 259-268.

75. Gmünder, F.K., Kiess, M., Sonnenfeld, G., Lee, J., and Cogoli, A. 1990. A ground-based model to study the effects of weightlessness on lymphocytes. Biol. Cell 70: 33-38.

76. Bechler, H., Cogoli, A., Cogoli-Greuter, M., Müller, O., and Hunzinger, E. 1992. Activation of microcarrier attached lymphocytes in microgravity. Biotechnol. Bioeng. 40: 991-996.

77. Cogoli, A. 1993. The effect of hypogravity and hypergravity on cells of the immune system. J. Leukocyte Biol. 54: 259-268.

78. Chapes, S.K., Morrison, D.R., Guikema, J.A., Lewis, M.L., and Spooner B.S. 1992. Cytokine secretion by immune cells in space. J. Leukocyte Biol. 52: 104-110.

79. Armstrong, J.W., Gerren, R.A., and Chapes S.K. 1995. The effect of space and parabolic flight on macrophage hematopoiesis and function. Exp. Cell Res. 216: 160-168.

80. Limouse, M., Manie, S., Konstantinova, I., Ferrua, B., and Schaffar, L. 1991. Inhibition of phorbol ester-induced activation in microgravity. Exp. Cell Res. 197: 82-86.

81. Schmitt, D.A., Hatton, J.P., Emond, C., Chaput, D., Paris, H., Levade, T., Cazenave, J.P., and Schaffar, L. 1996. The distribution of protein kinase C in human leukocytes is altered in microgravity. FASEB J. 10: 1627-1634.

82. Fleming, S.D., Edelman, L.S., and Chapes, S.K. 1991. Effects of corticosterone and microgravity on inflammatory cell production of superoxide. J. Leukocyte Biol. 50: 69-76.

83. Woods, K.M., and Chapes, S.K. 1994. Abrogation of TNF-mediated cytotoxicity by spaceflight involves protein kinase C. Exp. Cell Res. 211: 171-174.

84. Konstantinova, I.V., Rykova, M.P., Lesnyak, A.T., and Antropova, E.A. 1993. Immune changes during long duration missions. J. Leukocyte Biol. 54: 189-201.

85. Konstantinova, I.V., and Fuchs, B.B. 1991. The Immune System in Space and Other Extreme Conditions. Harwood Academic Publishers, Chur, Switzerland.

86. Cogoli, A., and Gmünder, F.K. 1991. Gravity effects on single cells: Techniques, findings and theories. Pp. 183-248 in Advances in Space Biology and Medicine, Vol. 1 (S.E. Bonting, ed.). JAI Press, Greenwich, Conn.

87. Cogoli, A., Valuchi-Morf, M., Müller, M., and Breigleb, W. 1980. The effect of hypogravity on human lymphocyte activation. Aviat. Space Environ. Med. 51: 29-34.

88. Cogoli, A., Tschopp, A., and Fuchs-Bislin, P. 1984. Cell sensitivity to gravity. Science 225: 228-230.

89. Cogoli, A. 1993. The effect of hypogravity and hypergravity on cells of the immune system. J. Leukocyte Biol. 54: 259-268.

90. Bechler, H., Cogoli, A., Cogoli-Greuter, M, Müller, O, and Hunzinger, E. 1992. Activation of microcarrier attached lymphocytes in microgravity. Biotechnol. Bioeng. 40: 991-996.

91. Konstantinova, I.V., and Fuchs, B.B. 1991. The Immune System in Space and Other Extreme Conditions. Harwood Academic Publishers, Chur, Switzerland.

92. LeBlanc, A.D., Schneider, V.S., Evans, H.J., Englebretson, D.A., and Krebs, J.M. 1992. Bone mineral loss and recovery after 17 weeks of bed-rest. J. Bone Miner. Res. 5: 843-850.

93. Cogoli, A., and Gmünder, F.K. 1991. Gravity effects on single cells: Techniques, findings and theories. Pp. 183-248 in Advances in Space Biology and Medicine, Vol. 1 (S.E. Bonting, ed.). JAI Press, Greenwich, Conn.

94. Hargens, A.R., Tipton, C.M., Gollnick, P.D., Mubarak, S.J. Tucker, B.J., and Akeson, W.H. 1983. Fluid shifts and muscle function in humans during acute simulated weightlessness. J. Appl. Physiol. 54: 1003-1009.

95. Konstantinova, I.V., and Fuchs, B.B. 1991. The Immune System in Space and Other Extreme Conditions. Harwood Academic Publishers, Chur, Switzerland.

96. LeBlanc, A.D., Schneider, V.S., Evans, H.J., Englebretson, D.A., and Krebs, J.M. 1992. Bone mineral loss and recovery after 17 weeks of bed-rest. J. Bone Miner. Res. 5: 843-850.

97. Albrecht-Buehler, G. 1992. The simulation of microgravity conditions on the ground. ASGSB Bull. 5: 3-10.

98. Chapes, S.K., Mastro, A.M., Sonnenfeld, G., and Berry, W.D. 1993. Antiorthostatic suspension as a model for the effects of spaceflight on the immune system. J. Leukocyte Biol. 54: 227-235.

99. Musacchia, X.J., Deavers, D., Meininger, G., and Davis, T. 1980. A new model for hypokinesia: Effects on muscle atrophy in the rat. J. Appl. Physiol. 48: 470-476.

100. Chapes, S.K., Mastro, A.M., Sonnenfeld, G., and Berry, W.D. 1993. Antiorthostatic suspension as a model for the effects of spaceflight on the immune system. J. Leukocyte Biol. 54: 227-235.

101. Berry, W.D., Murphy, J., Smith, B., Taylor, G.R., and Sonnenfeld, G. 1991. Effect of microgravity modeling on interferon and interleukin responses in the rat. J. Interferon Res. 11: 243-249.

102. Caren, L., Mandel, A.D., and Nunes, J. 1980. Effect of simulated weightlessness on the immune system in rats. Aviat. Space Environ. Med. 51: 251-255.

103. Rose, A., Steffen, J.M., Musacchia, X.J., Mandel, A.D., and Sonnenfeld, G. 1984. Effect of antiorthostatic suspension on interferon-alpha/beta production by the mouse. Proc. Soc. Exp. Biol. Med. 177: 254-256.

104. Sonnenfeld, G., Mandel, A.D., Konstantinova, I.V., Berry, W.D., Taylor, G.R., Lesnyak, A.D., Fuchs, B.B., and Rakhmilevich, A. 1992. Spaceflight alters immune cell function and distribution. J. Appl. Physiol. 73: 191S-195S.

105. Sonnenfeld, G., Morey, E.R., Williams, J.A., and Mandel, A.D. 1982. Effect of a simulated weightlessness model on the production of rat interferon. J. Interferon Res. 2: 267-270.

106. Chapes, S.K., Mastro, A.M., Sonnenfeld, G., and Berry, W.D. 1993. Antiorthostatic suspension as a model for the effects of spaceflight on the immune system. J. Leukocyte Biol. 54: 227-235.

107. Berry, W.D., Murphy, J., Smith, B., Taylor, G.R., and Sonnenfeld, G. 1991. Effect of microgravity modeling on interferon and interleukin responses in the rat. J. Interferon Res. 11: 243-249.

108. Chapes, S.K., Mastro, A.M., Sonnenfeld, G., and Berry, W.D. 1993. Antiorthostatic suspension as a model for the effects of spaceflight on the immune system. J. Leukocyte Biol. 54: 227-235.

109. Berry, W.D., Murphy, J., Smith, B., Taylor, G.R., and Sonnenfeld, G. 1991. Effect of microgravity modeling on interferon and interleukin responses in the rat. J. Interferon Res. 11: 243-249.

110. Kopydlowski, K.M., McVey, D.S., Woods, K.M., Landolo, J.J., and Chapes, S.K. 1992. Effects of antiorthostatic suspension and corticosterone on macrophage and spleen cell function. J. Leukocyte Biol. 52: 202-208.

111. Nash, P., Bour, B., and Mastro, A. 1991. Effect of hindlimb suspension simulation of microgravity on in vitro immunological responses. Exp. Cell Res. 195: 353-360.

112. Kopydlowski, K.M., McVey, D.S., Woods, K.M., Landolo, J.J., and Chapes, S.K. 1992. Effects of antiorthostatic suspension and corticosterone on macrophage and spleen cell function. J. Leukocyte Biol. 52: 202-208.

113. Fleming, S.D., Rosenkrans, C., and Chapes, S.K. 1990. Test of the antiorthostatic suspension model on the inflammatory cell response. Aviat. Space Environ. Med. 61: 327-332.

114. Miller, E.S., Koebel, D.A., Davis, S.A., Klein, J.B., McLeish, K.R., Goldwater D., and Sonnenfeld, G. 1994. Influence of suspension on the oxidative burst by rat neutrophils. J. Appl. Physiol. 76: 387-390.

115. Berry, W.D., Murphy, J., Smith, B., Taylor, G.R., and Sonnenfeld, G. 1991. Effect of microgravity modeling on interferon and interleukin responses in the rat. J. Interferon Res. 11: 243-249.

116. Caren, L., Mandel, A.D., and Nunes, J. 1980. Effect of simulated weightlessness on the immune system in rats. Aviat. Space Environ. Med. 51: 251-255.

117. Nash, P., Bour, B., and Mastro, A. 1991. Effect of hindlimb suspension simulation of microgravity on in vitro immunological responses. Exp. Cell Res. 195: 353-360.

118. Gould, C.L., and Sonnenfeld, G. 1987. Enhancement of viral pathogenesis in mice maintained in an antiorthostatic model: Coordination with effects on interferon production. J. Biol. Regul. Homeost. Agent. 1: 33-36.

119. Gould, C.L., and Sonnenfeld, G. 1987. Enhancement of viral pathogenesis in mice maintained in an antiorthostatic model: Coordination with effects on interferon production. J. Biol. Regul. Homeost. Agent. 1: 33-36.

120. Miller, E.S., and Sonnenfeld, G. 1993. Influence of suspension on the expression of protective immunological memory to murine Listeria monocytogenes infection. J. Leukocyte Biol. 54: 378-383.

121. Miller, E.S., and Sonnenfeld, G. 1994. Influence of antiorthostatic suspension on resistance to murine Listeria monocytogenes infection. J. Leukocyte Biol. 55: 371-378.

122. Gould, C.L., and Sonnenfeld, G. 1987. Enhancement of viral pathogenesis in mice maintained in an antiorthostatic model: Coordination with effects on interferon production. J. Biol. Regul. Homeost. Agent. 1: 33-36.

123. Miller, E.S., and Sonnenfeld, G. 1993. Influence of suspension on the expression of protective immunological memory to murine Listeria monocytogenes infection. J. Leukocyte Biol. 54: 378-383.

124. Miller, E.S., and Sonnenfeld, G. 1994. Influence of antiorthostatic suspension on resistance to murine Listeria monocytogenes infection. J. Leukocyte Biol. 55: 371-378.

125. Konstantinova, I.V., and Fuchs, B.B. 1991. The Immune System in Space and Other Extreme Conditions. Harwood Academic Publishers, Chur, Switzerland.

126. Dick, E.C., Mandel A.D., Warshaver, D.M., Conklin, S.C., and Jerde, R.S. 1977. Respiratory virus transmission at McMurdo Station. Antarctic J. 12: 2-3.

127. Meehan, R.T., Duncan, U., Neale, L., Taylor, G., Muchmore, H., Scott, N., Ramsey, K., Smith, E., Rock, P., Goldblum, R., and Houston, C. 1988. Operation Everest II: Alterations in the immune system at high altitudes. J. Clin. Immunol. 8: 397-403.

128. Schmitt, D.A., and Schaffar, L. 1993. Isolation and confinement as a model for spaceflight immune changes. J. Leukocyte Biol. 54: 209-213.

129. Schmitt, D.A., Peres, C., Sonnenfeld, G., Tkackzuk, J., Arquier, M., Mauco, G., and Ohayon, E. 1995. Immune responses in humans after 60 days of confinement. Brain Behav. Immunol. 9: 70-77.

130. Schmitt, D.A., Schaffar, L., Taylor, G.R., Loftin, K.C., Schneider, V.S., Koebel, A., Abbal, M., Sonnenfeld, G., Lewis, D.E., Reuben, J.R., and Ferebee, R. 1996. Use of bed rest and head-down tilt to simulate spaceflight-induced immune system changes. J. Interferon Cytokine Res. 15: 151-157.

131. Sonnenfeld, G., Measel, J., Loken, M.R., Degioanni, J., Follini, S., Galvagno, A., and Montalbini, M. 1992. Effect of isolation on interferon production and hematological parameters. J. Interferon Res. 12: 75-81.

132. Hargens, A.R., Tipton, C.M., Gollnick, P.D., Mubarak, S.J. Tucker, B.J., and Akeson, W.H. 983. Fluid shifts and muscle function in humans during acute simulated weightlessness. J. Appl. Physiol. 54: 1003-1009.

133. Albrecht-Buehler, G. 1992. The simulation of microgravity conditions on the ground. ASGSB Bull. 5: 3-10.

134. Cogoli, A., and Gmünder, F.K. 1991. Gravity effects on single cells: Techniques, findings and theories. Pp. 183-248 in Advances in Space Biology and Medicine, Vol. 1 (S.E. Bonting, ed.). JAI Press, Greenwich, Conn.

135. Pellis, N.R., Goodwin, T.J., Risin, D., McIntyre, B.W., Pizzini, R.P., Cooper, D., Baker, T.L., and Spaulding, G.F. 1997. Changes in gravity inhibit lymphocyte locomotion through type I collagen. In Vitro Cell Dev. Biol. Anim. 33: 398-405.

PART III

Additional Space Environment Issues

11

Radiation Hazards

INTRODUCTION

The long-range plans of NASA include construction and occupation of a space station in low Earth orbit, and possible human exploratory missions to the Moon and Mars within the next 25 years. The biological effects of exposure to radiation in space pose potentially serious health risks that have to be controlled or mitigated before such relatively long-term missions beyond low Earth orbit can be initiated.[1] The levels of radiation in space are high enough and the missions are long enough that adequate shielding is necessary to minimize carcinogenic, cataractogenic, and possible neurologic effects on crew members. The question to be answered is, What will provide the necessary protection, for the extent of a mission, against the biological effects of high-energy galactic cosmic-ray particles ranging from energetic protons with low mean linear energy transfer (LET) to nuclei of high atomic number and very high energies (HZE) with high LET and against the effects of transient radiation in solar particle events?[*]

This chapter summarizes a recent National Research Council (NRC) report, *Radiation Hazards to Crews of Interplanetary Missions: Biological Issues and Research Strategies,*[2] on current knowledge of the types and levels of radiation to which crews will be exposed in space, and also discusses the range of possible human health effects that must be protected against. It suggests steps to be taken and the types of experiments needed to reduce significantly the level of uncertainty regarding health risks to human crews in space, and it recommends priorities for research from which NASA can obtain the information required to evaluate the biological risks faced by humans exposed to radiation in space and to mitigate such risks. The chapter outlines, in general terms, the commitment of resources that NASA

---

[*]For example, with substantial uncertainty, the annual estimated equivalent dose behind ~7.5 cm of aluminum for galactic cosmic rays during the 1977 solar minimum (high fluence level of galactic cosmic rays) would have exceeded the current equivalent dose limit of 0.5 sievert per year.

should make to carrying out these experiments in order to accomplish an effective shielding design in time for a possible mission launch to Mars.

It is recommended that, if the goal of safe interplanetary missions with human crews is sought, NASA explore various possibilities, including the construction of new facilities, to increase the research time available for experiments with HZE particles.

A recent comprehensive report to NASA[3] focuses on the extent to which animal experiments are required to assess the health risks to humans exposed to the radiation in space. It examines carefully the molecular, cellular, and animal background information that will be needed to extrapolate these radiation effects to humans and concludes that animal experimentation should be pursued selectively, but decisively. This conclusion is strong support for the priority research recommendations and strategies given in this chapter.

The characteristics of the radiation field are more complicated in low Earth orbit (LEO) than in interplanetary space because the number of particles depends on altitude and on the magnitude and direction of Earth's magnetic field.[4] Since the orientation of the space station will be fixed and its orbit will be at an inclination of 51.6°, the radiation dose rate depends on orbit position and also on location within the station. The latter asymmetry is about 2-fold under identical shielding. The three principal radiation sources of concern are galactic cosmic rays, protons trapped by Earth's field, and solar particle events (SPEs), whose frequencies are a function of solar activity, being highest during solar maxima. "Currently there are not techniques available for forecasting either the occurrence or the magnitude of such events."[5] The peak of the next solar cycle is expected in spring of the year 2000. During an SPE the proton flux may increase 10-fold in less than 10 hours. A radiobiological problem in LEO is that the mix of high- and low-LET particles changes with altitude (more protons at higher altitudes) and with solar cycle (more high-LET particles during solar minimum).[6] Clearly, real-time monitoring of dose rates and the spectra of particles contributing to the dose are important because the dose equivalent, in sieverts (Sv), depends on the relative biological effectiveness (RBE), which varies with LET and, for end points of interest, is currently not well known for HZE particles. Thus, the research strategies outlined below for space missions are also necessary for understanding the radiation effects in LEO.

Determination of the frequencies of chromosomal aberrations in lymphocytes irradiated in vitro versus gamma ray dose has been used to estimate the dose equivalent for chromosomal aberrations in the lymphocytes of two astronauts on the Mir-18 115-day spaceflight.[7] The differences in postflight minus preflight frequencies are calibrated to give the dose equivalent from the results of acute exposures of preflight lymphocyte samples to gamma rays. The average dose equivalent was 0.15 Sv. The technique looks promising for retrospective dosimetry, but additional data are required if the method is to be applied to chronic and nonuniform body exposures.[8] If the RBEs for cancer induction, for example, differ from those for chromosomal aberrations, aberrations would not be a useful biodosimeter for cancer without knowledge of appropriate correction factors.

STATEMENT OF THE PROBLEM

The knowledge needed to design adequate shielding has both physical and biological components: (1) the distribution and energies of radiation particles present behind a given shielding material as a result of the shield being struck by a given type and level of incident radiation and (2) the effects of a given dose on relevant biological systems for different radiation types. Each component involves significant uncertainty that must be reduced to permit effective design of shielding, given that the level of uncertainty governs the amount of shielding. It is only prudent to design shielding that will protect space crew members from the predicted, but uncertain, high levels of biological effects resulting from their exposure to radiation. At the same time, excess shielding, based on current cost estimates, would impose an excess expenditure on the order of tens of billions of dollars.[9]

HZE particles impinging on shielding, or on human tissue, result in very dense ionization tracks (high LET) with numerous fragments that produce a spectrum of other energetic nuclei, protons, neutrons, and heavy fragments. The numbers of these other nuclei depend on the nature of the shielding and its mass per unit area. The energy loss of the individual particles depends on their types and energies. Thus, each particle contributes to the radiation dose and biological response, which are dependent on the number of particles, their types, and their energies. The theoretical calculations of doses per particle type obtained thus far for relevant shielding materials must be verified by ground-based experiments, because the radiation field rate in space is too complex for sufficient experimental analysis. Ionizing radiation either affects cellular macromolecules directly or reacts with water to produce free radicals that affect these macromolecules by so-called indirect effects. These effects are mitigated somewhat by the presence of free-radical scavengers in the surrounding medium. The scavengers are useful in significantly reducing the effects of low-LET radiation but, with some exceptions,[10] do not seem to result in any such decreases in the damage caused by high-LET radiation.

The biological effects of fast, charged particles depend on the nature of the particle (its charge and velocity) and on the specific biological end point under observation (e.g., cell killing, mutation at a specific genetic locus, chromosomal alterations, cell transformation in vitro, and tumor induction). The RBE is taken as the ratio of the dose of gamma rays required to produce a specific effect to the dose of particle radiation required to produce the same level of effect. The RBE depends on the type of particle and the biological effect under consideration and may vary with the magnitude of the biological effect. More importantly, RBE varies greatly with the LET of the particle. For example, high-energy protons may have an RBE value approaching 1.0, whereas high-energy iron nuclei may have an RBE approaching 40. For tumor induction in animals exposed at lower doses, the relationship between RBE and LET is known for only one tumor site—the Harderian gland in mice. There are no equivalent data for tumor induction in humans. Hence, there is a great uncertainty in extrapolating from cell and scanty mouse data to evaluate human risk. Moreover, there is also great uncertainty in the extrapolation from the analysis of cancer induction in Japanese individuals exposed to acute doses of radiation resulting from the atomic bombs.[11] These doses are not known precisely. Since this radiation was primarily low LET, in order to estimate risks to humans in spaceflight conditions one must extrapolate from the RBE-versus-LET data for cells in culture and from small mammals to humans. In addition, one must extrapolate from the risks due to acute exposures of humans to the low-dose-rate chronic exposures involved in space missions (except for the relatively acute exposures from solar particle events). As a general rule, as the dose rate decreases, the biological effect from a given dose also decreases. This dose reduction, in going from acute to chronic exposure, also depends on the biological system and may range from a factor of 2 to 10.[12] The dose rate reduction factor for HZE particles is not well known but is probably closer to 1.[13] Two other factors that must be considered, but whose impacts are currently unknown, are the effects of biochemical or cellular repair reactions following exposure to HZE particles and the effects of microgravity on such reactions. Thus, in estimating the risks to humans exposed to radiation in space, the uncertain factors are the radiation fields behind the shielding and the extrapolation, via cell culture and animal experiments, from the uncertain risks posed by acute low-LET exposure to risks posed by chronic high-LET exposure.

To quote Curtis et al.,[14] "Uncertainties in these numbers are difficult to estimate but a rough analysis leads to a 10-15% uncertainty in the initial charged particle spectra, a 50% uncertainty in the radiation transport calculation, a factor of 2-3 uncertainty in the risk coefficients for low-LET radiation (most of which is due to uncertainty in the dose and dose rate effectiveness factor) and perhaps a factor of 2-5 uncertainty in the risk cross sections at high-LET. Thus, an overall uncertainty in the risk of radiation-induced cancer of a factor of 4-15 for a space crew in the galactic cosmic ray environment appears to exist at the present state of our knowledge." Obviously, these uncertainties have, themselves, large uncertainties.

In 1989, the National Council on Radiation Protection and Measurements (NCRP) issued its report *Guidance on Radiation Received in Space Activities*, which had been requested by NASA.[15] That report introduced different career limits depending on gender and age of onset of exposure to radiation in space. The career equivalent dose limits of 1 to 4 Sv that were recommended were based on a lifetime excess risk of cancer mortality of 3×10^{-2} per gray (Gy) of low-LET, acute radiation. The report's scope was limited to low-Earth-orbit missions, although it considered the radiation environment in deep space and the biological effects of HZE particles. Recommendations for protection against deterministic effects were also made in 1989. The career limit of 12 Sv recommended in 1979 by the National Research Council for skin was reduced by the NCRP in 1989 to 6 Sv; the limit for the lens of the eye was reduced from 6 to 4 Sv; and the limit for blood-forming organs was set at 1 to 4 Sv, depending on age and gender. Since 1989, estimates of cancer risk based on studies of atomic bomb survivors have been increased significantly, and the NCRP will issue new recommendations in the near future.

CURRENT UNDERSTANDING OF BIOLOGICAL EFFECTS OF RADIATION

Types of Effects

For the purpose of settling radiation protection standards, it has been useful to divide the biological effects that are important for human health into stochastic and deterministic effects.

Stochastic effects are considered to be those due to radiation-induced changes randomly distributed in the DNA of single cells that may lead to cancer or genetically transmissible effects, depending on the target cells. Cancer occurs after a long latency period: after 2 or more years in the case of leukemia and from 2 years to decades in the case of solid cancers. It is assumed that the frequency of such effects increases with dose without a threshold and that the severity of the effect is independent of dose. Stochastic effects are the most important consideration in setting protection limits for human populations exposed to radiation at low doses. It is important to note that radiation weighting factors or quality factors apply only to stochastic effects in the dose range pertinent to radiation protection. Based on studies of atomic bomb survivors of Hiroshima who were exposed to acute levels mainly of gamma rays but also of fission neutrons at very high dose rates, estimates for the risk of contracting leukemia have been refined,[16] and there are also data on mortality and the incidence of solid cancers at more than 20 sites in the human body.[17,18] (The precise contribution of the neutrons from the fission reactions to the total dose at Hiroshima is poorly known but is not considered a major contributor to the risk of cancer in those exposed at Hiroshima.) In 1991, the International Commission on Radiological Protection (ICRP)[19] included leukemia and eight specific sites of solid cancers in its estimates of the probability of an individual contracting a fatal cancer after whole-body exposure to low-LET radiation at 1 Gy and at a high dose rate. The estimated probabilities were 7.12×10^{-2} per person based on a multiplicative projection model and 4.16×10^{-2} per person using an additive model.[20] A similar number, 3.7×10^{-2} per adult worker, using an additive model, was given in 1997 by the NCRP.[21]

Deterministic effects, previously termed nonstochastic effects, occur only after exposure to relatively high doses and affect cell populations to the detriment of specific organs or whole organisms. These effects can range from acute radiation sickness to hair loss or nausea. In contrast to stochastic effects, deterministic effects are dose dependent in both frequency and severity. Deterministic effects may occur early (in a matter of hours or days) or late (after many months or even longer). Radiation protection standards are set to prevent deterministic effects, whereas standards to protect against stochastic effects are selected to limit these effects to an "acceptable" level.

Effects Induced by Protons and Heavy Ions

In deep space, the radiation environment consists mainly of galactic cosmic radiation (GCR) at a low fluence rate. In the energy range from 100 MeV (100×10^6 electron volts [eV]) per nucleon to 10 GeV (10×10^9 eV) per nucleon, the GCR consists of 87 percent protons, 12 percent helium ions, and 1 percent heavier ions.[22] Protons are also the major component of solar particle events (SPEs), with a smaller contribution by helium and heavier ions emitted from the Sun. A major difference between SPE radiation and GCR is the much greater transient fluence in the former, which in very large solar particle events can be 10^{10} protons/cm^2 with energies greater than 10 MeV.[23]

No data are available for most of the deterministic effects induced in humans by exposure to protons, and very limited data are available from studies of animals. The latter[24] indicate that the biological effects of the higher-energy protons (>138 MeV) are similar to those caused by 2-MeV x rays and ^{60}Co gamma rays. The RBE for acute mortality was about 1.0 to 1.1. In the case of 160-MeV protons, Urano et al. found RBE values ranging between 0.8 and 1.3 for killing of jejunal crypt cells, skin damage, and effects on the lens of the eye in exposed mice.[25] An exception to these RBE values of about 1 was the indirect finding by Storer et al. of higher RBE values, namely 2.4 and 4.9 for 30-day lethality and testicular atrophy, respectively, in mice.[26] In the treatment of human cancer by proton irradiation, an RBE of 1.1 has been used for planning purposes, and this value does not appear to underestimate the effectiveness of the protons.

In the case of stochastic effects, there are no data for cancer induction in humans exposed to protons. In a study of the induction of tumors in mice exposed to 60-MeV protons, Clapp et al. found no RBE values greater than about 1.0.[27] Burns et al. reported an RBE of about 3 for the induction of skin tumors in rats exposed to 10-MeV protons compared to electrons.[28] The studies discussed above were carried out with single high-dose-rate exposures that are very different from the exposures occurring in space. Burns et al. noted a reduction in tumorogenic effect with the increasing fractionation of 10-MeV proton irradiation, an indication of recovery,[29] and the curvilinear response to single doses was similar to the response to low-LET radiation, indicating that the protons have attributes both of low-LET radiation and, because the RBE is about 3, of high-LET radiation. Obviously, more animal data, using protons of several energy ranges, are needed to estimate human cancer risks from galactic protons.

The deterministic effects of exposure to heavy ions have been studied in experimental animals. The RBE of various ions was determined for effects that result from cell killing in the gut, testes, and bone marrow, and in vitro systems.[30] As Figure 11.1 indicates, the RBE increases with increasing LET, reaches a maximum between 100 and 200 keV/μm, and decreases rapidly at higher LET values. The maximum RBE values for effects involving cell killing, such as shown in Figure 11.1, are in the range of 2.0 to 3.0.[31] However, for cataract induction in rats and mice, the RBE values obtained by Merriam are in the range of 40 to 50 at low doses.[32] Although the work of Lett and co-workers and of Worgul and co-workers suggests that it may be possible, with further data, to extrapolate across species to obtain RBE values for cataract induction, current data do not allow reliable estimation of the risk of cataract induction occurring in humans as a result of exposure to radiation in deep space.[33,34] Observations on radiotherapy patients indicate that very high doses of low-LET radiation give rise to deterministic-type damage. HZE particles produce high-dose ionization tracks and kill the cells they traverse. The concern about such microlesions in the central nervous system resulting from traversal of cells by heavy charged

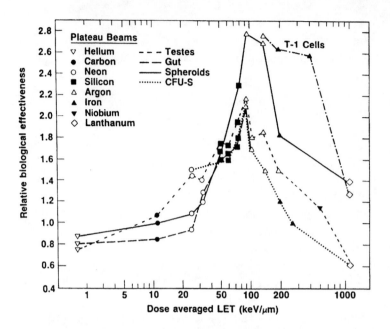

FIGURE 11.1 RBE-LET relationship for inactivation of different cells in vitro (CFU-S, intestinal crypt clono-genic cells, cells in spheroids, human T-1 cells) and loss of testis weight. SOURCE: National Council on Radiation Protection and Measurements (NCRP). 1989. Guidance on Radiation Received in Space Activities. Recommendations of the National Council on Radiation Protection and Measurements. NCRP Report No. 98. National Council on Radiation Protection and Measurements, Bethesda, Md.

particles, such as iron, has not been eliminated, nor has any evidence been produced to show that the concern is justified.

The main concern about stochastic effects is the risk of cancer induction. It is agreed that the RBE for carcinogenesis increases with the increasing LET of the radiation. The evidence comes largely from animal experiments with fission neutrons but also from data on induction of lung tumors in humans exposed to alpha particles from radon.[35] There has been only one systematic study of the relationship between the LET of heavy ions and the RBE values of the ions for tumor induction, which was carried out on the Harderian gland of mice.[36] Although this gland is a suitable epithelial system, it is the *only* tumor model that has been examined over the range of LET values encountered in space. Hence, it is not possible to generalize on the basis of these data about RBE values for induction of cancer in important sites of the human body such as the breast, lung, and bone marrow. The data on Harderian gland tumors show a rise in RBE with increasing LET, reaching a maximum and a plateau of about 30 at approximately 100 to 200 keV/μm. [37] However, unlike the case of the RBE-LET relationship for cell killing and mutation (as in Figure 11.1),[38] there is no evidence of a rapid decrease in RBE at higher LET values. Data on induction of skin tumors by argon ions[39] support the expectation of a high RBE for induction of tumors by heavy ions but do not allow any more precise determination of what value should be used to estimate the risk of cancer induction in humans. Because of the importance of establishing the precise RBE values for cancer induction, more experimental studies are required.

First, a pragmatic set of studies is needed to provide data necessary for the determination of appropriate factors that should be used in making risk calculations. These should be systematic studies of RBE as a function of particle type and energy for a select number of heavy ions and for protons, using well-defined animal models for tumorigenesis. In addition, information on dose rate and fractionation effects for protons is also needed. Improvements in risk estimates beyond those attainable with these data require a more complete understanding of the mechanisms of tumor induction and of principles that will aid in using data from experimental systems subjected to relatively high radiation doses, to estimate effects on humans exposed to low, protracted doses, and to estimate risks across populations. These kinds of studies will require the development and exploration of new model systems and the application of developing technologies in cell and molecular biology.

PRIORITY RESEARCH RECOMMENDATIONS AND STRATEGIES

An estimate of the risk of adverse biological effects due to irradiation during a space mission corresponds to measurements of the relevant deterministic and stochastic biological consequences of exposure to radiation as a function of (1) dose, (2) dose rate, and (3) radiation quality. Such measurements should also be made as a function of radiation type and the shielding thickness for each type of spacecraft material. The risk estimates should include their uncertainties.

Given in the two subsections that follow, and summarized from the 1996 NRC report *Radiation Hazards to Crews of Interplanetary Missions*,[40] are the priority research recommendations and the issues that must be addressed in any endeavor to significantly reduce the risk and uncertainty of radiation hazards to the crews of interplanetary missions.* The research strategy presented to address each recommendation is narrowly focused on the recommendation and describes the minimum research model likely to provide the necessary data. The development of such a narrow set of strategies should not necessarily be taken as a recommendation to limit the scope of studies to those outlined below. If funds become available, many of these studies could be usefully expanded to provide additional relevant information.

In accordance with current understanding of the risks and uncertainties, the research recommendations are separated into higher- and lower-priority groups. As more data become available, some questions may shift in priority. Some strategies can be carried out independently, whereas others will be influenced by the outcome of research on the other recommended strategies and should be scheduled accordingly. The reasoning that forms the basis of these recommendations is discussed in greater detail in the 1996 report.[41]

Higher-Priority Research Recommendations

The higher-priority research recommendations, and strategies to answer them, follow. Research recommendations deemed important but of a lower priority are given in the next subsection.

1. Determine the carcinogenic risks following irradiation by protons and HZE particles.

Strategy

This key recommendation requires that two related questions be addressed: (1) Can the risk due to irradiation by protons in the energy range of the space environment be predicted on the basis of the risk posed by exposure to low-LET radiation, such as gamma rays, and is there evidence for the repair of

---

*Although the recommendations of the Task Group on the Biological Effects of Space Radiation were published independently in the 1996 report, it was the intent of CSBM that the task group's report would also constitute the basis of the material on radiation hazards presented in this CSBM strategy for research in space biology and medicine into the new century.

damage in cells following fractionated exposure to protons and HZE particles? (2) What are the appropriate RBEs for HZE particles? The answers to these questions are fundamental to understanding the risk of contracting cancer as a result of travel in deep space. These important questions can be addressed using solely ground-based studies if appropriate funding and additional radiation resources such as accelerator time are made available.

Initial studies should focus on cellular effects that are relevant to cancer. Research with cells would provide a more rapid resolution than would tumor induction studies with animals of whether effects of exposure to high-energy protons are similar to those arising from low-LET radiation. Theoretical models of radiation effects, as well as currently available data for cellular and tumorigenic effects of exposure to protons (mostly at energies lower than those encountered in space), would argue that risks due to proton irradiation are similar to those from low-LET irradiation. Although considerable data are already available for protons in this energy range, these data are not satisfactory to answer questions related to high-energy protons in the 1-GeV energy range. To verify such a prediction for higher-energy protons, a series of studies should be conducted in several cellular systems, including human fibroblasts and lymphocytes, to examine the effects of protons in the 1-GeV energy range on cell killing, induction of chromosomal aberrations, and induction of gene mutations and transformation. Chromosomal aberrations could also be studied in lymphocytes from animals irradiated in vivo. By conducting such studies with both acute and fractionated exposure regimens, it would be possible to determine whether fractionation effects (sparing of radiation response by allowing for DNA repair between fractions) similar to those for low-LET radiation exist. Animal carcinogenesis experiments with protons should be conducted only if the results of cellular studies indicate discrepancies from the predictions. If tumorigenesis studies are warranted, the same animal models recommended for the study of tumorigenesis following exposure to HZE particles (described below) should be employed.

To obtain more reliable risk coefficients for HZE particles, a systematic series of studies of RBE-LET relationships for a select number of heavy ions—with emphasis on iron particles—should be conducted using well-defined animal models for tumorigenesis. Adequately defining these relationships requires that the dose-response relationship be determined for these particles in the dose range lower than 20 to 30 cGy, because at higher doses of high-LET radiation, the response appears to reach a maximum followed by a decrease. The model systems chosen should be those for which substantial dose-response data for other high- and low-LET radiation are already available. Model systems that are particularly amenable to concomitant cellular and molecular studies should receive priority. Given the fact that mice have been used more extensively than other mammalian species in studies of carcinogenesis,[42] the use of the murine models for radiation-induced myelogenous leukemia and mammary gland cancer is appropriate. The conduct of these studies would require considerable commitment of time (beam time) at an HZE accelerator in an appropriate facility. Under the assumption that 3 months of beam time would be available per year, it is estimated that these studies would take approximately 6 years to complete. This estimate assumes that the irradiation of sufficient numbers of animals would require 2 years. Following irradiation, the completion of animal studies can be expected to require approximately 4 years. Under current conditions, which provide only 2 weeks of beam time each year, it would be almost impossible to complete a meaningful series of animal studies, because the period of time between the first set of animal irradiations and the last would probably be on the order of 6 years (assuming that half of the beam time were devoted to animal irradiation). This long temporal separation of experimental groups makes comparisons more difficult even with well-defined systems. Under these conditions, completion of the carcinogenesis studies would require a minimum of 10 years after the first irradiation.

Improvements in risk estimates beyond those attained with such data would require a more complete understanding of mechanisms and of principles that will aid in the direct extrapolation of results from experimental systems to astronauts. These kinds of studies will likely require the development and exploration of new model systems and the application of developing technologies in cell and molecular biology.

2. Determine how cell killing and induction of chromosomal aberrations vary as a function of the thickness and composition of shielding.

Strategy

The data obtained from studies conducted in response to recommendation 1 are necessary background for determining the biological effects of the specific radiation qualities and fluences in a spacecraft. The quality and dosimetry of radiation produced by HZE particles traversing shielding of different thicknesses and composition would be assessed from studies that address recommendation 6. Cellular studies should not be initiated for any particular energy (of HZE particle) and shielding until physical characterization of that radiation is completed. At a minimum, studies using cell-killing chromosomal aberrations and mutations and/or transformation as end points should be conducted using radiation qualities defined in dosimetry studies. Only the effects from acute exposures should be measured. Ground-based HZE particle sources, used with appropriate shielding to simulate in-flight conditions, should be suitable for such experiments. In-flight studies would be prohibitively difficult to conduct, and little gain in information would be realized.

It would be appropriate to conduct the initial studies in vitro using the same cell lines employed to address recommendation 1 (i.e., both rodent and human cell lines). Subsequent cytogenetic studies would have to be conducted in vivo to develop a more appropriate database for use in risk assessment calculations. It is recommended that bone marrow cells and peripheral lymphocytes, which are easily analyzed cytogenetically, be examined for chromosomal aberrations. Based on the information obtained in these cytogenetic studies, it would then be feasible to design a study to assess the induction of leukemia and breast cancer in mice exposed, behind shielding, to acute doses of HZE-particle radiation incident on the shielding (if cellular studies indicate this is necessary). Irradiation for in vitro studies could be accomplished in a relatively short time (i.e., about 2 days for each radiation type and energy). Typically, about 6 months would be required for analysis. If six radiation types were examined in consecutive order, as might be expected for a single research team, then such a study would require on the order of 3 years. Similarly, in vivo cytogenetic studies also require about 2 days of irradiation time for each radiation quality. If the in vivo and in vitro studies were performed in parallel, then both might reasonably be completed during the 3-year period.

The in vitro studies would allow the comparison of animal and human sensitivities to changes in shielding parameters. The in vivo experiments would provide data for in vitro and in vivo comparisons of cytogenic responses for the mouse. This parallelogram approach would provide estimates of the chromosomal aberration frequencies induced in humans in vivo (for bone marrow cells and lymphocytes). The mouse cancer studies (leukemia and breast cancer) can then be extrapolated in terms of human tumors by assuming that chromosomal aberration sensitivity factors apply. This approach would seem to be particularly reasonable for leukemias since chromosomal alterations are involved in the etiology of this cancer.

3. Determine whether there are studies that can be conducted to increase the confidence of extrapo-

lation from rodents to humans of radiation-induced genetic alterations that in turn could enhance similar extrapolations for cancer.

Strategy

The studies proposed for addressing recommendation 2 would yield relative sensitivity factors for mutations, chromosomal aberrations, and cell killing in rodent and human cells in vitro, and the in vivo cytogenetic studies would allow comparison of in vitro and in vivo responses in a single species, most likely the mouse. The sensitivity factors and other comparative data could then be used to provide an estimate of the responses mentioned above in humans by using cancer induction data obtained in rodents. However, the reliability of a relative sensitivity factor for this use must first be established. Chromosomal aberration and mutation frequencies induced by exposure to radiation are influenced to a great extent by the kinetics and fidelity of DNA repair processes. The relative values of these parameters for mice and humans are not well known. Therefore, a secondary measure of relative sensitivity pertinent to cancer risk assessment would be a comparison of the features of DNA repair in human and rodent cells in vitro following acute exposure to protons and HZE particles. Techniques based on pulsed-field gel electrophoresis have been developed that can measure DNA strand breaks at very low exposure levels (<10 cGy). If these experiments had to compete with other high-priority items for beam time within the total of ~100 hours currently available per year, they would probably extend over 3 years.

4. Determine if exposure to heavy ions at the level that would occur during deep-space missions of long duration poses a risk to the integrity and function of the central nervous system.

Strategy

A multifaceted research approach is required to address recommendation 4 so as to relate molecular changes to alterations in function. Some of the necessary experiments could take a long time, and since a few definitive answers must be obtained before final decisions about shielding and mission planning can be made, it is essential to ensure coordination of the strategy for this field of research. The studies range from the induction of DNA damage, repair, and maintenance of the fidelity of DNA into old age, to studies of the heavy-ion-induced morphological and functional changes in the central nervous system. The time necessary would vary from about 2 years for the DNA studies to perhaps 10 or more years for studies of functional changes, depending on the species required for a definitive assessment.

The scope of this research should be agreed upon by representatives of the disciplines involved, including both experimental and clinical neurologists. One essential study that could be started now is confirmation of the findings of Lett et al. on retinal cells—that late breakdown of DNA exposed to heavy ions occurs and that age at exposure is important.[43] New sensitive techniques for assessing DNA damage can be applied to the problem and also to the determination of dose-response relationships and the influence of LET. The studies described for this question could not be performed at all with only 2 weeks of beam time per year. If 3 months were available per year, then experience with similar studies[44-47] suggests a rough time estimate for the performance of all the required studies of from 5 to 7 years because of the long time interval required to observe late effects. This estimate is based on the assumption that sufficient animal facilities and staff will be available at the beam site and that rodents will be used as the animal model. If rodents are not used, then another way to complete these studies in the time frame proposed might be to repeat, confirm, and extend the work of Lett et al. A minimum of three ions with a spread of LET values would have to be examined.

5. Determine if better error analyses can be performed of all factors contributing to the estimation of risk by a particular method, and determine the types and magnitude of uncertainty associated with each method. What alternate methods for calculation of risk can be used for comparison with conventional predictions in order to assess absolute uncertainties? How can these methods be used to better determine how the uncertainties in the methods affect estimates of risk to humans and estimates of mission costs?

Strategy

The relative significance of uncertainties in risk assessments must be adequately established, and the impact of reductions in the level of these various uncertainties must be determined. The conventional approach for the assessment of risks is initially to calculate a dose, defined as the equivalent dose for the radiation field of interest corresponding to the dose of low-LET radiation that produces the same level of risk. The simplest method for obtaining the equivalent dose is to multiply the physical dose by a quality factor for the radiation field, but there are several other approaches, including models for normal-tissue responses, microdosimetric methods, and fluence-based techniques. In all of these cases, there is uncertainty associated with the method itself and additional uncertainty associated with each of the input quantities used to calculate the risk. In the former case, each of the quantities (e.g., physical dose or quality factors) required as input to establish the risk has a level of uncertainty associated with it. Reductions in the uncertainties in the values of the specific input quantities have differing effects on the magnitude of the uncertainty of the total risk, depending on the method chosen. Currently, the lack of knowledge concerning the uncertainties in the values of the quantities needed to assess risks is a major limitation in establishing realistic design requirements for a planetary mission.

In addition to uncertainties in the values of the input quantities, there is an intrinsic uncertainty associated with the method used. Since the use of only one method with a possible large uncertainty is at best questionable, it is recommended that risk estimates be determined by different, independent methods as a means of determining the overall uncertainty from input quantities and methods. The results of an error analysis (i.e., an analysis of the relative and absolute uncertainties) should be used to evaluate which methods will most effectively reduce the uncertainties in risk estimates and, therefore, uncertainties in the cost of shielding.

An analysis of the uncertainties in risk based on present data and methods could be achieved within about 1 year with proper support. Such analyses, however, should be updated routinely as part of a continuing effort throughout the entire project, and all investigators should be required to provide error analyses of their results. Lack of knowledge of cross sections for producing secondary nuclear particles in the materials used to construct a spacecraft represents a source of uncertainty that might be reduced, with a consequent potential for cost savings. However, variability in the types and energies of incident particles resulting from variation in the number and quality of solar events is not representative of an error in the input data used to calculate risk.

6. Determine how the selection and design of the space vehicle affect the radiation environment in which the crew has to exist.

Strategy

The capability for determining effects of space vehicle selection and design is based in part on having accurate knowledge of the incident radiation field, the reaction cross sections for the incident particles reacting with vehicle materials, and the fragmentation or recoil products that such reactions produce. Current knowledge of the fragmentation products of HZE particles is limited to only a few particles in a few materials. For knowledgeable shielding design, the initial radiation fields, reaction probabilities, and secondary particles produced as a function of angle must be determined through physical measurements, at an HZE-particle accelerator, of the particle types and energies resulting behind different compositions and thicknesses of shielding.

Based on the predictions of current transport codes, hydrogen-containing materials are preferred for shielding because they offer better shielding than other materials on a per-unit-mass basis. To properly assess the accuracy of these predictions, the transport codes used to calculate shielding efficiency have to be benchmarked against measured data for elemental (aluminum, iron, etc.) and composite shields.

Complementary measurements should be made with a microdosimetric detector of the type currently being flown in space. The absorbed dose as a function of depth should likewise be measured along the axis of the beam at selected positions along the axial plane. The measured data should be compared with predictions by the Langley Research Center transport code and/or a Monte Carlo transport code. Similar measurements as a function of depth should be made for the simplest possible geometry in space, and these results should be compared with calculations of the dose, radiation quality, and particle spectra.

Engineering of the storage for a spacecraft's supplies so as to form an enhanced "storm shelter" against transient high levels of radiation would be subject to the same verification of the accuracy of data and calculations. At the current level of availability of heavy-ion particle accelerator time, it is estimated that more than 10 years will be required to collect the necessary data.[48] With increased availability of accelerator time and other resources, data collection and analysis could be compressed into a time frame of about 4 years.

7. Determine if solar particle events can be predicted with sufficient advance warning to allow crew members to return to the safety of a shielded storm shelter.

Strategy

The ability to predict the time of occurrence and/or the magnitude of solar particle events is currently an inexact science at best. Protecting a mission crew from SPE radiation requires improving the capability to accurately predict solar events. This effort requires that information on the status of the total solar surface be continually available. One mechanism for accomplishing this would be a series of space platform monitoring stations. Given the necessary information on the status of the Sun's surface, the science and models that interpret these data must be enhanced, with the goal of achieving accurate forecasts 8 hours in advance of a spacecraft encountering an SPE. Prediction of the resources and time required to achieve this capability is beyond the expertise of the authors of this report. However, the ability to predict solar events 8 hours in advance of their occurrence is thought to be an operational requirement for a safe interplanetary mission.

Lower-Priority Research Recommendations

1. Estimate the risks of reduced fertility and sterility as a result of exposure to radiation on missions of long duration in deep space.

Strategy

• *Female:* Studies of women receiving pelvic and abdominal radiotherapy in which there is good dosimetry could provide useful information on the effects of radiation on ovarian function. It is probable that prospective studies of women treated with cytotoxic drugs at young enough ages, in whom ovarian function is compromised, could provide valuable information when combined with modeling. Complementary studies of both normal and radiation-induced loss of ovarian follicles, preferably in a nonhuman primate, will be required.

• *Male:* An assessment of the effect of dose rate and protraction of radiation on spermatogenesis is essential. The study should be carried out on a primate, but studies done previously on other mammals could be extended to include low dose rate or fractionated proton exposures. Sperm counts are an easy and economical assay of the effect of exposure. However, histological studies of the testes are required, especially in cases of azoospermia (total loss of sperm). The stem cells may not be the most sensitive target, because loss of the ability of the supporting tissue to enable differentiation in the spermatogenic process may determine the probability of sterility. Paracrine mechanisms, which release locally acting substances from cells directly into intracellular space, are involved in the differentiation process, but little is known about either their role or the effects of radiation on them. Studies of men receiving cyclophosphamide (a cell-killing chemical) could provide some help in comparing the relative effects of acute and chronic administration of radiation doses on sperm production.

To improve understanding of the effects of radiation on fertility, pragmatic studies of the loss of ova or sperm and studies of the basic aspects of ovarian function or spermatogenesis should be carried out hand in hand. As much clinical data as can be obtained and are relevant should be collected, and priority should be given to animal experiments designed to answer the questions that cannot be answered from clinical data. Ideally, the studies should include the effects of repeated exposure to protons and heavy ions at low fluences. However, protracted exposure to gamma rays may be the most practical approach, and gamma rays should be an adequate surrogate for protons. Since there are several sources available for both gamma rays and protons, beam time is probably not a limiting factor for conducting this study. If the group conducting the study were co-located with the source, and the appropriate support staff and animal care facilities were available, such a study might be completed in as little as 4 years. Currently, however, such a resource does not exist.

2. Estimate the risks of clinically significant cataracts being induced by exposure to radiation at the levels that will occur on extended spaceflights.

Strategy

A considerable body of data provides information about the induction of cataracts in different species by different types of radiation. There is, however, no consensus on how to collate the data and use them to estimate risk to humans. This objective, however, appears to be within reach and should be pursued. Another approach is to determine experimentally the relationship of RBE for cataractogenesis to LET and to apply the RBE value to the data for cataract induction in humans by low-LET radiation. A better understanding is needed of the effects of protracted exposure at low dose rates for both low- and high-LET radiation, because the data currently available for humans are for high-dose-rate, low-LET radiation.

Since research efforts on atomic bomb survivors (who were exposed to low-LET radiation) are already under way, the results of which could readily be applied to address recommendation 2, the most cost- and time-effective approach to this issue would be to ensure that current work on the survivors receives continued support.[49] The cataractogenic effects of protons, the most prevalent particles in galactic cosmic rays, can be estimated directly with reasonable confidence from data on the effects of low-LET radiation. Moreover, estimating the risk from exposure to HZE particles by any of the methods suggested so far in this report depends on the use of data obtained from humans exposed to low-LET radiation, and the major source of such data is atomic bomb survivors. Under these circumstances a sufficient answer to the question of the magnitude of the risk for cataractogenesis owing to long-duration spaceflight might be obtained in a time frame of 4 or 5 years.

3. Determine whether drugs can be used to protect against the acute or carcinogenic effects of exposure to radiation in space.

Strategy

A program to develop drugs capable of protecting humans against the acute toxic effects of radiation has been conducted for many years under the auspices of the Department of Defense. These efforts have yielded a number of drugs that are moderately protective against the effects of low-LET radiation because they are free-radical scavengers. Such scavengers are relatively less effective against high-LET radiation because ionizations are produced more frequently as a result of direct effects rather than indirectly through the products of water radiolysis. At present, the effectiveness of such agents against acute high-dose exposure to protons, such as might be experienced during an SPE, is not known. Studies should be conducted in animal models to determine the efficacy of single doses of such drugs in protecting against the damaging effects of protons, similar to those associated with an SPE, on blood-forming cells.

More recently, studies have suggested that agents related to the compound WR-2721 may be efficacious at relatively low doses in protecting against the mutagenic and carcinogenic effects of radiation through a mechanism independent of the drug's activity as a free-radical scavenger. It is recommended that studies be pursued to determine whether such protective effects can be obtained after exposure to HZE particles. Such studies could concentrate on radiation-induced somatic cell mutagenesis, since these effects are likely to be reasonably predictive of protective effects observed for carcinogenesis in animals. Additional mechanistic studies would allow the possibility of the development of more effective and less toxic agents that might be useful for protection against late effects associated with doses resulting from SPEs. It is unlikely that a strategy of daily doses of such agents is warranted as a means of modifying the risks from daily exposure to cosmic radiation, given the relatively low risks associated with exposure at these levels, although such a strategy might be useful at the time of an SPE.

These studies would require access to appropriate facilities for irradiation with HZE particles. Under current conditions (2 weeks' available beam time per year), it is estimated that such studies would require approximately 4 years to complete, assuming that cells for these studies could be "piggybacked" with those of other cellular studies. Under more ideal conditions, with 3 months of beam time available each year, these studies would require approximately 2 years to complete.

4. Determine if there is an assay that can provide information on an individual's sensitivity to radiation-induced mutagenicity and that also can be predictive of a predisposition for susceptibility to cancer.

Strategy

For at least 10 years, Sanford and colleagues have reported on the use of a G_2 chromatid aberration assay for detecting individuals with a predisposition for cancer.[50-52] In this assay, human lymphocytes or cultured skin fibroblasts are irradiated with x rays, and metaphase cells arrested with the chemical colcemid between 0.5 and 1.5 hours after exposure represent cells irradiated in the G_2 phase of the cell cycle. The analysis of chromatid aberrations gives a measure of chromosomal damage induced in G_2; comparison of this aberration frequency with that for metaphase cells collected in the first 0.5 hour after exposure provides an estimate of DNA repair capacity. It has been reported that individuals designated as cancer prone, irrespective of the tumor type, show an enhanced frequency of aberrations and a reduced repair capacity. Attempts to duplicate the assay in other laboratories have proved unsuccessful.[53] Scott et al., for example, found no difference in sensitivity between controls and lymphocytes from individuals who were homozygous or heterozygous for xeroderma pigmentosum (a DNA-repair-deficiency disease), or who had the cancer-predisposing disease familial adenomatous polyposis, or who had the cancer-predisposing syndromes Li-Fraumeni, basal cell nevus, Down's, or Fanconi's.[54] They were able to show an enhancement with ataxia telangiectasia homozygotes, a very predictable result. It remains to be determined if this modified assay can detect all individuals with a cancer predisposition or at least a predisposition in specific cases. To validate the assay, it is also of considerable importance that it be conducted in several different laboratories and that an extensive sample from the general population be assessed to obtain an estimate of the range of sensitivities. Since the assay can be validated with low-LET x rays or gamma rays, no beam time is required. The analysis of at least 100 individuals in the general population and at least 10 cancer-prone families (in several laboratories simultaneously) would take approximately 2 years to complete.

5. Determine if there are differences in biological response arising from exposure to particles with similar LET, but with different atomic numbers and energies.

Strategy

There is experimental evidence to suggest that differences in both the energy and the track structure of particles may lead to differing biological effects of exposure to radiation that are independent of LET.[55-57] The differences in observed RBE values generally have been in the range of 2 to 3. However, the available data are derived from various sources that utilize different models and experimental conditions, thus making comparison among them difficult. Carefully designed experiments should be carried out under controlled dosimetric conditions, such that the effect of factors such as atomic number, track structure, and energy can be specifically compared in the same system. It would seem reasonable to employ well-defined experimental systems such as those proposed to address the higher-priority recommendations 1 and 2, for which substantial data are already available for various types of radiation. These would include cellular systems to examine effects on cell killing, mutagenesis, and chromosomal aberrations. If the differences observed are restricted to a factor of 2 to 3 or less, as predicted from currently available data, conducting additional experiments in animal models for tumorigenesis would not be warranted.

Based on the assumptions that the appropriate heavy ions are available, that a dedicated facility is used to minimize tuning time, and that 2 weeks of beam time per year are set aside for this strategy, the in vitro experiments could be completed in 2 years, particularly if they were carried out in parallel with those intended to address the higher-priority recommendation 2. This estimate is based on the use of three biological end points and three different LET ranges, with three particles in each range. Since it is doubtful, however, that the necessary level of resources would be reserved for lower-priority projects such as this, an estimate of 3 years is probably more realistic at current levels of availability for HZE-particle accelerator time.

If the annual available beam time were increased to 3 months, then it should still be possible to carry out this strategy in 2 years.

TIME SCALE OF RESEARCH

To carry out the research necessary to reduce the physical and biological uncertainties inherent in estimating risk and to design shielding to protect against a credible maximum risk, approximately 3,000 hours of beam time are required for experiments with HZE particles and energetic protons. At the present utilization rate of approximately 100 hours per year at the Brookhaven Alternating Gradient Synchrotron, the research could take more than 20 years—an unacceptably long time.

Figures 11.2 and 11.3 show potential research time lines based on the assumption that the currently available beam time at a heavy-ion accelerator of about 2 weeks per year (Figure 11.2) remains unchanged or that 3 months become available per year (Figure 11.3).

It is recommended that, if the goal of safe interplanetary missions with human crews is sought, NASA explore various possibilities, including the construction of new facilities, to increase the research time available for experiments with HZE particles.

NEED FOR ANIMAL USE

There are no estimates for the risk of cancer induction in humans exposed to protons, the major component of galactic cosmic radiation and solar radiation, or to heavy ions such as iron. Therefore, risk estimates currently must be based on either (1) information about the risks incurred by exposure of animals to low-LET radiation modified to allow for the different RBE values for the different types of radiation involved or (2) data from animal experiments used in conjunction with some method of extrapolating the risk estimates to humans. Both approaches are hampered by insufficient experimental data. For example, it is essential to have adequate data on the induction of cancer by radiation at a sufficient range of LET values to obtain the RBE values to estimate the risk from exposure to GCR in deep space. Obtaining such data involves the use of animals and, to a lesser extent, in vitro studies on human chromosomes and cells. Specific deterministic effects, such as reduction in fertility, cataractogenesis, and damage to the central nervous system, are important in assessing the total risk posed by prolonged sojourns in the radiation environments in space. The effects of heavy ions on the central nervous system are of particular importance. Although no information about such effects on humans is available that is suitable for setting radiation limits, it is essential that the possibility of effects on the central nervous system be adequately assessed. Because the ideal of obtaining data from primates exposed to heavy ions is unlikely to be realized, critical animal experiments must be carefully crafted and executed.

EXPERIMENTAL TECHNIQUES AND NEW DATA REQUIRED

This section touches on new techniques being used for the qualitative assessment of mutations and chromosomal aberrations and for the characterization of molecular events involved in tumor development. It is assumed that significant progress in the next few years will be made in these broad areas.

Estimates of cancer risk posed by low-LET radiation are quite well founded and are based on fairly extensive animal, but limited human, studies (mostly those of atomic bomb survivors). Testing the reliability of the extrapolation of results from rodent studies to humans would require a better understanding of the mechanism of formation of specific tumor types, both background and x ray induced, for both human and animal models (with the same tumor type). Although little information is available on the genetic alterations associated with radiation-induced tumors, the methods exist to determine such alterations, and candidate genes such as the tumor suppressor gene p53 have been proposed.[58] What remains to be developed are sufficiently sensitive assays for detecting mutations in nonselectable genes that could be markers of early stages in tumor development. Although specific polymerase chain reaction (PCR) methods are becoming more sensitive, they are still 1 or 2 orders of magnitude away from being able to detect induced mutations at the frequencies of occurrence required, typically mutation frequencies of 1 in 10^7 cells.

Limited data are available on cancer induction in rodents exposed to high-LET radiation; information on other biological effects is also sparse. It will be necessary to conduct additional cancer studies in rodents exposed to different types of high-LET radiation and to characterize the resulting tumors at the molecular level. In fact, for high-LET radiation, the conversion of DNA lesions into mutations is not well understood. To better simulate conditions of exposure during spaceflight, it is necessary to consider the effectiveness of induction of mutations by low-dose-rate exposure to both high- and low-LET radiation. The use of fluorescence in situ hybridization allows reciprocal translocations to be assessed following protracted exposure. A translocation is a significant chromosomal end point when considering genomic alterations that are associated with adverse health effects. Assays are also under development for detecting low-frequency aberrations in genes above background. Although currently available only for selectable genes such as that for hypoxanthine phosphoribosyl transferase, for which mutants have a growth advantage (i.e., they are selected for their ability to grow faster than nonmutants), it is anticipated that new assays will be available for nonselectable tumor genes and genes such as p53 and other tumor suppressor genes in the future.

The identification of populations that are genetically susceptible to cancer development is also of considerable importance. Uncovering the mechanisms involved in tumor formation is critical for this purpose but, despite considerable progress, is still a distant goal. A more attainable goal may be the development of surrogate assays for predicting increased sensitivity for tumor induction. The G_2 chromosomal aberration assay described by Jones et al. is promising.[59] It appears to be able to identify individuals who have at least increased radiosensitivity of lymphocytes; in one instance, this increase was quite marked in about 40 percent of breast cancer patients. More work and probably a number of modifications to the technique are in order before it can be used as a predictor of radiation sensitivity.

FIGURES 11.2 and 11.3   The designation HP indicates strategies to address higher-priority research recommendations 1 through 7 and LP indicates strategies to address lower-priority recommendations 1 through 5, given 2 weeks of beam time per year and 3 months of beam time per year, respectively. In both figures, the bottom axis indicates the estimated amount of time required to carry out the various strategies; the top axis indicates the general dependence of the mission time line on the research. Strategies that can be carried out independently are separated by dotted lines. The length of the time bar associated with each strategy is based on the approximate amount of time estimated for the required research, except in the case of HP 5 and HP 7. HP 5 (the shaded box) is not a laboratory research strategy, but rather a computational methodology, and the amount of time reserved for it may be flexible. It is therefore set to end when the construction of shielding begins. The amount of time devoted to HP 7 will depend on the time available between the initiation of research and flight. Since HP1(b) utilizes protons and therefore does not require a heavy-ion accelerator, it is not affected by the given restrictions on beam time. NOTE: 1(a), cell studies; 1(b), animal carcinogenesis studies with protons; and 1(c), RBE-LET relationship studies.

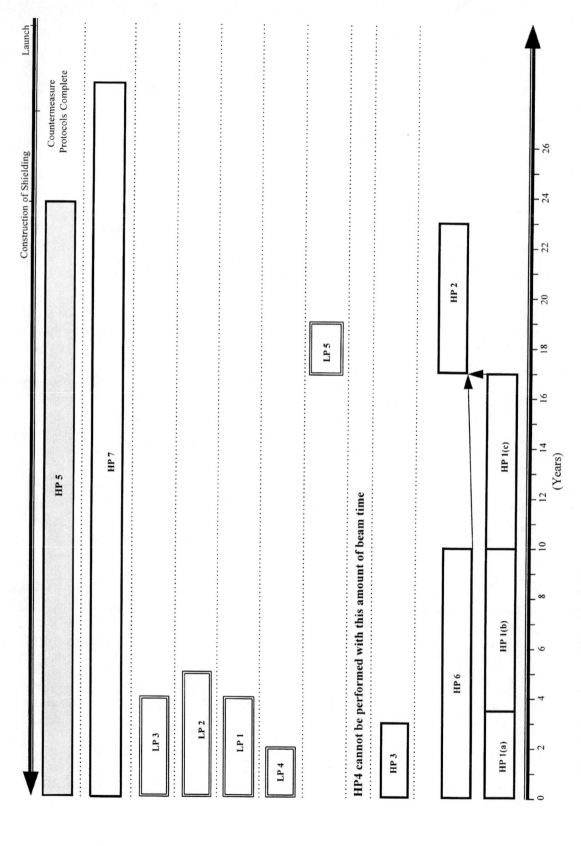

FIGURE 11.2   Potential research time line, given beam time of about 2 weeks per year.

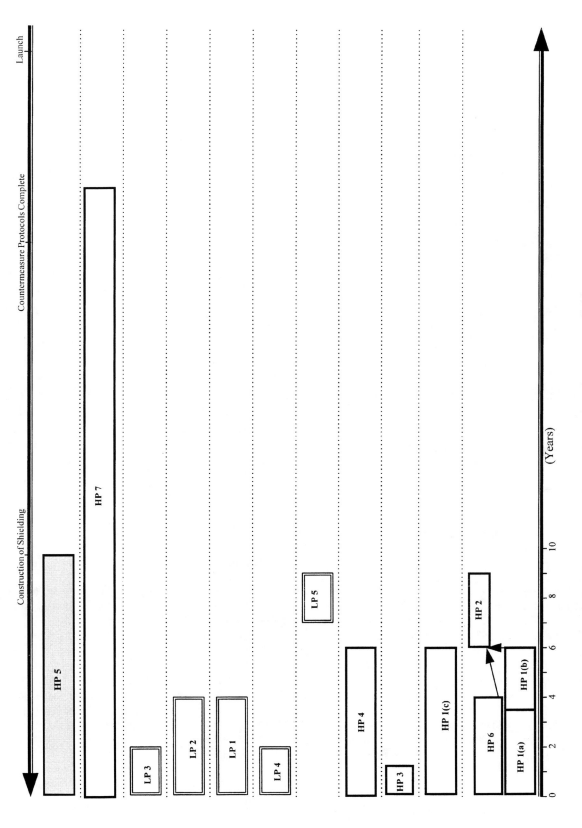

FIGURE 11.3   Potential research time line if beam time of 3 months per year becomes available.

GROUND- VERSUS SPACE-BASED RESEARCH

The influence of microgravity on the effects of low-LET radiation has been reviewed by Horneck[60] and was discussed by Nelson.[61] Most experiments showed negligible or small effects of microgravity on radiation-induced changes. Typical changes observed had to do with increased chromosomal alterations in fruit flies and in *Tradescantia* (the spiderwort plant) following irradiation before liftoff. Horneck suggests that changes in chromosomal structure or position in microgravity could have prevented effective rejoining of chromosomes. On the other hand, there was no control in such experiments for vibration or acceleration during liftoff or return of the satellites. In another example cited in these reviews—an experiment measuring viability in yeast—survival was lower for microorganisms irradiated before liftoff compared to ground-based controls treated in the same way. It was noted that the difference in survival did not seem to be dose dependent. These results were interpreted as indicating that DNA repair was less efficient in microgravity. No experiments were carried out in space using a 1-*g* centrifuge for controls.

Recent ground-based experiments, summarized by Kronenberg, on radiation-induced DNA fragmentation, neoplastic transformation of cells plated 24 hours after irradiation, and the effects of a chemical radioprotector on mutation induction showed that DNA repair and cell recovery take place readily after low-LET irradiation, but not after exposure to HZE particles.[62] Since the only reported significant effect of microgravity may be on DNA repair and cell recovery following low-LET exposure, and there seems to be no DNA repair or cell recovery following high-LET exposure, microgravity should not be important for HZE particle effects.

The above considerations indicate that HZE particles are a very important factor in the damage resulting from long space missions and that the effects of microgravity probably will not alter the cellular response to HZE particles but might actually increase the effect of low-LET radiation.

Hence, it is concluded that the majority of the useful information on radiation effects and risks will come from ground-based experiments and that radiation experiments in space, with all their logistical difficulties, will not be rewarding and may not be worth the effort.

REFERENCES

1. Space Studies Board, National Research Council. 1993. Scientific Prerequisites for the Human Exploration of Space. National Academy Press, Washington, D.C.
2. Space Studies Board, National Research Council. 1996. Radiation Hazards to Crews of Interplanetary Missions: Biological Issues and Research Strategies. National Academy Press, Washington, D.C.
3. Lawrence Berkeley Laboratory. 1997. Modeling Human Risk: Cell and Molecular Biology in Context. LBNL Report 40278. U.S. Department of Energy, Washington, D.C.
4. Badhwar, G.D. 1997. The radiation environment in low-Earth-orbit. Radiat. Res. 148: S3-S10.
5. Badhwar, G.D. 1997. The radiation environment in low-Earth-orbit. Radiat. Res. 148: S3-S10.
6. Badhwar, G.D. 1997. The radiation environment in low-Earth-orbit. Radiat. Res. 148: S3-S10.
7. Yang, T.C., George, K., Johnson, A.S., Durante, M., and Federenko, B.S. 1997. Biodosimetry results from spaceflight Mir-18. Radiat. Res. 148: S17-S23.
8. Straume, T., and Bender, M.A. Issues in biological dosimetry: Emphasis on radiation environments in space. Radiat. Res. 148: S60-S70.
9. Wilson, J.W., Cucinotta, F.A., Shinn, J.L., Kim, M.H., and Badavi, F.F. 1997. Shielding strategies for human space exploration: Introduction. Chapter 1 in Shielding Strategies for Human Space Exploration: A Workshop (J.W. Wilson, J. Miller, and A. Konradi, eds.). National Aeronautics and Space Administration.
10. Malgorzata, A.W., Adelstein, S.J., and Kassis, A.I. 1998. Indirect mechanisms contribute to the biological effects produced by decay of DNA-incorporated iodine-125 in mammalian cells in vitro: Double-strand breaks. Radiat. Res. 149:134-141.

11. Board on Radiation Effects Research, National Research Council. 1990. Health Effects of Exposure to Low Levels of Ionizing Radiation: BIER V. National Academy Press, Washington, D.C.

12. Board on Radiation Effects Research, National Research Council. 1990. Health Effects of Exposure to Low Levels of Ionizing Radiation: BIER V. National Academy Press, Washington, D.C.

13. Blakely, E.A., Ngo, F.Q.H., Curtis, S.B., and Tobias, C.A. 1984. Heavy-ion radiobiology: Cellular studies. Adv. Radiat. Biol. 11: 295-389.

14. Curtis, S.B., Nealy, J.E., and Wilson, J.W. 1995. Risk cross sections and their application to risk estimation in the galactic cosmic-ray environment. Radiat. Res. 141: 57-65.

15. National Council on Radiation Protection and Measurements (NCRP). 1989. Guidance on Radiation Received in Space Activities. Recommendations of the National Council on Radiation Protection and Measurements. NCRP Report No. 98. National Council on Radiation Protection and Measurements, Bethesda, Md.

16. Preston, D.L., Kusumi, S., Tomonaga, M., Izumi, S., Ron, E., Kuramato, A., Kamada, N., Dohy, H., Matsuo, T., Nonaka, H., Thompson, D.E., Soda, M., and Mabuchi, K. 1994. Cancer incidence in atomic bomb survivors. Part III. Leukemia, lymphoma and multiple myeloma, 1950-1987. Radiat. Res. 137: S68-S97.

17. Pierce, D.A., Shimizu, Y., Preston, D.L., Vaeth, M., and Mabuchi, K. 1996. Studies of the mortality of atomic bomb survivors. Report 12, Part I. Cancer: 1950-1990. Radiat. Res. 146: 1-27.

18. Thompson, D.E., Mabuchi, K., Ron, E., Soda, M., Tokunaga, M., Ochikubo, S., Sugimoto, S., Ikeda, T., Terasaki, M., Izumi, S., and Preston, D.L. 1994. Cancer incidence in atomic bomb survivors. Part II: Solid tumors, 1958-1987. Radiat. Res. 137: S17-S67.

19. International Commission on Radiological Protection (ICRP). 1991. 1990 Recommendations of the International Commission on Radiological Protection. ICRP Publication 60. Annals of the ICRP 21. Pergamon Press, Elmsford, N.Y.

20. International Commission on Radiological Protection (ICRP), 1991. 1990 Recommendations of the International Commission on Radiological Protection. ICRC Publication 60. Annals of the ICRP 21. Pergamon Press, Elmsford, N.Y.

21. National Council on Radiation Protection and Measurement (NCRP). 1997. Uncertainties in the Fatal Cancer Risk Estimates Used in Radiation Protection. Recommendations of the National Council on Radiation Protection and Measurements. NCRP Report No. 126. NCRP, Bethesda, Md., p. 73.

22. Simpson, J.A. 1983. Introduction to the Galactic Cosmic Radiation. Composition and Origin of Cosmic Rays (M.M. Shapiro, ed.). Reidel Publishing, Dordrecht, The Netherlands.

23. Simpson, J.A. 1983. Introduction to the Galactic Cosmic Radiation. Composition and Origin of Cosmic Rays (M.M. Shapiro, ed.). Reidel Publishing, Dordrecht, The Netherlands.

24. Dalrymple, G.V., Lindsay, J.R., Mitchell, J.C., and Hardy, K.A. 1991. A review of the USAF/NASA proton bioeffects project: Rationale and acute effects. Radiat. Res. 126: 117-119.

25. Urano, M., Verkey, L.J., Guitein, M., Lepper, J.E., Suit, H.D., Mendrondo, O., Gragoudos, E., and Koehler, A. 1984. Relative biological effectiveness of modulated proton beams in various murine tissues. Int. J. Oncol. Biol. Phys. 10: 509-514.

26. Storer, J.B., Harris, P.S., Furchner, J.E., and Langham, W.H. 1957. The relative biological effectiveness of various ionizing radiations in mammalian systems. Radiat. Res. 6: 188-288.

27. Clapp, N.K., Darden, D.B., Jr., and Jernigan, M.C. 1974. Relative effects of whole-body sublethal doses of 60-MeV protons and 300-kVp x rays on disease incidence in RF mice. Radiat. Res. 57: 158-186.

28. Burns, F.J., Hosselet, S., and Garte, S.J. 1989. Extrapolations of rat skin tumor incidence: Dose, fractionation and linear energy transfer. Pp. 571-582 in Low Dose Radiation: Biological Bases of Risk Assessment (K.F. Baverstock and J.W. Stather, eds.). Taylor and Francis, London.

29. Burns, F.J., Albert, R.E., Vanderlaan, M., and Strickland, P. 1975. The dose-response curve for tumor induction with single and split doses of 10 MeV protons. Radiat. Res. 62: 598 (abstract).

30. National Council on Radiation Protection and Measurements (NCRP). 1989. Guidance on Radiation Received in Space Activities. Recommendations of the National Council on Radiation Protection and Measurements. NCRP Report No. 98. National Council on Radiation Protection and Measurements, Bethesda, Md.

31. Ainsworth, E.J. 1986. Early and late mammalian responses to heavy charged particles. Adv. Space Res. 6: 153-165.

32. Merriam, G.R., Jr., Worgul, B.V., Medvedovsky, C., Zaider, M., and Rossi, H.H. 1984. Accelerated heavy particles and the lens 1 cataractogenic potential. Radiat. Res. 98: 129-140.

33. Lett, J.T., Lee, A.C., and Cox, A.B. 1991. Late cataractogenesis in rhesus monkeys irradiated with protons and radiogenic cataract in other species. Radiat. Res. 126: 147-156.

34. Worgul, B.V., Medvedovsky, C., Huang, Y., Marino, S.A., Randers-Pehrson, G., and Brenner, D.J. 1996. Quantitative assessment of the cataractogenic potential of very low doses of neutrons. Radiat. Res. 145: 343-349.

35. Committee on the Biological Effects of Ionizing Radiation, National Research Council. 1988. Health Effects of Radon and Other Internally Deposited Alpha-Emitters: BEIR IV. National Academy Press, Washington, D.C.

36. Alpen, E.L., Power-Risius, P., Curtis, S.B., DeGuzman, R., and Fry, R.J.M. 1994. Fluence-based relative biological effectiveness for charged particle carcinogenesis in mouse harderian gland. Adv. Space Res. 14: 573-581.

37. Alpen, E.L., Power-Risius, P., Curtis, S.B., DeGuzman, R., and Fry, R.J.M. 1994. Fluence-based relative biological effectiveness for charged particle carcinogenesis in mouse harderian gland. Adv. Space Res. 14: 573-581.

38. National Council on Radiation Protection and Measurements (NCRP). 1989. Guidance on Radiation Received in Space Activities. Recommendations of the National Council on Radiation Protection and Measurements. NCRP Report No. 98. National Council on Radiation Protection and Measurements, Bethesda, Md.

39. Burns, F.J., Hosselet, S., and Garte, S.J. 1989. Extrapolations of rat skin tumor incidence: Dose, fractionation and linear energy transfer. Pp. 571-582 in Low Dose Radiation: Biological Bases of Risk Assessment (K.F. Baverstock and J.W. Stather, eds.). Taylor and Francis, London.

40. Space Studies Board, National Research Council. 1996. Radiation Hazards to Crews of Interplanetary Missions: Biological Issues and Research Strategies. National Academy Press, Washington, D.C.

41. Space Studies Board, National Research Council. 1996. Radiation Hazards to Crews of Interplanetary Missions: Biological Issues and Research Strategies. National Academy Press, Washington, D.C.

42. Fry, R.J.M. 1981. Experimental radiation carcinogenesis: What have we learned? Radiat. Res. 87: 224-239.

43. Lett, J.T., Keng, P.C., Bergtold, D.S., and Howard, J. 1987. Effects of heavy ions on rabbit tissues: Induction of DNA strand breaks in retinal photoreceptor cells by high doses of radiation. Radiat. Environ. Biophys. 26: 23-36.

44. Lett, J.T., Cox, A.B., Keng, P.C., Lee, A.C., Su, C.M., and Bergtold, D.S. 1980. Late degeneration in rabbit tissues after irradiation by heavy ions. Pp. 131-142 in Life Sciences and Space Research, Vol. XVIII (R. Holmquist, ed.). Pergamon Press, Oxford, England.

45. Lett, J.T., Keng, P.C., Bergtold, D.S., and Howard J. 1987. Effects of heavy ions on rabbit tissues: Induction of DNA strand breaks in retinal photoreceptor cells by high doses of radiation. Radiat. Environ. Biophys. 26: 23-36.

46. Space Studies Board, National Research Council. 1996. Radiation Hazards to Crews of Interplanetary Missions: Biological Issues and Research Strategies. National Academy Press, Washington, D.C.

47. Williams, G.R., and Lett, J.T. 1995. Damage to the photoreceptor cells of the rabbit retina from 56Fe ions: Effect of age at exposure. Adv. Space Res. 18: 55-58.

48. Space Studies Board, National Research Council. 1996. Radiation Hazards to Crews of Interplanetary Missions: Biological Issues and Research Strategies. National Academy Press, Washington, D.C.

49. Wu, B., Medvedovsky, C., and Worgul, B.V. 1994. Non-subjective cataract analysis and its application in space radiation risk assessment. Adv. Space Res. 14: 493-500.

50. Parshad, R., Sanford, K.K., and Jones, G.M. 1983. Chromatid damage after G2 phase X-irradiation of cells from cancer-prone individuals implicates deficiency in DNA repair. Proc. Natl. Acad. Sci. U.S.A. 80: 5612-5616.

51. Sanford, K.K., Parshad, R., Gantt, R., Tarone, R.E., Jones, G.M., and Price, F.M. 1989. Factors affecting and significance of G2 chromatin radiosensitivity in predisposition to cancer. Int. J. Radiat. Biol. 55: 963-981.

52. Parshad, R., Price, F.M., Pirollo, K.F., Chang, E.H., and Sanford, K.K. 1993. Cytogenetic response to G2-phase X-irradiation in relation to DNA repair and radiosensitivity in a cancer-prone family with Li-Fraumeni syndrome. Radiat. Res. 136: 236-240.

53. Bender, M.A., Viola, M.V., Riore, J., Thompson, M.H., and Leonard, R.C. 1988. Normal G2 chromosomal radiosensitivity and cell survival in the cancer family syndrome. Cancer Res. 48: 2579-2584.

54. Scott, D., Spreadborough, A.R., Jones, L.A., Robert, S.A., and Moor, C.J. 1996. Chromosomal radiosensitivity in G2-phase lymphocytes as an indicator of cancer predisposition. Radiat. Res. 145: 3-16.

55. Kranert, T., Schneider, E., and Kiefer, J. 1990. Mutation induction in V79 Chinese hamster cells by very heavy ions. Int. J. Radiat. Biol. 58: 975-987.

56. Belli, M., Cera, F., Cherubini, R., Haque, A.M., Ianzini, F., Moschini, G., Sapora, O., Simone, G., Tabocchini, M.A., and Tiveron, P. 1993. Inactivation and mutation induction in V79 cells by low energy protons: Re-evaluation of the results at the LNL facility. Int. J. Radiat. Biol. 63: 331-337.

57. Stoll, U., Schmidt, A., Schneider, E., and Kiefer, J. 1995. Killing and mutation of Chinese hamster V79 cells exposed to accelerated oxygen and neon ions. Radiat. Res. 142: 288-294.

58. Culotta, E. and Koshland, R.D.E. 1993. p53 sweeps through cancer research. Science 262: 1958-1961.

59. Jones, L.A., Scott, D., Cowan, R., and Roberts, S.A. 1995. Abnormal radiosensitivity of lymphocyte from breast cancer patients with excessive normal tissue damage after radiotherapy: Chromosome aberrations after low-dose-rate irradiation. Int. J. Radiat. Biol. 67: 519-528.

60. Horneck, G. 1992. Radiobiological experiments in space: A review. Int. J. Radiat. Appl. Instrum. 20: 82-205.
61. Nelson, G., Space-based radiation biology, presentation to the Task Group on the Biological Effects of Space Radiation, November 13, 1995, Washington, D.C.
62. Kronenberg, A., NASA space radiation health program: Ground-based radiobiology research program, presentation to the Task Group on the Biological Effects of Space Radiation, Washington, D.C.

12

Behavioral Issues

INTRODUCTION

Astronauts on long-duration missions are subjected to many factors that may affect their health, well-being, and performance of mission-related duties. Some of these factors are unique to the space environment (e.g., prolonged periods of microgravity), whereas others are also present in other environments (e.g., confinement, isolation, exposure to physical hazards, altered work or rest schedules). Still others are characteristic of the individuals, groups, and organizations involved in crewed space missions. This chapter discusses the factors that affect behavior and performance during the preflight, inflight, and postflight phases of manned space missions, and it makes recommendations for research and changes in operations to ensure crew members' safety, well-being, and productivity, along with mission success.[1] As such, it is much broader in scope than the study of space human factors, which focuses on the role of humans in complex systems, the design of equipment and facilities for human use, and the development of environments for comfort and safety.[2] Although these issues are not addressed here, the committee's objective is nevertheless to provide a comprehensive assessment of the current status and future direction of research in human behavior and performance in space.

Program History

Over the past 30 years, a number of reports and publications of scientific panels, working groups, and scientific conferences have identified various requirements for conducting research on human behavior and provided NASA with several recommendations. Those recommendations have been based on a similar set of conclusions as summarized in the 1987 Goldberg report:[3]

> There is not enough objective data to determine the seriousness of behavioral impairments in past spaceflight missions. Nevertheless, there is reason to suppose that pyschological problems have already

occurred in spaceflights and that these will increase in frequency and severity as missions become longer and more complex, as crews become larger and more heterogeneous, and as the dangers of spaceflight become more fully appreciated.

Despite this assessment of the importance of behavioral issues, little progress had been made in transforming the recommendations for research on human behavior and performance in space into action.[4] In contrast to the routine collection of data on cardiovascular, neurological, and musculoskeletal changes in flight since the early days of the U.S. space program, there has been relatively little effort to collect data on behavior and performance in a systematic fashion. Nevertheless, as missions have become longer in duration, issues relating to human behavior and performance have gained increasing prominence. This prominence is reflected in studies of individuals and groups in analogue environments as well as the largely anecdotal accounts of long-duration missions of the Russian space program and the Shuttle-Mir Space Program (SMSP). As could be predicted from controlled simulation studies,[5-7] the history of space exploration has seen many instances of reduced energy levels, mood changes, poor interpersonal relations, faulty decision making, and lapses in memory and attention. Although these negative psychological reactions have yet to result in a disaster, this is no justification for ignoring problems that may have disastrous consequences. Furthermore, there are degrees of failure short of disaster and degrees of success short of perfection; if favorable organizational and environmental conditions can increase the level and probability of success, they are worthy of consideration.

Statement of Goals

The 1987 NRC report stated: "The overall goal for the study of human behavior in space is the development of empirically based scientific principles that identify the environmental, individual, group, and organizational requirements for the long-term occupancy of space by humans."[8] This goal remains as important today as it was 11 years ago. However, it is important to acknowledge three interrelated elements of this overall goal. The first is the identification of operational requirements for the facilitation and support of optimal individual, group, and organizational performance and the prevention and treatment of performance decrements during long-duration missions. Although priority should be given to the operational relevance of this research, the second element of research on behavior and performance in space is the advancement of fundamental knowledge of human behavior that has relevance beyond long-duration missions. Both of these elements, in turn, are related to a third—promotion of the physical and mental well-being of those directly and indirectly involved in long-duration missions, including flight crews, their families, and their ground support colleagues. All three objectives presume the existence of certain environmental and organizational constraints that will give priority to operational over basic science issues as well as observational over experimental research designs. Nevertheless, the committee acknowledges the importance of utilizing approaches to the study of behavior and performance in space that seek to integrate these diverse sets of issues and designs.

Definition and Assessment of Behavior and Performance in Space

Research in human behavior and performance during long-duration missions is concerned with individual crew members and ground support personnel; groups of individuals who comprise the flight crews and ground support teams of specific missions; and the organizations that recruit, train, and support these individuals. Individual performance has traditionally been assessed using measures of task productivity (ability), emotional health and well-being (stability), and the quality and quantity of social interactions with other crew members (compatibility).[9,10] Group performance has traditionally been assessed using measures of cooperation and conflict.[11,12] Organizational performance has traditionally been assessed using measures of the costs and benefits of accomplishing mission, organizational, and overarching national political goals and objectives.

In the past, emphasis has been placed on human behavior and performance research that is conducted in flight. However, research conducted during the extended period of training and preparation prior to the mission (preflight) and during the aftermath of a crewed mission (postflight) is also viewed as essential in accomplishing the overall scientific goals described above.

Research in Analogue Environments

Much of our understanding of human behavior and performance in space has been obtained from the study of "analogues" such as Antarctic research stations, polar expeditions, nuclear submarines, isolated military outposts and national parks, undersea habitats, oil drilling rigs, small rural communities, and space simulator experiments.[13-16] These analogues serve as "model systems" of behavior in space in much the same way that hind-limb unloading and bed-rest studies described elsewhere in this report serve as model systems of physiological changes related to microgravity. In many instances, a considerable amount of data that are relevant to long-duration space missions has already been collected. These data sets also offer a larger sample size than is typically available from either the U.S. or the Russian space programs. The collection of new data in analogue environments may also be cheaper than collecting similar data in space for logistical reasons. Thus, analogue studies are more cost-effective than new research conducted in space and provide an opportunity to identify and resolve problems on the ground before they occur in space. Simulator studies can be especially useful, since they lead to the characterization of important space-related psychosocial factors under controlled conditions, expose possible relationships between these factors, and allow variables to be manipulated in a systematic fashion.

However, analogue settings vary with respect to their relevance to long-duration space missions.[17] Such missions and each of the analogue environments exhibit important differences with respect to characteristics of crew members, procedures for screening and selection, crew size, mission objectives and duration, and the nature of the physical environment. Patterns of behavior associated with adapting to the extreme cold and extended periods of darkness during the Antarctic winter, for instance, may be very different from patterns of behavior associated with adapting to the microgravity environment of space capsules. Similarly, lessons learned about the effects of duration of exposure to isolation and confinement from studies of submarine crews may not be relevant to astronaut crews that are much smaller in size. In general, analogue studies are believed to be necessary and suitable for examining many of the issues described in this chapter, including: environmental conditions common to isolated and confined environments, circadian rhythms and sleep, the psychophysiology of emotion and stress, stress and coping, cognition and perception, emotion, personality, crew tension and conflict, crew cohesion, ground-crew interaction, leadership, psychosocial countermeasures, organizational culture, mission duration, and mission management. Other issues can be examined only under actual conditions of spaceflight. These include the effects of microgravity and the absence of a normal diurnal cycle on sleep and circadian rhythms, cognition and perception, affect and the hypothalamic-pituitary-adrenal axis; psychopharmacology; and the requirements for conducting valid and reliable psychophysiological measures in space. There is a need to determine how to connect data collected in analogue settings to the characteristics of the space program. In addition, greater emphasis is required on determining how knowledge obtained from analogue environments can be incorporated into the National Aeronautics and

Space Administration (NASA) activities with respect to the screening and selection of astronaut personnel, development and implementation of training programs and psychological countermeasures, and development and testing of data collection procedures that meet the requirements listed above.

Integration of Research and Operations

Research on behavior and performance in space presents numerous opportunities to integrate the goals and objectives of researchers and operations managers and to simultaneously address issues of theoretical and operational importance. There is a need to bring researchers, clinicians, and managers involved in operations together to identify and prioritize the needs for research from the operational perspective and the resources available to conduct behavior and performance research. There is also a need to conduct research that demonstrates a relationship between issues with operational relevance and the more fundamental questions related to human behavior and performance in general. For instance, research conducted on a long-duration mission or in analogue environments offers the potential for determining the extent to which various personality characteristics such as neuroticism, extraversion, and openness influence cognitive performance, patterns of stress and coping, social dynamics, or measures of health and well-being. Similarly, the extent to which crew member autonomy or control over workload influences health, well-being, and performance could provide greater insight into the role of self-efficacy and personal autonomy in general.[18] Ultimately, research on behavior and performance in long-duration space missions that meets the needs of the larger society, as well as the more specific programmatic needs of NASA and other space agencies, should be designed and developed.

Organizational Support of Research

Research on human behavior and performance in space requires first the development of an organizational climate that is supportive of such research. The principal task in creating such a climate is providing access to the objects of study (i.e., space crews and ground control personnel). Previous NRC recommendations and other reviews of the need for research on behavior and performance in space have emphasized the importance of gaining access to those involved in crewed space missions.[19] In the past, there has been a great reluctance on the part of the astronaut community to participate in such research because of concerns over the lack of confidentiality and the potential for inappropriate use of information that could threaten assignments to specific missions and jeopardize careers. Even if astronauts are willing to participate in such research, they are unlikely to be in a position to provide information on behavior and performance if they operate in an organizational environment that does not support the collection and analysis of such information and that provides no safeguards against its misuse. The ability to convince the individuals, groups, and organizations who are the objects of this research to participate is critical to its success. However, access implies some form of reciprocity between behavioral scientists and their subjects. Although previous reports have implied that it is the primary responsibility of NASA to provide access to study subjects, behavioral scientists must share responsibility in this regard. This responsibility includes clear explanation of the importance of such research, demonstration of its relevance to mission operations, and implementation of procedures that will ensure the confidentiality of data collected and prevent the misuse of these data. It may also be worth considering an alternative strategy in which researchers and study participants collaborate in refining the design and execution of studies that are of interest and importance to both parties. This kind of "action research" in operational settings often can generate more trustworthy findings (and therefore better science) than studies using methods that have been optimized for laboratory conditions. Moreover, a more collaborative approach can often avoid the "us-versus-them" dynamic that not infrequently develops between researchers who need compliance with their procedures and subjects who understandably may object to invasive questions, instruments, or protocols for research whose purpose is not fully understood or for which they view themselves merely as objects under study.

Research in behavior and performance also requires an increase in the level of institutional support for psychologists, psychiatrists, and other behavioral scientists within the respective organizations involved in long-duration missions. Such support may take several different forms—from increased funding for NASA scientists to interact with colleagues at professional meetings and conferences, to a greater role in management decision making. Increased visibility of the behavioral sciences within the management framework of NASA would contribute substantially to creating a climate that is favorable to research in behavior and performance.

A third requirement is a commitment to ongoing data collection for operational and research purposes. As with any data collection activity, it is much easier and more cost-effective to maintain an ongoing data collection system than to intermittently initiate and discontinue data collection efforts.

A fourth requirement for the organizational support of research in human behavior and performance in space is improvement of the peer review process for grant proposals and expansion of opportunities for the dissemination of study results to the wider research community. At present, proposals for research in behavior and performance are reviewed by panels that are composed almost exclusively of investigators with expertise in space human factors, with almost no input from investigators with expertise in the social and behavioral sciences. In contrast to the recommendations of the Committee on Advanced Technology for Human Support in Space (1997),[20] this committee believes that NASA should continue to separate behavior and performance from space human factors at the managerial level. Although there are admittedly certain advantages to the integration of management efforts in these two interrelated areas, one of the problems with conducting behavior and performance research in the past has been the low priority accorded to such research in favor of work conducted in areas such as ergonomics, biomechanics, anthropometrics, and workload. In part, this low priority has been the result of failure to understand the different focus of behavior and performance research and space human factors research. Greater efforts are required on the part of NASA to identify and recruit experts in human behavior and performance in space and analogue environments, as well as experts in disciplines that have relevance to the issues described below, to participate in such research, review grant proposals, and evaluate study findings. Greater efforts are also required on the part of investigators to publish many of the results that currently exist only in anecdotal form, abstracts of conference proceedings, or the "gray" literature of technical reports.

ENVIRONMENTAL FACTORS

Living in a small capsule for a prolonged period exposes the group to chronic, cumulative stress from a variety of environmental sources. This is an increasingly serious issue as missions become longer, because stressors and vulnerability to stress accumulate over time.[21] Stress effects are also exacerbated by unexpected reductions in living space, habitability, or safety, which may become even more serious as missions involve longer duration and lower accessibility.

Environmental Conditions Unique to Spaceflight

As described in Chapter 5, microgravity significantly disrupts such basic aspects of perception and behavior as vertical orientation and energy expenditure, and is causally linked to space motion sickness.

Microgravity has important implications for ergonomic design, and its possible effects on task performance, visual orientation, and emotional well-being during long-duration missions clearly require intensive research. Prolonged periods in small enclosed spaces, combined with microgravity, have effects on perception and motor behavior.[22] A recent report that astronauts cannot accurately recall the position of an unseen target has implications for safe manipulation of controls when visual contact is lost.[23] As noted in Chapter 5, depth perception, eye-hand coordination, and visual constancies may also be affected during flight. Systematic in-flight research on how design features might remedy these problems (e.g., by establishing a consistent local vertical)[24] is needed.

A host of minor discomforts peculiar to spaceflight may add up to the most stressful aspects of capsule life. Such chronic stressors or "daily hassles" include uncomfortable levels of temperature and humidity, limited light (to conserve power), noise and vibration from machinery, constant vigilance involved in monitoring instruments, the need to put on and take off clumsy protective clothing for every venture outside the capsule, the growing accumulation of garbage and floating particles, and limited facilities for sanitation and refreshment. Environmental pollution, in particular, may have different emotional and performance effects on different individuals, and tolerance for such conditions may be worth testing as an aspect of astronaut selection and training. Environmental pollution may also cause olfactory distress, and although psychological adaptation and physiological habituation to unpleasant olfactory stimuli have been studied, they have not been investigated during prolonged capsule living. Analogue studies typically have not involved the deliberate induction of discomfort, but future studies should do so because such conditions are very likely to have emotional consequences that would affect individual performance and group dynamics.

Some unique environmental features are not only unpleasant but also dangerous. A good example is the degradation of the internal atmosphere through increased carbon dioxide levels or through fuel or gas leaks. For example, glycol was released into the atmosphere on board Mir in April 1997; the crew had to don protective gear for extended periods of time, which further increased discomfort and inconvenience. These and other potential dangers, such as radiation and loss of pressure, make spaceflight intrinsically stressful. It is important to establish how space crews assess and react to the continuous possibility and occasional occurrence of hazards and whether these processes differ in short-duration and long-duration missions. There is a critical need to examine the perceived likelihood and severity of risk factors, how the perceived risk influences the individual and group performance of the crew, and what countermeasures would minimize their fears.

Environmental Conditions Common to Isolated, Confined Environments

Space vehicles are part of the class of isolated, confined environments (ICEs) that have a number of shared characteristics. Some of these pose a challenge for the crew and therefore must be considered in physical and behavioral design processes. For instance, several ICE stressors are related to physical and social density: crowded, minimally partitioned spaces violate the fundamental need to maintain control over other people's access to oneself. In normal environments, this control is exercised by means of interpersonal distance, territoriality, and privacy. When control over these factors is eroded, stress results.[25] Violations of interpersonal distance, territoriality, and privacy are common in ICEs and can result in irritability, tension, personal conflicts, withdrawal, and performance decrements (especially in tasks that require cooperation). These problems may be exacerbated when the crew simultaneously has to cope with other demands, when cohesion is already low, or when the group includes people from cultures that follow different norms. For instance, preferred interpersonal distances are greater in North America and Western Europe than in the Middle East and some parts of Asia and Latin America, and a distance that appears comfortable to one participant in a cross-cultural interaction may signal aggression or rejection to the other.[26]

Research is needed to understand how stressors associated with the physical and social density of ICEs are likely to be affected by the unique characteristics of spaceflight (e.g., the peculiarity of microgravity makes possible some new and strange interaction orientations)[27] and how they influence specific aspects of performance and adjustment in missions of various types and durations. Space simulators could be used to assess how interior design (e.g., movable bulkheads; improved sound insulation; curtains around sanitary facilities; seats that can be arranged to promote or discourage social, visual, and auditory interaction) can optimize privacy, interpersonal distance, and territoriality within the limitations of the vehicle's payload and construction. Research should also be devoted to the development and evaluation of training programs and other countermeasures that promote individual privacy and territoriality and interpersonal distance. Particular attention should be devoted to the development and evaluation of programs that provide instruction on individual and cultural differences in the need for distance, privacy, and territory and efforts to maintain them.

Monotony is another characteristic of isolated and confined environments that will likely influence behavior and performance on long-duration spaceflights. Previous research in analogue settings indicates that the lack of variation in the physical and social environment can result in boredom, loss of energy and concentration, and interpersonal friction. Because it is difficult to change the environment, perception of reduced control may also occur, a situation that is very disturbing to individuals such as astronauts who view themselves as effective agents in controlling their own lives.[28] However, additional research is required to determine whether physical monotony intensifies the desire to explore and manipulate elements of the environment during long-duration missions and the conditions under which the desire to test one's abilities can result in the manufacturing of artificial challenges, sometimes by taking unnecessary risks and ignoring familiar safety rules.[29]

Research on the effects of physical monotony can also contribute to the development of effective countermeasures. This might include, for instance, the design of activities that offer challenge and perhaps excitement without jeopardizing the crew or the mission. "Surprises" could be delivered by resupply flights, or the capsule itself (e.g., on a Mars voyage) might carry research or other tasks, music, reading matter, videotapes, family packages, and so on, to be opened at different points along the mission. The efficacy of such countermeasures should be evaluated in analogue studies before the policy is implemented.

There is also a need to develop and evaluate countermeasures that maximize the crew's control over the environment.[30] Nonstandard items of clothing, food, recreational materials, and reminders of home are already in use (although their effect has not been systematically studied); in addition it might be possible to let astronauts change color schemes within the capsule (perhaps through projection[31]), move some pieces of furniture or equipment, and attach items of personal significance or pleasure to the interior. News from and communication with home, and stimulating multimedia resources, can relieve the unchanging ambient stimulus array.[32] Natural images—projected scenes, audiotapes, released odors[33]—might be especially useful, and windows have been found to relieve visual monotony and maintain a sense of contact with Earth.[34] We need to study what happens when the view from the window is essentially unchanging and when Earth is not visible. In such cases, the window may become a source of more boredom, and both it and the natural scenes can become sources of loneliness.

Long stretches of monotonous unstructured time are likely to be a problem on long-duration missions. Crew members in such environments (e.g., polar crews,[35] prisoners in solitary confinement[36]) can suffer from too much free time. When the team spends many months or years in the environment, leisure periods can become a time of boredom, lassitude, neglected hygiene, and apathy.[37] Research is necessary to develop countermeasures for this problem, such as systematic programs of some sort, whether fitness exercises, study, or hobbies. Group activities of the same sort can also be developed.

Recommendations

Research on environmental factors should include an examination of the following topics, listed in order of priority:

1. The effects of the physical and psychosocial environment of spacecraft on cognitive, psychophysiological, and affective measures of behavior and performance:

- *Affective and cognitive responses to microgravity-related changes in perceptual and physiological systems;*
- *Behavioral responses to perceived risks associated with the space environment (e.g., radiation, contamination of the ambient atmosphere, buildup of debris, use of breathing apparatus and space suits); and*
- *Psychosocial predictors of the use and perceived importance of "personal" territories and individual strategies for coping with physical and social monotony.*

2. The development and evaluation of countermeasures for mitigating adverse effects of the physical and social environments on individual and group performance:

- *Alternative methods for filling unstructured time such as organized group and individual recreation and leisure activities;*
- *Use of novelty or "surprises" in terms of work assignments, recreational materials, and messages/gifts from home to reduce boredom and monotony;*
- *Crew training in communication and acceptance of different levels of desired privacy, interpersonal distance, territoriality, and personal disclosure; and*
- *Design of spacecraft interiors and amenities to maximize control over the physical environment and reduce impacts of physical monotony on behavior and performance.*

PSYCHOPHYSIOLOGICAL ISSUES

Circadian Rhythms and Sleep

Sleep is a basic human need that normally occurs in a temporally regular pattern, beginning at night and ceasing with the day. Alteration or disruption of this normal circadian pattern affects sleep duration and may induce a sleep deficit. Lack of sleep contributes to stress and to inefficiency in cognitive and psychomotor performance. Poor sleep and fatigue have been reported on a number of short-duration Shuttle missions as well as long-duration Mir missions. About 30 percent of U.S. Shuttle astronauts have requested sleep medication in-flight, although none had a history of such usage on Earth.[38]

Under natural conditions, the sleep-wake cycle and other circadian rhythms are synchronized with the period of Earth's rotation by means of periodic factors in Earth's environment. The most important of these *zeitgebers* are day-night changes in light and temperature.[39] Current plans for the International Space Station call for illumination levels ranging from 108 to 538 lux. This is much lower than the 2,500 lux necessary to entrain circadian rhythms in humans.[40] Hence, some desynchronization of sleep-wake and other circadian cycles would be expected to occur.

There are wide individual performance differences that are dictated by circadian rhythm patterns, ranging from those whose optimal work physiology peaks early in the day to those who can work efficiently long into the night. These patterns are consistent and are likely to persist in the space environment despite the imposition of a common work schedule. On the other hand, groups in temporal isolation have been found to display synchronous circadian rhythms.[41,42] Research is needed to determine whether it is important to match work groups in circadian pattern for greater crew harmony or whether, perhaps, different types should be distributed in a crew so that someone is always at optimal physiological efficiency.

Sleep deprivation has also been commonly reported by personnel working in other ICEs. However, research is required to identify the consequences of sleep disruption. For instance, the inability to concentrate, intellectual inertia, an increase in suggestibility and hypnotizability, and sensory hallucinations reported in laboratory and antarctic studies[43] may be a consequence of sleep disruption or of reduced social and environmental stimulation and increased monotony. Studies of night-shift workers and individuals in temporal isolation have found sleep disruptions to be associated with significant decrements in cognitive performance and increases in tension, depression, anger, and confusion.[44,45] A number of theories have been proposed to explain the association between sleep, cognitive performance, and mood, including the phase-advance or phase-shift hypothesis (which postulates that during depression the strong circadian X oscillator—which governs core body temperature, rapid eye movement, sleep propensity, and plasma cortisol secretion—is advanced in its timing relative to the weaker Y oscillator, which modulates the rest-activity cycle, propensity for slow-wave sleep, and plasma growth hormone levels),[46] the circadian dysregulation or attenuation hypothesis (which postulates that circadian rhythms in depression are characterized by a loss of power or rhymicity),[47] and the entrainment error hypothesis (which postulates that depression is the result of a lack of temporal stability in the circadian cycle resulting from weakened entrainment to the 24-hour day).[48] However, the conditions under which sleep affects or is affected by mood and cognition during a long-duration mission remain to be identified. Studies conducted in flight, along with ground-based simulations of long-duration missions, offer a unique opportunity to examine the merits of these competing theories and to determine how different circadian systems mediate the association between sleep and performance or well-being. Similarly, research is required to determine whether changes in sleep that occur during long-duration missions are a function of stress, workload, isolation, confinement, the absence of social *zeitgebers* or other environmental cues (e.g., a 24-hour day-night cycle) that influence circadian rhythms, or other features of the physical environment (microgravity, noise, light, atmosphere).

Another issue that should be investigated in the next decade is whether members of space crews experience improved sleep when they are allowed to set their own work-rest schedules or when they are required to adhere to a fixed schedule. During the Soyuz program, when sleep schedules were set counter to the local time of the launch site, cosmonauts experienced some degradation of performance and disturbed sleep.[49] Early research under conditions of isolation revealed that human circadian rhythms tend to "free-run" for a period of approximately 25 hours.[50] However, further research is required to determine the probability of major dissociations and desynchronizations in circadian systems under uniform or individualized work schedules in an isolated and confined environment and the long-term consequences of such desynchronizations. The environment used in most ground-based research is contrived, work activities set for subjects are often purposeless, and the experimental designs invariably involve solitary confinement without knowledge of time.[51] Evidence from analogue settings as to the costs and benefits of allowing personnel to "free-cycle" and set their own schedule versus maintaining a fixed work-rest schedule remains inconclusive.[52-54]

A third issue related to sleep during long-duration missions concerns the long-term effects of sleep deprivation. Evidence from the Antarctic suggests that the restoration of normal sleep patterns, especially Stage IV sleep, may take as long as 2 years for individuals in an isolated and confined environment.[55] Similarly, laboratory studies suggest that individuals may unknowingly accumulate very large sleep debts with an associated vulnerability to a decline in performance during periods of monotony and reduced stimulation.[56] However, the extent to which differences in individual characteristics (e.g., age, gender, levels of stress, coping styles, personality) or environmental context (e.g., workload, light-darkness, noise) are associated with such accumulation has yet to be determined.

The Psychophysiology of Emotion and Stress

The data on human emotional reactions involve three primary response systems: (1) language responses, including reports of feelings, evaluative judgments, and expressive communications; (2) behavioral acts such as avoidance, attack, and performance deficits; and (3) alterations of the sympathetic-adrenal-medullary (SAM) system and the hypothalamic-pituitary-adrenal (HPA) axis. These different outputs have unique sensitivities and temporal parameters and can be separately shaped by the environment. Cross-correlations between systems are often low, and a single channel of information on emotion may grossly mislead. It is not uncommon for some individuals, reporting no fear or anxiety when under palpable stress, to nevertheless show marked, sustained sympathetic activation of the HPA axis. In this context, the physiological marker may be the better predictor of future behavioral disturbances and performance inefficiency.

Furthermore, it is important to consider that the psychological meaning of physiological changes is not obvious or always consistent with common assumptions. For example, heart rate (a measure often taken during spaceflight) decreases (not increases) when organisms orient to threatening stimuli. Physiological data must always be interpreted in the context of their collection, with reference to focal stimuli, tasks being performed, social and physical environment, and related verbal and behavioral responses. Understanding the emotional significance of physiological data depends on careful research in the applied settings of interest.

Research on the psychophysiology of stress and emotion in the general population suggests that the alterations in the HPA axis likely to occur during long-duration missions, as described in Chapter 9, may also have significant impacts on mood and memory.[57] However, the precise nature of these mechanisms and the extent to which they affect other aspects of performance, including cognition, emotion, and social interactions, remain to be determined. For instance, relatively pronounced HPA activation is common in depression, but it is unclear whether HPA activation causes or results from depressed affect. Studies in flight will be required to identify the cognitive and affective correlates of hormonal, cardiovascular, respiratory, and other physiological changes that represent alterations of the HPA axis and the direction of causality of such associations.

Considerable research has been conducted on the use of flight simulators for training and evaluation,[58] but there is a dearth of psychophysiological data from this setting. It has been shown that media representations and simulated stress situations involving anger, fear, or anxiety generate physiological patterns of response similar to those occasioned in the actual context.[59] This suggests that much could be learned from psychophysiological studies of simulated stressful missions that might then be used in the development of effective countermeasures. For example, does the crew's preflight physiological state predict the probability of crew conflict, social interaction problems, and poor performance? Are there physiological markers during the mission that augur reduction in crew efficiency (error proneness, inattention, etc.) that could prompt early preventive remediation?

Most psychophysiological research uses average responses of many subjects and many trials with the same stimulus event within the experimental session to isolate weak but important signals in naturally noisy biological systems. Although this technique is suitable for defining general psychological principles, it is less useful in telling us about persisting individual characteristics. Lacey observed that physiological stress response patterns differ among individuals.[60] That is, one subject's arousal reaction may involve strong cardiac rate increases, whereas the cardiovascular reaction of another subject may be more strongly reflected in blood pressure. Similarly, some subjects show clear changes in sudomotor systems in response to any stimulus input; for others, the skin conductance response may be totally absent. These idiosyncratic patterns, particularly their stability over time, have not received extensive study with modern methods of measurement and analysis. Where the performance of specific crew members is the concern, the understanding of these unique signatures is a key goal.

To fill in this data gap, we need more studies that look at the organization and stability of response patterns *in individuals* over periods of weeks or months. It is particularly important to evaluate physiological patterns that are reactions to defined, repeatable stressors (e.g., physical threat, social stress) and to meaningful tasks. Problem features of the space environment will need to be included in the stimulus array being tested (e.g., confined environment, small-crew interactions). Eventually, these experiments should be conducted in space so that the physiological effects of microgravity can be made part of the equation.

As detailed in Chapters 5 and 8, weightlessness involves gross changes in cardiovascular and neurological functioning. Many crew members report that these physiological events are highly unpleasant, and they show related, significant performance deficits. An effort should be made to assess individual differences in reactions to these physical stressors prior to the actual experience of space. Stern and colleagues, for example, have investigated reactions to bodily rotation and the reactions of a stationary subject within a rotating optokinetic drum (which fills the visual field).[61] The latter produces symptoms of motion sickness like those found in space. The electrogastrogram—a noninvasive method for measuring changes in stomach and intestinal motility—can be used to measure smaller and anticipatory autonomic nervous system (ANS) reactions that predict more severe motion sickness. It is important to study parallels in cardiovascular and psychomotor patterns and to see if there are common characteristics of emotional temperament that correlate with this form of distress.[62] The effectiveness of biofeedback and other behavioral coping strategies in reducing these patterns and their effect on performance should also be explored.

Psychophysiological Measurement in Space

Psychophysiological measurement is most meaningful in the context of a clearly defined stimulus and a focused task demand. For the busy astronaut, mission projects need to be analyzed to identify natural assessment tasks. This might be instrument monitoring or motor control requirements embedded in regular maintenance activities, or in recreational interactions with the computer (e.g., competitive games), by individuals or groups.

The following measures are potentially sensitive to psychological stress, mood, motivation, and attention.[63] The current technology is adaptable for use outside the laboratory. However, a continuous effort should be made to develop instrumentation that is more portable, nonintrusive and natural on the body, and robust in the space environment. This is an important engineering goal that NASA should support.

• *Heart rate and variability, electrocardiograph (EKG) waveform, pulse volume.* Heart rate tends to decrease in characteristic ways with attention to external stimuli. Heart rate changes are also associated with emotional reactivity, being modulated differently by affective valence and motivational intensity. Spectral analyses of interbeat interval distributions permit an estimation of vagal tone and autonomic balance (sympathetic versus parasympathetic) that may help predict stress vulnerability. Sympathetic cardiac control varies with t-wave amplitude, providing a collateral estimate of ANS arousal.

• *Facial muscle action.* Facial expression defines emotion for some theorists. Formal coding of facial changes is done in social psychological studies of emotion, but it is slow and labor intensive. The development of computer analytic methods for on-line measurement of facial expression from video observation is an important goal for instrumentation development. This could be used in conjunction with voice frequency analysis (employed in the Russian space program) to provide a powerful, unobtrusive method for assessing emotional interactions. Facial muscle action potentials can also be monitored directly. This has the advantage of providing data on expressions that are below the threshold of visual detection, but has the disadvantage of requiring that electrodes be placed on the face, with a resulting intrusiveness of measurement.

• *Blink rate and blink magnitude (probe induced).* Blink rates vary with attention and anxiety, and can be monitored by simple electrodes, strain gauges, or high-speed cameras. Furthermore, brief acoustic probes (e.g., 50 ms, 90-decibel noise) can be presented, prompting mild reflexive blink-startle reactions. During perception, the magnitude of such noise-induced blinks has been shown to reflect emotional state—larger startle during unpleasant states and smaller in pleasant states.

• *Temperature (ear canal and skin surface) measurement.* Facial temperature varies with emotional reactivity. Tonic changes in temperature are valuable in assessing circadian rhythms.

• *Electroencephalograph (EEG).* The EEG is particularly useful in evaluating sleep patterns and in assessing attention and arousal. Currently, the simultaneous measurement of up to 128 electrode sites is practical.

In recent years, a variety of neural imaging methods have been developed that add significantly to the psychophysiologist's armamentarium. For example, as noted in Chapter 5, magnetic resonance imaging (MRI) can be used to acquire a detailed picture of the living brain in which individual structures are readily discerned. Furthermore, by relying on the detection of changes in the brain's regional blood flow, the functional activity of specific brain areas can be imaged as subjects perform cognitive tasks or are exposed to emotional stimuli. Although these methodologies cannot be taken into space, they can be used before and after a flight and thus allow evaluation of changes in brain structure and function that could result from extended stress, isolation, confinement, and microgravity. One might speculate, for example, that the reported impairment of short-term memory in some crew members[64] might be secondary to actual changes in the hippocampus. In conjunction with its use to assess changes in cortical maps associated with changes in sensorimotor integration, as described in Chapter 5, MRI would provide a method to assess these and similar questions about changes in the brain that may be attendant on prolonged exposure to the unique physical and psychosocial stressors of long-duration space missions.

Recommendations

Research on psychophysiological factors of behavior and performance during long-duration missions should include an examination of the following issues in order of priority:

1. In-flight studies of the characteristics of sleep during long-duration missions, including predictors of change in sleep quality and quantity; whether sleep deprivation is cumulative; how much sleep debt is necessary to produce an overall impairment of cognitive performance, mood, and interpersonal behavior; and whether reductions in sleep debt are associated with improved performance;

2. Ground-based studies of change and stability in individual physiological patterns (e.g., cardiovascular, neuroendocrine, and immune system changes related to alterations in the hypothalamic-

pituitary-adrenal axis) in response to psychosocial and environmental stress and their applicability to measures of behavior and performance in flight; and

3. Development of psychophysiological instrumentation that is highly portable, nonintrusive, and robust in the space environment.

INDIVIDUAL ISSUES

Psychological Issues

Stress and Coping

Cumulative stress has certain reliable effects, including psychophysiological changes related to alterations in the sympathetic-adrenal-medullary system and the hypothalamic-pituitary-adrenal axis (hormonal secretions, muscle tension, heart and respiration rate, gastrointestinal symptoms), subjective discomfort (anxiety; depression; changes in sleeping, eating, and hygiene), interpersonal friction, and impairment of sustained cognitive functioning. The person's appraisal of a feature of the environment as stressful and the extent to which he or she can cope with it are often more important than the objective characteristics of the threat.[65] Several behavioral and psychophysiological measures of responses to various major and minor stressors have been identified and used in studies of the general population. However, as noted in the preceding discussion of psychophysiological issues, more information is needed on the relationship between self-perceived, performance-related, and physiological signs of stress within a highly selected and motivated group such as space crews because these measures do not consistently correlate with one another. Use of these behavioral and physiological measures to assess the efficacy of coping responses should be examined critically for the same reasons.

Given that spaceflight is intrinsically and perhaps chronically stressful, coping is a crucial area of research. Selye's model of the General Adaptation Syndrome, widely accepted as the basic description of the stress and coping process, indicates that as a stressor appears and continues, the individual's coping resources are first mobilized (the *alarm* stage), then deployed (*resistance*), and eventually, if the situation is not resolved, depleted (*exhaustion*).[66] Among highly competent and self-confident people such as space crews, resistance is likely to be effective and last a long time. However, prolonged severe stress or the impact of several simultaneous stressors will eventually reduce the coping resources of even the strongest organisms. Where one stage ends and the next begins—especially the threshold between resistance and exhaustion—can be crucial to the success of a mission and the survival of the crew, and both behavioral and physiological markers of the stage transition must be established through studies that compare baseline and follow-up measures of these stress responses with mission duration, workload, changes in the physical and social environment (e.g., increase in environmental pollutants, interpersonal tension), and scheduled and unscheduled events.

The specific coping strategies of spacefarers should also be studied. Research has identified two major categories of strategies, problem oriented and emotion oriented.[67] The former include methods for solving or avoiding a problem; the latter are appropriate when neither a solution nor escape is feasible, and the only available strategy is to try to endure the situation. The selection of one or both types of coping strategies in responding to a specific stressor depends, in part, on the characteristics of both the stressor and the individual. Differences in demands placed on the individual by the physical and social environment and the organization of the mission may require different coping strategies. An understanding of what coping strategies are employed under what situations is critical to the development of effective countermeasures designed to support effective coping responses to the chronic and acute stressors likely to occur during long-duration missions.

Although inter- and intrapersonal problems arising from environmental stressors may occur during long missions, measures of both positive and negative reactions are needed for a complete picture of what one may expect from an extended mission. Understanding and predicting beneficial and enjoyable characteristics of the environment, as well as successful coping, adaptation, and positive long-term outcomes among crew members, are just as important and informative as studying environmental stressors and human failure, maladaptation, and postflight problems.[68] Such information is essential to the development of effective countermeasures during all phases of a mission.

Cognition and Perception

Although the perceptual changes associated with microgravity and disruption of circadian rhythms have been thoroughly examined in both in-flight and ground-based studies, there is very little systematic knowledge of changes in cognition and perception associated with long-duration spaceflight. Information on cognitive and perceptual distortion during short-duration missions is of little help because of potential qualitative as well as quantitative differences associated with prolonged exposure to the space environment. Decrements in cognitive performance are of particular concern because of the likelihood of emergency situations. Therefore, additional research is required to determine what cognitive changes occur during long-duration missions, what individual and environmental characteristics are associated with such changes, and whether preflight training is effective as a countermeasure to any performance decrements.

A particular problem requiring research in the next decade is the increased likelihood of errors as missions grow longer in duration. These can involve something as simple as disconnecting the wrong cable (which happened aboard Mir in July 1997) or flicking the wrong switch. An even more likely outcome is the inability to break away from one's training and invent ways of solving unexpected problems when an unanticipated situation calls for a careful information search and a complex, creative decision strategy.[69] The complex of abilities generally referred to as "fluid intelligence" is most likely to be influenced by hostile or restrictive environments. This includes problem solving in situations that require novel solutions, decision making in situations in which multiple aspects of the situation must be considered, and dealing with situations in which it is important to keep track of several things at one time. These cognitive processes are tied to the status of the forebrain, and this area is especially subject to insult from environmental damage (e.g., carbon monoxide poisoning, excessive alcohol consumption). It is also possible to take a more psychological approach and observe that all the fluid intelligence functions require tracking many variables in working memory. Therefore, if one's working memory space is being taken up by distractions (including social tensions, depressed affect, or simply discomfort due to impoverished environments), there is less "mental space" available to deal with the problem at hand. Alternatively, the absence of certain distractions under conditions of social and physical monotony during long-duration flights may also produce a decline in cognitive skills.

Environmental impacts on either simple, overlearned performance or on the solving of complex, novel problems have not been studied with highly selected, trained, and motivated subjects such as astronauts. As a result, it is not known what levels of stress and mission duration are likely to impair crew responses to different kinds of problems. Simulations of both ongoing stressors and acute events are needed to predict how crews would cope with them. It is important to note that the types of cognitive skills that are at risk due to the sorts of conditions expected on long-duration spaceflight are also the kinds of skills that are most subject to deterioration with age. Therefore, there is a difficult trade-off between a wealth of knowledge and experience and a potentially diminished cognitive ability to utilize this knowledge, both of which are age-related, that may have important implications for the screening and selection of personnel for such missions.

The possibility of brief periods of anterograde amnesia (i.e., failure to consolidate information) in-flight must be considered. Functionally, the appearance of mild anterograde and/or brief retrograde amnesias also can be produced by environmental agents, including some drugs used to combat motion sickness. Retrograde amnesias can be produced by striking emotional events, so that individuals may forget what they were doing just prior to a dramatic incident. Research is required to determine the incidence of low-level amnesias during long-duration missions and to develop effective countermeasures.

Emotion

Emotions are valenced reactions to personally significant events, including physiological reactions, behavioral reactions, cognitive reactions, and subjective feelings of pleasure and displeasure. Both positive and negative mood can take on added significance during long-duration space missions because trivial issues are often exaggerated by people living and working in isolated and confined environments.[70] However, we do not know whether this exaggeration is the result of changes in physiological processes, emotional behavior, or cognitive activity. Whether some or all such sources contribute to emotional feelings in any environment is a central issue in the theory of emotional experience. Investigation of the emotional experiences of astronauts during long-duration missions might enable us to advance our understanding of the foundations of emotional experience by contributing to the development of models that link characteristics of the individual (e.g., concerns, values, goals) and the environment (e.g., limited resources, information, alternatives) to physiological (e.g., facial expressions, autonomic arousal, HPA axis alteration) and cognitive (appraisal of issues or events as trivial or important, self-perception, interpretation of a situation as being "emotional") changes.

The underlying causes of expressions of negative emotion in ICE environments include factors such as fatigue and disruptions of circadian rhythms, difficulty in adjusting to microgravity-related physiological changes and alterations in the HPA axis, learned patterns of coping with stress (e.g., avoidance, expressiveness), perception of the risks associated with the physical environment, loneliness and separation from family and friends, receipt of unexpected and distressing news, and interpersonal conflicts.[71-73] However, the specific contribution of each of these factors to the expression of specific emotions in specific individuals remains to be determined. A greater understanding of individual, social, and environmental predictors of negative emotions in long-duration missions is essential to the development of effective countermeasures. Although the probability that the expression of these negative emotions will escalate into episodes of psychopathology during long-duration space missions is quite low, they have been associated with decline in task performance and motivation; disruption of attention, short-term memory, and other cognitive processes; increased interpersonal conflict, leading to both voluntary and involuntary isolation from other crew members; and the occurrence of various psychosomatic or psychophysiological symptoms.[74,75]

Equally important is an understanding of the factors that contribute to the expression of positive emotions. For instance, research from analogue settings suggests that long-duration isolation and confinement in an extreme environment are likely to produce personally rewarding experiences for certain crew members.[76] Such optimal experiences are characterized by feelings of being strong, alert, in effortless control, unself-conscious, and at the peak of one's abilities.[77] Investigation of such experiences during actual spaceflight conditions may help us to understand why certain environments perceived by some to be stressful, with adverse effects on health, well-being, and performance, can be perceived by others as challenging, with positive outcomes. However, additional research is required to determine whether individual optimal experiences can lead to enhanced task performance and the likelihood of overall mission success, which individuals are likely to have these experiences, and which aspects of long-duration space missions are likely to produce such experiences for individuals. Additional research is also required to determine whether preflight training can increase the probability that such experiences will occur.

Personality

In space, personality traits interact with the various stressors and unique aspects of the physical and social environment to determine how crew members behave and perform. Several traits relevant to spaceflight are essentially motivational, such as risk-taking and telic dominance (focusing narrowly on reaching one's goal as opposed to valuing the challenging process of trying to reach it). Others affect one's reactions to stressful situations. These include optimism, hardiness (a composite of more specific characteristics, such as a feeling that one can understand and control events), resilience (the ability to "get past" setbacks and continue on), and resourcefulness (being able to look for and find alternative ways of solving a problem). Still other traits define social tendencies (e.g., sociability, dominance, aggressiveness, affiliative needs, and dependency).

In the next decade, research on human behavior in space should focus on the extent to which personality characteristics vary among astronaut personnel and whether these variations are associated with differences in behavior and performance preflight, in flight, and postflight. Such research could focus on single traits believed to be specifically relevant to spaceflight, such as those described above. For example, it may be hypothesized that individuals who are high in thrill-seeking and dominance, and low in resourcefulness, will engage in aggressive competition with their crewmates to relieve boredom in a long-duration mission. However, because of the vast number of known personality traits, it is difficult to determine a priori which should be studied; at this point, perhaps the most fruitful approach is to adopt a multitrait approach and look at the traits that in the past decade of research have been noted as powerful correlates of many kinds of behavior, known as "the Big Five."[78] These cover underlying tendencies across the contexts of mental health and emotional stability (*neuroticism*), social interaction styles (*extraversion, agreeableness*), and cognitive or performance orientations (*openness, conscientiousness*). How they relate specifically to the ICE performance measures of ability, stability, and compatibility and how they influence behavior in all of the contexts within the domain of spaceflight are important questions for basic research that also offer potential contributions to improving individual selection and crew composition. In addition to their operational relevance, comparisons of astronaut personality traits and behavior preflight and in the controlled social and physical environment of spaceflight offer the potential for addressing some of the fundamental questions underlying the fields of social and personality psychology, such as the extent to which personality is stable and consistent over time, is a social construct, or is a function of the situation.

Personality traits are stable but not unchangeable; major life events have been known to cause long-term changes such as increased spirituality and religiousness, greater sensitivity to others, and a new set of motivational priorities. A substantial personality change as a result of spaceflight may have a significant impact on the life of the astronaut and his or her family. Longitudinal studies of future long-duration missions are needed to establish the probability of such changes and their possible consequences, to identify individual and mission-related characteristics that are associated with such changes, to prepare the individuals likely to be affected by these changes, and to design strategies for possible intervention.

Psychiatric Issues

Psychopathology

No selection program—no matter how thorough—can completely prevent the development of psychiatric problems in space. There have been several anecdotal accounts of episodes of depression, bereavement, anger, and anxiety among astronauts in long-duration missions where clinical intervention was necessary or advisable.[79,80] Studies of submariners[81] and antarctic winter-over personnel[82] have indicated that between 5 and 10 percent of individuals who are psychologically screened and selected for assignments in isolated and confined environments experience clinically significant psychopathology. The Russians also have described a syndrome called asthenia that frequently results from the hypostimulation that may affect cosmonauts during long monotonous periods in space.[83] Elements of this syndrome include irritability, emotional lability, poor appetite, and sleep disruption. In some cases, negative personality changes have occurred that have led to marital problems and severe psychiatric difficulties.[84] Spouses also can be affected. For example, some investigators have coined the term "submariners' wives syndrome" for the depression and stress experienced by spouses who are trying to adjust to the reintegration into the family of their husbands who have just returned from sea patrol.[85,86]

Although the incidence of severe psychopathology is assumed to be low during long-duration missions, ongoing studies of the etiology and incidence of psychiatric disorders preflight, in flight, and postflight will be necessary to validate this assumption and to develop effective countermeasures in the hopefully rare instances in which such disorders do occur. The collection of baseline measures of personality during the screening and selection phase and the ongoing monitoring of affective, cognitive, and psychophysiological responses to discrete stressors during preflight training and in-flight operations are particularly relevant to this objective.

Psychopharmacology

Medical kits on U.S. and Russian crewed space missions have included a variety of psychoactive medications: anxiolytics such as diazepam, sleeping pills such as flurazepam, antipsychotics such as haloperidol, and intramuscular promethazine for space motion sickness.[87] Seventy-eight percent of shuttle crew members have taken medications in space, primarily for space motion sickness, headache, sleeplessness, and back pain.[88]

Physiological changes due to spaceflight may change the pharmacokinetic behavior of psychoactive drugs, thus influencing their dosage and route of administration.[89] For example, microgravity can increase blood flow in the upper part of the body and decrease it in the lower part. Thus, an intramuscular injection to the arm rather than the hip might alter the bioavailability of the medication. Also, gastric emptying, intestinal absorption, first-pass effects through the liver, and metabolism and secretion rates may be influenced by fluid shifts and other effects of microgravity. This can lead to medication dosages and direct and side effects that differ from what is predicted from experiences on Earth. Like other medications, psychoactive drugs may be sensitive to such changes,[90] and empirical study in the space environment is required to fully understand these effects.

Countermeasures

Psychological and psychiatric countermeasures have two primary objectives: (1) Measures such as procedures for the screening and selection of astronauts are intended to prevent the occurrence of decrements in performance that may negatively affect individual health and well-being and mission success. (2) Measures such as the monitoring of flight crews, training programs, in-flight services, and clinical interventions are intended to support individual crew members and ensure optimal performance during a mission.

Screening and Selection

In the history of space exploration, there have been no descriptions in the scientific literature of either the rationale used for screening and selection of American astronauts or the reliability and validity of that rationale. The Soviet scientific literature has been less reticent in documenting the process of screening and selection, but these methods also have not been subjected to any assessment of validity and reliability.

The screening and selection of astronaut personnel in the U.S. space program has traditionally emphasized a "select-out" approach in which candidates who have a diagnosable psychiatric disorder, or are considered to be at risk for developing such a disorder, are identified and not selected for astronaut training.[91] Screening procedures used in this approach have relied on standardized assessments and an examination of relevant biographical data. Santy reported the incidence of psychiatric disorders in a study of 223 astronaut candidates to be 7.7 percent.[92] Furthermore, this approach has been credited for the relatively low risk of astronaut personnel developing psychiatric disorders in-flight. Nevertheless, it is important to continually review this process and update it when necessary, using the latest state-of-the-art psychiatric screening techniques. For instance, the impact of revisions in standardized psychiatric classification systems on screening and selection of astronaut personnel has yet to be determined.

In recent years, greater emphasis has been placed on a "select-in" approach to identify candidates whose character traits enhance the likelihood of being able to perform under stress and interact productively as a member of a crew. Research on astronaut personnel and individuals in analogue settings has identified a number of characteristics as predictors of successful performance in extreme, isolated environments. McFadden and colleagues found that expressive traits are significant predictors of astronaut effectiveness in interpersonal domains.[93] A similar study by Rose and colleagues found that astronaut professional effectiveness was associated with high negative expressivity and "communion" (subordinate and gullible), low openness, low negative instrumentality (egotism), and high agreeableness.[94] A study by Ursin and colleagues hypothesizes that a moderate level of aggressiveness or vitality would be appropriate for a short-duration mission but not for a long-duration mission.[95] The latter would require selecting-in characteristics such as readiness to bear privation and emotional stability. Data on these select-in psychological characteristics are very promising, but require validation against in-flight performance measures. This task is somewhat challenging because it means that specific select-in criteria must not be used as the initial basis of astronaut selection until they have been found to predict astronaut performance. Only after research demonstrates the validity of a criterion should it become part of the operational selection process.[96] Furthermore, screening and selection of multinational crews for long-duration missions must take into account the fact that preferences for certain personality traits or characteristics linked to a specific job description (e.g., pilot or astronaut) are likely to vary among the different cultural groups represented in such missions.[97,98]

Support of Individuals and Families

Research should be conducted to determine which forms of psychological and psychiatric support are most appropriate for astronaut and ground support personnel during the preflight, in-flight, and postflight phases of a mission. Data from the Mir incidents in 1997 may be especially relevant in this regard, because of the frequency and severity of the problems encountered on this space vehicle. Research has consistently pointed to the direct and moderating influence of social support on physical health and emotional well-being. However, under conditions of prolonged isolation and confinement, influences may be minor or inconsequential. In part, these conditions place certain restraints on the transmission of support to astronauts in long-duration missions. They also serve as stressors to family members who would otherwise provide support to the astronauts. Both limit the capacity of family units to provide support as well as the capacity of astronauts to utilize such support to maximize their performance and well-being. Research should also be conducted to identify the types of support necessary and effective in reducing stress and enhancing the ability to cope with prolonged separation in families of astronauts during training and during the flight.

Preflight training of astronauts should include familiarizing the astronauts themselves with possible psychological problems that may arise, ranging from alterations in perception to the complexity of group interactions. Training should include accommodation to potential cognitive problems (e.g., taking special precautions against inattention when one is forced to go through long periods of sleeplessness) as well as training in how to deal with changes in muscle coordination or perception as described in Chapter 5.

In-flight psychosocial support programs have been lauded by Russian crews and favorably commented on by astronauts participating in the SMSP.[99] This support is provided by ground-based teams who maintain crew mental health by providing them with reminders of connections to Earth, such as music, audio tapes, and other personal items. The support team can also provide valuable "coaching" in the event of an on-board psychosocial problem. Although most of these will be in the normal range (e.g., interpersonal friction, being informed about a family crisis), serious psychiatric problems could occur, as noted above. Records of the effectiveness of such support should be subjected to the same sorts of debriefing and analysis as mission performance data to identify areas of potential improvement.

Psychosocial support also should be extended into the postflight period. As noted earlier, studies of personnel in submarines and polar stations indicate that the returning crew member may have difficulty dealing with normal levels of physical and social stimulation. When problems do occur a careful analysis should be made of the effectiveness of the countermeasures, in order to improve the process.

Recommendations

Research on the individual factors of behavior and performance during long-duration missions should include an examination of the following topics, listed in order of priority:

1. The effects of individual characteristics of crew members on cognitive, psychophysiological, and affective measures of behavior and performance:

- *The relationship between self-reports and external (i.e., performance-related, physiological) symptoms of stress;*
- *The use of specific coping strategies and behavioral and physiological indicators of coping stage transitions during long-duration missions;*

> • *The effect of physical and psychosocial stressors on problem-solving tasks and fluid intelligence;*
> • *Associations between general and mission-relevant personality characteristics and performance criteria of ability, stability, and compatibility; and*
> • *Individual and mission-related predictors of postflight changes in personality and behavior.*

2. Development and evaluation of countermeasures to performance decrements during the preflight, in-flight, and postflight phases of a long-duration mission:
> • *The validity and reliability of current and alternative screening and selection procedures based on performance criteria;*
> • *The effectiveness of psychological and psychiatric countermeasures that have been employed in past flights, using data collected from observations, postmission briefings, and interviews;*
> • *The influence of spaceflight on the pharmacology, pharmacokinetics, efficacy, and side effects of psychoactive medications; and*
> • *The effectiveness of countermeasures that provide psychosocial support to family members on the performance and well-being of astronauts.*

INTERPERSONAL ISSUES

Crew Tension and Conflict

Several factors affect crew tension or conflict. The extent to which heterogeneity (i.e., differences in the psychological, social, and demographic characteristics of individuals who comprise a flight crew) is a risk for interpersonal conflict in ICE settings remains largely unresolved and requires additional research. Although analogue studies have indicated that heterogeneous crews are more likely to experience interpersonal conflicts under conditions of isolation, studies in space simulation environments have found that crewmates who share certain personality characteristics such as a high need for dominance do not work well together,[100,101] whereas people who are compatible and sensitive to each other in a complementary manner do much better. However, little is known about specific personality traits that enhance compatibility and reduce interpersonal conflict in space.

Gender factors also are important. Instances of sexual stereotyping have been reported, both in space and in Earth analogues.[102] Although it is not unusual for such behavior to take place in the general population, research in analogue environments has demonstrated that it often takes on added significance in isolated and confined environments, resulting in misunderstandings and increased tension between men and women who must learn to work together.

Another factor affecting crew conflict relates to differences in career motivation between crew members. In space analogue environments, people with different training backgrounds and career objectives have responded in different ways to missions involving prolonged isolation.[103] In some cases, conflicts have erupted between groups that have compromised mission goals.[104] Similarly, operationally oriented pilots and engineers in space may view the mission objectives differently from scientifically oriented payload specialists or "guests" who have nonoperational duties. Tensions also can occur when some crew members view their roles as more important than those of other crew members.

Finally, cultural and language differences may affect space crews by producing intracrew friction and ineffective responses to danger, both of which can have a negative impact on the success of a mission. Reports from long-duration Russian space missions involving people from other nations have highlighted conflicts among crew members based on differences in language competency and culturally determined expectations, values, attitudes, and patterns of behavior.[105,106] On the other hand, one may argue that cross-cultural issues will have a minimal impact on crew behavior and performance since, as members of a common profession, astronauts share a body of knowledge, a set of expectations, and common skills that comprise the "microculture" of the space crew.[107] However, as crews become larger and include individuals with a diverse set of backgrounds, skills, and responsibilities, it may be increasingly difficult to produce an allegiance to a common professional culture.

A common language is also essential to maintaining optimal levels of interpersonal coordination and cooperation. A survey of 54 astronauts and cosmonauts found that all of the respondents believed that it was important for members of a space crew to be fluent in a common language.[108] Interestingly, U.S. and Russian space travelers rated the importance of having a common language significantly higher than did astronauts from other countries, possibly reflecting the concern the former groups felt for the operational aspects of the missions for which they were responsible. Limited fluency in the language of the host crew may also slow down communication[109] and might prove life-threatening during an emergency where timing and accuracy of communication are critical.

Crew Cohesion

As with conflict, factors that potentially affect crew cohesion require further investigation. Dysfunctional crews may scapegoat one of their members and blame him or her for the inability of group members to interact more productively. In isolated and confined groups, such scapegoating may actually serve to unify all but the ostracized member.[110] Lack of cohesion also may lead to subgrouping, where the crew splits into factions along social or task-oriented lines (e.g., scientists versus pilots).[111]

Social monotony may also contribute to reduced crew cohesion. During antarctic and submarine missions, a long monotonous period has been reported where tasks become routine and where crew members frequently experience boredom, depression, homesickness, and withdrawal.[112] This phase also has been described during long-duration Russian space missions,[113,114] and it has resulted in decreased cohesion and increased interpersonal problems.[115] These social ramifications of monotony may be relieved by increasing audio-video communications with family, friends, interesting strangers, and ground control on Earth. Different problems arise when social monotony is not absolute. For example, in a space station, unlike a long-duration mission to Mars, people would occasionally leave and new ones arrive. Such changes relieve monotony but can produce their own stressors, since the group has to get used to new arrivals and socialize them into its accustomed ways of doing things.[116] The microcultures of crews and missions have to be better understood in order to ease this process. The effects of periods of social monotony on space crews should be addressed further in order to maintain morale, provide meaningful use of leisure time, and prevent negative consequences of low stimulation (e.g., asthenia, crew member withdrawal).

Crew cohesion seems to decrease over time during the course of long-duration space missions, especially as the effects of monotony and familiarity begin to take their toll. However, common interests and activities can help to bond crew members together. For example, a study of astronauts and cosmonauts who had flown in space found that the shared experience and excitement of spaceflight contributed significantly to enhancing crew member communication.[117] Interestingly, these same two factors were the only ones that were rated as significantly enhancing communication between space crews and monitoring personnel on the ground.

Similarly, efforts are required to identify the factors that are likely to promote and maintain crew cohesion during the preflight and in-flight phases of long-duration missions. Although heterogeneity in crew composition may contribute to increased levels of tension and conflict as noted earlier, variety in background and experience may also minimize social monotony and lead to rewarding interpersonal experiences that contribute to enhanced group performance and individual well-being. For example, during leisure periods, different hobbies and interests may be shared among crew members that could stimulate learning and lead to the formation of new social bonds. The extent to which crew cohesion is influenced by the characteristics of individual members may also vary with mission duration such that cohesion may be enhanced by homogeneous features of crew composition early in the mission and heterogeneous features later on. The impact of homogeneity or heterogeneity and mission phase on crew cohesion should be studied empirically under actual spaceflight conditions. Such research would help to advance our understanding of the extent to which the interactions of space crews and other small groups can be explained by processes of interpersonal attraction (e.g., similarity, cooperation, interpersonal acceptance, shared threat), as predicted by attraction models of cohesion, or in conjunction with other effects (e.g., conformity, in-group differentiation, stereotyping, in-group solidarity) produced by the process of self-categorization specified in self-categorization theory.[118]

Ground-Crew Interaction

Tension involving a confined group of people may be displaced to outsiders who are monitoring their activities, since it is easier to express anger and anxiety toward more remote individuals than toward people with whom one must frequently interact. Such displacement has been reported during both Russian and U.S. space missions[119-121] and during previous ground-based simulation studies.[122,123] Astronaut perceptions of the demands placed on them by ground-control as excessive, unreasonable, or unclear have led to expressions of hostility and conflict in the past. For their part, ground control personnel have complained of the failure of astronauts to adhere to schedules or follow directions, raising concerns of an increased risk of accidents and mission failure. More often, degradation in ground-crew interactions has led to instances of miscommunication.[124]

On the other hand, these apparent degradations in ground-crew interactions may actually have an adaptive function. Because they are remote from the crew in a physical sense, ground control personnel may serve as an outlet for crew aggression and irritability that is the result of factors external to ground-crew relations.[125] The direction of anger and hostility toward external authorities and individuals may also serve to unite astronaut crews, thereby facilitating cooperation and enhancing performance. However, no attempt has been made to study the benefits of such alterations in ground-crew interactions relative to costs in a quantitative or qualitative fashion.

Another important element of ground-crew interaction that requires further investigation is the interaction between crew members in flight and family and friends on the ground. As noted earlier, such interaction is critical to reducing social monotony and individual feelings of isolation and confinement. However, as studies of personnel in analogue settings suggest, contact with family members may lead to problems.[126] One important issue is under what conditions bad news should be communicated to the crew. This is a complex issue for study, whose parameters (e.g., seriousness of the event, who is involved, whether the astronaut can help solve the problem) need to be manipulated systematically in simulations. Studies conducted to date have provided no consensus on the question of whether negative personal information should be withheld from crew members until the end of a mission.[127] Information also is lacking on the optimal duration and timing of family calls.

Leadership Role

To date, there have been no studies of crew leadership conducted during actual spaceflight.[128] Research on small groups in other isolated and confined environments suggests that effective leadership of such groups in general is a function of certain characteristics of the leader (i.e., hard-working, optimistic, sensitive to needs of the crew), forms of leadership behavior (i.e., democratic-participative rather than authoritarian decision making), and situations (democratic in response to task and crew maintenance and authoritarian in response to emergencies and unexpected situations).[129-132] However, recent advances in the theory of leadership suggest that greater attention should be devoted to understanding the context and consequences of leadership during spaceflight. For instance, the appropriate exercise of different leadership roles and lines of authority at different times in the mission has to be studied in order to understand factors that lead to status leveling, leadership competition, and role confusion. Research is required to determine whether the characteristics of effective leadership during long-duration missions differ from those of short-duration missions. During short-term spaceflights, the identified leader is the mission commander, the lines of authority are clear, and activities are task-oriented. However, studies of polar expeditions indicate that task leadership is more important during the early phases of a mission, but less important than supportive leadership and shared decision making as the mission progresses.[133,134] In a European Space Agency (ESA)-sponsored study of a three-man crew isolated for 135 days in the Mir space station simulator in Moscow, both the leader control measure (which addressed the task-oriented, instrumental characteristics of the identified leader) and the leader support measure (which addressed his supportive, expressive qualities) correlated positively with group cohesion.[135] However, the commanders of long-duration missions may not be able to provide this social or emotional support as well as some other member of the crew. As a result, lines of authority may alter, and the identified leader may experience status leveling, leading to interpersonal conflict and the breakdown of command structure. In addition to its operational relevance, research on these issues would also serve to advance our fundamental understanding of leadership in small groups by enabling us to test the hypothesis that relationship-oriented leadership is more successful than task-oriented leadership in situations moderately favorable to the leader, or in very favorable or unfavorable situations,[136] and the hypothesis that the individual traits and social situations necessary for the exercise of transformational leadership (e.g., nonconformity) are incompatible with those necessary for transactional leadership.[137,138]

Countermeasures

Selection

NASA has developed a number of strategies to deal with the interpersonal issues that could potentially lead to a degradation in individual and group performance.[139] Procedures for the selection of flight crews are less formal than procedures for the screening and selection of individuals into the astronaut corps described above. The mission commander, who is picked first, has some input into who his or her fellow crew members will be; however, the final decision is the responsibility of management.

In contrast to U.S. selection procedures, the Russian space program has expended considerable effort on the construction of compatible crews. When forming space crews, Russian psychologists take into account the similarity of values among potential crew members; their social and motivational attitudes toward the job; the presence of complementary personality and character traits and cognitive styles; and the ability to learn rapidly and efficiently. One of the compatibility testing methods used by

Russian psychologists when forming a crew is the Mutual Talking Test in which candidates are required to cooperate in completing a computer-simulated task. Cosmonauts participate in a number of other experiments that simulate interdependent activities. In addition, Russian psychologists believe that biorhythms are useful in selecting cosmonauts for specific missions. Crew members with similar biorhythms are often assigned to the same crew. Despite a strong conviction on the part of the Russian space program about the effectiveness of these procedures, they have been subjected to little objective evaluation.

In the future, interpersonally oriented testing methods should be used along with more individually oriented psychological tests preflight to assess the compatibility of crew members. Formal testing procedures have correctly predicted crew incompatibilities during the Ben Franklin submersible simulator mission,[140] and interpersonally oriented psychological tests, such as the FIRO-B and sociometric questionnaires, have shown promise in crew selection in several simulation studies.[141-146]

Training

In the future, more didactic and experiential training similar to the programs involving participants in the SMSP should be provided concerning the influences of specific sociocultural factors (e.g., personality, compatibility, gender bias, cultural and language differences, career motivation) in order to minimize crew tension, sustain cohesion, and prevent subgrouping and scapegoating. Astronauts who have flown on previous international space missions have endorsed the importance of receiving cultural and language training that is related to the crew composition of their missions.[147] In addition, preflight activities involving team building and conflict resolution should be incorporated into the training of space crews.[148,149] Experiential sensitivity training before a mission has been found to be useful in helping individuals interact better during Earth-bound confinement experiments.[150,151] Also, crew interaction training modules developed for use with airline cockpit crews (e.g., crew resource management [CRM] and line-oriented flight training [LOFT])[152-154] should be modified for use in the space program. Crews preparing for long-duration space missions should spend some time together confined in an Earth-bound simulator so that potential interpersonal conflicts can be observed and dealt with prior to the actual mission. Finally, both crew members and monitoring personnel on Earth should be briefed on the psychological phenomenon of displacement in order to counter the effects of ground-crew miscommunication and interpersonal conflicts.

Issues that cause tensions between crew members and outside monitoring personnel have been ameliorated in the past through "bull sessions," both in simulations[155] and during actual space missions.[156] Future crews and their ground control personnel should be trained together preflight to use interactive techniques, and experts in small-group dynamics who work with and are trusted by the crews should be available on the ground to assist in conducting such sessions during the mission if the need occurs.

In-flight Monitoring

Current NASA in-flight monitoring activities consist of daily communications between crew members and flight surgeons and, less frequently, with psychological support personnel. However, these communications are not confidential and provide little or no opportunity to address and resolve sensitive issues related to interpersonal conflicts. The Russians find ongoing monitoring of voice communications to have some operational value in assessing crew tension, cohesion, and morale,[157] but this approach has not been fully explored by NASA. A 23-item Crew Status and Support Tracker (CSST) questionnaire is completed weekly by U.S. astronauts on the Mir space station and down-linked to the ground, where crew members write in responses to questions that are related to mood, morale, personal privacy, interactions with each other and with people on Earth, sources of stress, and physical state. However, the psychometric properties of this instrument have not been assessed.

In the future, new attempts should be made to monitor crew member interactions in-flight to look for potential interpersonal problems. For example, NASA should explore further the voice analysis techniques used by the Russians in monitoring their space crews during long-duration space missions. The usefulness of the CCST and other tracking instruments should be formally validated. Finally, additional efforts are required to provide a secure communication link.

In-flight Support

The in-flight support activities of the United States are modeled after the Russian system[158] and include surprise presents and favorite foods sent up on resupply rockets; two-way communication with family and friends on the ground via audio-video links, e-mail, or ham radio; and on-board recreational software and videocassettes for leisure time use. At times, crew members are encouraged to talk with one another to resolve interpersonal difficulties. A computer-based family picture album of spouses, children, friends, and co-workers is currently provided and has been well received by crew members. Although these support activities possess face validity and have been enthusiastically endorsed by participants in long-duration missions to date, their effectiveness in reducing intragroup, individual-family, and crew-ground tension or conflict and in promoting crew cohesion has not been systematically investigated.

Postflight Debriefing

Finally, research is required to evaluate the effectiveness of postflight debriefings in resolving issues of interpersonal conflict that occur in flight. In the past decade, debriefing protocols have been developed and implemented to help crew members readjust to their life on Earth. Although the focus of these protocols has been on individual well-being, as described above, supportive attention also is given to the reintegration of these individuals with their families. These debriefings offer a potentially effective countermeasure for addressing residual intracrew, crew-family, and ground-crew tensions that may develop during long-duration missions. An evaluation of the effectiveness of these debriefings in addressing such tensions will require systematic efforts to collect information on the issues in flight and postflight, and the randomized assignment of crew members to an experimental or control condition.

Recommendations

To enhance data collection and validate the research described above, the following operational and policy changes should be implemented:

1. Valid and reliable interpersonally oriented testing methods, along with more individually oriented psychological preflight tests, should be used to assess the compatibility of crew members.

2. More didactic and experiential training should be provided to crews on the following:
  • Specific sociocultural factors in order to minimize crew tension, sustain cohesion, and prevent subgrouping and scapegoating;
  • Preflight activities involving team building and conflict resolution; and

• The psychological phenomenon of displacement in order to counter the effects of ground-crew miscommunication and interpersonal conflicts.

3. Monitoring of crew member interactions in flight should be expanded as a tool for supporting crew cohesion and resolving interpersonal conflict.

Research on the interpersonal factors of behavior and performance during long-duration missions should include an examination of the following issues in order of priority:

1. The influence of different crew compositions (i.e., personality types, gender, culture, language, occupation, career motivation) on crew tension, cohesion, and performance during the mission;

2. Factors affecting ground-crew interactions, including the impact of crew tension and unhappiness on crew-ground communication; impact of ground-crew communication on crew cohesion and task performance; and conditions that affect the distribution of authority, decision making, and task assignments between space crews and members of ground control; and

3. The development of new countermeasures and assessment of the effectiveness of existing countermeasures, including the following:

• *Interpersonally oriented psychological tests in selecting crews;*

• *Team-oriented training modules in reducing crew tension and enhancing cohesion and performance during the mission; and*

• *Evaluation of Russian techniques such as voice analysis for monitoring the interpersonal performance of crews.*

ORGANIZATIONAL ISSUES

Organizational Culture

Each of the major space agencies participating in the International Space Station (NASA, the Russian Space Agency [RSA], ESA, the Canadian Space Agency [CSA], and the National Space Development Agency of Japan [NASDA]) possesses different "macrocultures," i.e., values, attitudes, and behavior as they relate to the principles and practices of management and organization that embody the cultural systems of the respective nations. They also have different experiences with the application of these values, attitudes, and behavior to the challenge of crewed spaceflight that may account for differences in expectations and operational procedures during long-duration missions. NASA and RSA, for instance, have been involved in manned spaceflight for a longer time than the other space agencies and are the only ones that currently possess crewed space vehicles. Furthermore, NASA and RSA are characterized by a number of operational features that reflect differences in their respective organizational cultures. These include differences in ground-crew interactions (e.g., Russian personnel have been reported to be more confrontational than Americans in their ground-crew interactions); extent of ground-crew communications (e.g., U.S. ground control personnel remain in contact with space crews for longer periods of time); the NASA emphasis on overtraining for missions versus the RSA emphasis on "on-the-job" training; and the structure of rewards and restraints (e.g., the Russian practice of docking the pay of cosmonauts who fail to perform prescribed tasks). These differences have been reported by astronauts and cosmonauts to significantly influence crew dynamics.[159] However, this impact has not been examined in a systematic manner. Similarly, there is a lack of data from analogue environments on this issue. Differences in organizational cultures have already had an impact on crew performance in the SMSP, and the committee believes that these impacts are likely to become more significant as future missions become longer in duration and involve more organizations.

Innovative qualitative data collection and analysis methods should be employed in examining the influence of organizational macrocultures and space crew microcultures on individual and group behavior and performance. Such methods have been found to be useful in studying microcultures in analogue settings[160] and include participant observation, nondirected interviews, pile-sort tasks, and cultural consensus modeling.

Mission Duration

Despite the general consensus that long-duration missions represent a qualitatively different experience in terms of behavior and performance from missions of short duration, it is unclear whether mission duration is a significant predictor of performance and behavior. For instance, several studies of small groups in isolated undersea research labs and space simulation studies have reported significant increases in symptoms of depression, anxiety, and group hostility over time.[161] These results have supported the hypothesis that ICE settings influence human behavior in a linear dose-response manner such that the longer the exposure, the more significant are the decrements. Other studies have been used to support the hypothesis that decrements in performance under these environmental conditions occur in stages.[162-163] Rohrer described three stages of reaction to isolation and confinement: (1) an initial period of heightened anxiety related to perceived danger during the first few days of a mission; (2) a prolonged period of boredom and depression; and (3) a period of anticipation during the end of a mission characterized by hypomanic affect and increased aggression and hostility.[164] Bechtel and Berning described the "third quarter phenomenon" in which performance is likely to decline during the third quarter of a mission in an isolated and confined environment regardless of the total duration of the mission itself.[165] Other studies have reported no significant decrements in behavior and performance during long-duration missions in analogue settings.[166] Still other studies have reported an improvement in performance over time.[167] In the Mir simulator study described previously, there was significantly less total mood disturbance and overall crew tension during the last half of the seclusion than during the first half, and a dramatic reduction in mood disturbance and anxiety occurred after a resupply event in which the crew received additional food and equipment as well as letters of support from family and friends.[169]

Management

There is a need to identify the managerial requirements for long-duration missions and the necessary steps for management structures to transition from short-duration to long-duration mission operations. One of the important management functions of the organizations involved in long-duration missions is the scheduling and monitoring of tasks performed in flight. Identifying the optimum amount of work that can and should be performed during long-duration missions is important for a number of reasons. Evidence from previous short-duration missions has pointed to the potentially adverse impacts of scheduling too many tasks within the time available. These impacts have included conflicts between astronauts and ground control personnel, refusal to perform assigned tasks, fatigue, sleep deprivation, a decline in cognitive performance, and an increase in negative affect.[169] On the other hand, evidence from long-duration missions and analogue environments suggests that a lack of sufficient amounts of meaningful and productive tasks can result in boredom, producing many of the same symptoms associated with overwork.[170] Individual and group performance may also be affected when disparities in workload occur among crew members, such that some are given too much to do and others are not given enough to do during a long-duration mission.

The microgravity environment is another key component of workload and the scheduling of tasks to be performed during long-duration missions. Despite the wealth of experience in working in such environments and the opportunities to develop task schedules based on their performance in simulators, the inability to perform tasks in the allotted time remains a chronic problem. Many astronauts attribute this to the inadequate consideration given by management to the challenge of performing these tasks in a microgravity environment and to individual variation in the time required to complete certain tasks. However, it remains unclear whether these concerns are justified or reflect the lack of communication and understanding between space crews and ground control described above. Similarly, space crews may estimate the time required to perform certain tasks in flight differently from estimates made on the ground because of distortions in temporal perception caused by characteristics of the environment.[171] Further research is required to determine the organizational requirements for developing realistic schedules for operations in space and on the ground, and the impact of such schedules on the management of long-duration missions. Further research is also required to determine the organizational impacts of giving space crews more control over the scheduling of tasks.

Another management issue relates to the formal distribution of authority and tasks during long-duration missions. Although issues relating to leadership, task assignment, and decision making have been addressed above in the section on interpersonal factors, they are also relevant here because NASA, other national space agencies, and other organizations involved in long-duration missions all possess formal definitions of organizational positions and specify the relationship of one position to another. Associated with each position is a set of expectations that define certain behaviors as either essential, admissible, or unacceptable.[172] The organizational structure of authority, decision making, and task assignment developed for short-duration missions may have little relevance for the organizational needs of long-duration missions. For instance, the traditional organizational structure that separated space crews into "pilots" and "scientists" or "payload specialists" and the distribution of authority, decision making, and task assignments based on this distinction may have little relevance for the effective management of long-duration missions. Similarly, the conditions under which authority structures, decision-making processes, and task assignments result in interpersonal and crew-ground control conflicts during long-duration missions may be quite different from the conditions leading to conflict during short-duration missions.

Recommendations

Research on the organizational influences on behavior and performance during long-duration missions should include an examination of the following issues:

1. The extent to which participating agencies and personnel representing these agencies share similar systems of knowledge, attitudes, and behavior regarding mission operations, goals, and objectives; how these systems are organized; and how they influence individual and group performance and behavior;

2. Whether changes in behavior and performance are associated with mission duration or specific periods during the course of a mission; what form this association assumes if it does occur; and whether the implementation of countermeasures should correspond to these specific periods; and

3. The requirements for effective management of long-duration missions as they relate to the following:

- *Task scheduling;*

• *Disparity between expected and observed time required for operational procedures, space-craft maintenance, scientific experimentation, and emergency response; and*
  • *Distribution of authority and decision making.*

REFERENCES

1. National Aeronautics and Space Administration. 1991. Space Human Factors Discipline Science Plan. Space Human Factors Program, Life Sciences Division. NASA, Washington, D.C.
2. Aeronautics and Space Engineering Board, National Research Council. 1997. Advanced Technology for Human Support in Space. National Academy Press, Washington, D.C.
3. Space Science Board, National Research Council. 1987. A Strategy for Space Biology and Medical Science for the 1980s and 1990s. National Academy Press, Washington, D.C., p. 168.
4. Helmreich, R.L. 1983. Applying psychology to outer space: Unfulfilled promises revisited. Am. Psychol. 38: 445-450.
5. Helmreich, R.L. 1973. Psychological research in Tektite 2. Man Environ. Syst. 3: 125-127.
6. Sandal G.M., Vaernes R., and Ursin, H. 1995. Interpersonal relations during simulated space missions. Aviat. Space Environ. Med. 66: 617-624.
7. Vinograd, S.P. 1974. Studies of Social Group Dynamics Under Isolated Conditions. Objective Summary of the Literature as It Relates to Potential Problems of Long Duration Space Flight. NASA-CR-2496. National Aeronautics and Space Administration, Washington, D.C.
8. Space Science Board, National Research Council. 1987. A Strategy for Space Biology and Medical Science for the 1980s and 1990s. National Academy Press, Washington, D.C., p. 169.
9. Gunderson, E.K.E. 1974. Psychological Studies in Antarctica. Human Adaptability to Antarctic Conditions (E.K.E. Gunderson, ed.). American Geophysical Union, Washington, D.C.
10. Taylor, A.J.W. 1991. The research program of the International Biomedical Expedition to the Antarctic (IBEA) and its implication for research in outer space. Pp. 43-56 in From Antarctica to Outer Space: Life in Isolation and Confinement (A.A. Harrison, Y.A. Clearwater, and C.P. McKay, eds.). Springer-Verlag, New York.
11. Sherif, M. 1966. Common Predicament: Social Psychology of Intergroup Conflict and Cooperation. Houghton Mifflin, Boston.
12. Austin, W.G., and Worchel, S., eds. 1979. The Social Psychology of Intergroup Relations. Brooks/Cole, Monterey, Calif.
13. Earls, J.H. 1969. Human adjustment in an exotic environment: The nuclear submarine. Arch. Gen. Psychiatry. 20: 117-123.
14. Palinkas, L.A. 1990. Psychosocial effects of adjustment in Antarctica: Lessons for long-duration spaceflight. J. Spacecraft Rockets 27: 471-477.
15. Sandal, G.M., Vaernes, R., and Ursin, H. 1995. Interpersonal relations during simulated space missions. Aviat. Space Environ. Med. 66: 617-24.
16. Weybrew, B. 1991. Thirty years of research on submarines. From Antarctica to Outer Space: Life in Isolation and Confinement (A.A. Harrison, Y.A. Clearwater, and C.P. McKay, eds.). Springer-Verlag, New York.
17. Stuster, J. 1986. Space Station Habitability Recommendations Based on a Systematic Comparative Analysis of Analogous Conditions. NASA-CR-3943. NASA Ames Research Center, Moffett Field, Calif.
18. National Aeronautics and Space Administration, Life Sciences Division. 1997. Task Force on Countermeasures Final Report. NASA, Washington, D.C.
19. Space Science Board, National Research Council. 1987. A Strategy for Space Biology and Medical Science for the 1980s and 1990s. National Academy Press, Washington, D.C.
20. Aeronautics and Space Engineering Board, National Research Council. 1997. Advanced Technology for Human Support in Space. National Academy Press, Washington, D.C.
21. Selye, H. 1978. The Stress of Life. McGraw-Hill, New York.
22. Bock, O. 1996. Grasping of virtual objects in changed gravity. Aviat. Space Environ. Med. 67: 1185-1189.
23. Watt, D.G.D. 1997. Pointing at memorized targets during prolonged microgravity. Aviat. Space Environ. Med. 68: 99-103.
24. Stuster, J. 1996. Bold Endeavors: Lessons from Polar and Space Exploration. Naval Institute Press, Annapolis, Md.
25. Carrere, S., and Evans, G.W. 1994. Life in an isolated and confined environment. Environ. Behav. 26: 707-741.
26. Aiello, J.R., and Thompson, D.E. 1980. Personal space, crowding, and spatial behavior in a cultural context. Pp. 107-178 in Human Behavior and Environment, Vol. 4 (I. Altman, J.F. Wohlwill, and A. Rapoport, eds.). Plenum, New York.

27. Summit, J.E., Westfall, S.C., Sommer, R., and Harrison, A.A. 1992. Weightlessness and interaction distance: A simulation of interpersonal contact in outer space. Environ. Behav. 24: 617-633.

28. Paterson, R.J., and Neufeld, R.W.J. 1995. What are my options? Influences of choice availability on stress and the perception of control. J. Res. Pers. 29: 145-167.

29. Kanas, N. 1990. Psychological, psychiatric, and interpersonal aspects of long-duration space missions. J. Spacecraft Rockets 27: 457-463.

30. Taylor, T.C., Spencer, J.S., and Rocha, C.J. 1986. Space Station Architectural Elements and Issues Definition Study. NASA-CR-3941. National Aeronautics and Space Administration, Washington, D.C.

31. Wise, J.A., and Rosenberg, E. 1988. The Effects of Interior Treatments on Performance Stress in Three Types of Mental Tasks. CIFR Tech. Rep. No. 002-02-1988. Grand Valley State University, Allendale, Mich.

32. Kelly, A.D., and Kanas, N. 1994. Leisure time activities in space: A survey of astronauts and cosmonauts. Acta Astronautica 6: 451-457.

33. Kelly, A.D., and Kanas, N. 1994. Leisure time activities in space: A survey of astronauts and cosmonauts. Acta Astronautica 6: 451-457.

34. Stuster, J.W. 1986. Space Station Habitability Recommendations Based on a Systematic Comparative Analysis of Analogous Conditions. NASA-CR-3943. National Aeronautics and Space Administration, Washington, D.C.

35. Taylor, A.J.W. 1987. Antarctic Psychology. Science Information Publishing Centre, Wellington, New Zealand.

36. Suedfeld, P., Ramirez, C., Deaton, J., and Baker-Brown, G. 1982. Reactions and attributes of prisoners in solitary confinement. Crim. Justice Behav. 9: 303-340.

37. Taylor, A.J.W. 1987. Antarctic psychology. Science Information Publishing Centre, Wellington, New Zealand.

38. National Aeronautics and Space Administration. 1991. Space Human Factors Discipline Science Plan. Space Human Factors Program, Life Sciences Division. NASA, Washington D.C.

39. Aschoff, J. 1965. Circadian rhythms in man. Science 148: 1427-1432.

40. Czeisler, C., Allan, J.S., Strogatz, S.H., Ronda, J.M., Sanchez, R., Rios, C.D., Freitag, W.O., Richardson, G.S., and Kronauer, R.E. 1986. Bright lights reset the human circadian pacemaker independent of the timing of the sleep wake cycle. Science 233: 667-670.

41. Aschoff, J., Fatranska, M., Giedke, H., Doerr, P., Stamm, D., and Wisser, H. 1971. Human circadian rhythms in continuous darkness: Entrainment by social cues. Science 171: 213-215.

42. Wever, R. 1979. The Circadian System in Man: Results of Experiments Under Temporal Isolation. Springer-Verlag, New York.

43. Barabasz, M., Barabasz, A.F., and Mullin, C. 1983. Effects of brief antarctic isolation on absorption and hypnotic susceptibility. Int. J. Clin. Exp. Hypn. 31: 235-238.

44. Bonnet, M.H. 1985. Effect of sleep disruption on sleep, performance, and mood. Sleep 8: 11-19.

45. Williams, H.L., Lubin, A., and Goodnow, J. 1959. Impaired performance with acute sleep loss. Psychol. Monogr. 73: 1-26.

46. Goodwin, F.K., Wirz-Justice, A., and Wehr, T.A. 1992. Evidence that the pathophysiology of depression and the mechanism of antidepressant drugs both involve alterations in circadian rhythms. Adv. Biochem. Psychopharmacol. 32: 1-11.

47. Siever, L.J., and Lund, R. 1985. Overview: Toward a dysregulation hypothesis of depression. Am. J. Psychiatry 142: 327-339.

48. Teicher, M.H., Glod, C.A., Magnus, E., Harper, D., Benson, G., Kruegar, K., and McGreenery, C.E. 1997. Circadian rest-activity disturbances in seasonal affective disorder. Arch. Gen. Psychiatry 54: 124-130.

49. National Aeronautics and Space Administration. 1991. Space Human Factors Discipline Science Plan. Space Human Factors Program, Life Sciences Division. NASA, Washington, D.C.

50. Aschoff, J. 1965. Circadian rhythms in man. Science 148: 1427-1432.

51. Kennaway, D.J., and Van Dorp, C.F. 1991. Free-running rhythms of melatonin, cortisol, electrolytes, and sleep in humans in Antarctica. Am. J. Physiol. 260: R1137-R1144.

52. Ross, J.K., Arendt, J., Horne, J., and Haston, W. 1995. Night-shift work in Antarctica: Sleep characteristics and bright light treatment. Physiol. Behav. 57: 1169-1174.

53. Kennaway, D.J., and Van Dorp, C.F. 1991. Free-running rhythms of melatonin, cortisol, electrolyes, and sleep in humans in Antarctica. Am. J. Physiol. 260: R1137-R1144.

54. Schaefer, K.E., Kerr, C.M., Buss, D, and Haus, E. 1979. Effect of 18-h watch schedules on circadian cycles of physiological functions during submarine patrols. Undersea Biomed. Res. 6: S81-S90.

55. Natani, K., Shurley, J.T., Pierce, C.M., and Brooks, R.E. 1970. Long-term changes in sleep patterns in men on the south polar plateau. Arch. Intern. Med. 125: 635-659.

56. May, J., and Kline, P. 1987. Measuring the effects upon cognitive abilities of sleep loss during continuous operations. Br. J. Psychol. 78: 800-809.

57. Baum, A., Grunberg, N.E., and Singer, J.E. 1982. The use of psychological and neuroendocrinological measurements in the study of stress. Health Psychol. 1: 217-236.

58. Helmreich, R.L. 1997. Managing error in aviation. Sci. Am. 276: 62-67.

59. Lang, P.J. 1984. Cognition and emotion: Concept and action. Pp. 192-226 in Emotion, Cognition, and Behavior (C.E. Izard, J. Kagan, and R.B. Zajonc, eds.). Cambridge University Press, Cambridge, Mass.

60. Lacey, J.I. 1967. Somatic response patterning and stress: Some revisions of activation theory. Psychological Stress: Issues in Research (M.N. Appleay and Trumbull, eds.). Appleton, New York

61. Andre, J.T., Muth, E.R, Stern, R.M., and Leibowitz, H.W. 1996 The effect of tilted stripes in an optokinetic drum on gastric myoelectric activity and subjective reports in motion sickness. Aviat. Space Environ. Med. 67: 30-33.

62. Stern, R.M., and Koch, K.L. 1996. Motion sickness and differential susceptibility. Psychol. Sci. 5: 115-120.

63. Cacioppo, J.T., and Tassinary, L.G. 1990. Principles of Psychophysiology: Physical, Social, and Inferential Elements. Cambridge University Press, New York.

64. Gushin, V.I., Efimov, V.A., and Smirnova, T.M. 1996. Work capability during isolation. Adv. Space Biol. Med. 5: 297-307.

65. Lazarus, R.S. 1966. Psychological Stress and the Coping Process. McGraw-Hill, New York.

66. Selye, H. 1978. The Stress of Life. McGraw-Hill, New York.

67. Lazarus, R.L., and Folkman, S. 1984. Stress, Appraisal, and Coping. Springer, New York.

68. Antonovsky, A. 1979. Stress, Health, and Coping. Jossey-Bass, San Francisco.

69. Suedfeld, P. 1987. Extreme and unusual environments. Pp. 863-886 in Handbook of Environmental Psychology, Vol. 1 (D. Stokols and I. Altman, eds.). Wiley, New York.

70. Stuster, J. 1996. Bold Endeavors: Lessons from Polar Exploration and Spaceflight. Naval Institute Press, Annapolis, Md.

71. Connors, M.M., Harrison, A.A., and Akins, F.R. 1986. Psychology and the resurgent space program. Am. Psychol. 41: 906-113.

72. Palinkas, L.A., Gunderson, E.K.E., and Burr, R.G. 1989. Social, psychological, and environmental influences on health and well-being of antarctic winter-over personnel. Antarct. J. 24: 210-212.

73. Palinkas, L.A. 1991. Effects of the physical and social environment on the health and well-being of antarctic winter-over personnel. Environ. Behav. 23: 782-799.

74. Kanas, N. 1987. Psychological and interpersonal issues in space. Am. J. Psychiatry 144: 703-709.

75. Santy, P. 1983. The journey out and in: Psychiatry and space exploration. Am. J. Psychiatry 140: 519-527.

76. Palinkas, L.A., Suedfeld, P., and Steel, G.D. 1995. Psychological functioning among members of a small polar expedition. Aviat. Space Environ. Med. 50: 1591-1596.

77. Csikszentmihalyi, M. 1990. Flow: The Psychology of Optimal Experience. Harper & Row, New York.

78. Costa, P.T., Jr., and McCrae, R.R. 1985. The NEO Personality Manual. Psychological Assessment Resources, Odessa, Fla.

79. Associated Press. 1997. Astronaut says he was feeling low. San Francisco Chronicle, January 22, p. A-6.

80. Carpenter, D. 1997. Are blunders on Mir signs the stress is too great? San Francisco Examiner, July 18, p. A-1.

81. Weybrew, B.B., and Noddin, E.M. 1979. Psychiatric aspects of adaptation to long submarine missions. Aviat. Space Environ. Med. 50: 575-580.

82. Palinkas, L.A., Cravalho, M.A., and Browner, D. 1995. Seasonal variation of depressive symptoms in Antarctica. Acta Psychiatr. Scand. 91: 423-429.

83. Kanas, N. 1991. Psychological support for cosmonauts. Aviat. Space Environ. Med. 62: 353-355.

84. Aldrin, E.E. 1973. Return to Earth. Random House, New York.

85. Isay, R.A. 1968. The submariners' wives syndrome. Psychiatric Quart. 42: 647-652.

86. Pearlman, C.A., Jr. 1970. Separation reactions of married women. Am. J. Psychiatry 126: 946-950.

87. Pavy-Le Traon, A., Saivin, S., Soulez-LaRiviere, C., Pujos, M., Guell, A., and Houin, G. 1997. Pharmacology in space: Pharmacotherapy. Adv. Space Biol. Med. 6: 107-121.

88. Aleksandrovsky, Y.A., and Novikov, M.A. 1996. Psychological prophylaxis and treatments for space crews. Space Biology and Medicine. III: Humans in Spaceflight, Book 2 (A.E. Nicogossian, S.R. Mohler, O.G. Gazenko, and A.I. Grigoriev, eds.). American Institute of Aeronautics and Astronautics, Reston, Va.

89. Saivin, S., Pavy-Le Traon, A., Soulez-LaRiviere, C., Guell, A., and Houin, G. 1997. Pharmacology in space: Pharmacokinetics. Adv. Space Biol. Med. 6: 107-121.

90. Saivin, S., Pavy-Le Traon, A., Soulez-LaRiviere, C., Guell, A., and Houin, G. 1997. Pharmacology in space: Pharmacokinetics. Adv. Space Biol. Med. 6: 107-121.

91. Santy, P.A. 1994. Choosing the Right Stuff: The Psychological Selection of Astronauts and Cosmonauts. Praeger Scientific, Westport, Conn.

92. Santy, P. 1997. Behavior and performance in the space environment. Fundamentals of Space Life Sciences, Vol. 2 (S.E. Church, ed.). Krieger, Malabar, Fla.

93. McFadden, T.J., Helmreich, R.L., Rose, R.M., and Fogg, L.F. 1994. Predicting astronaut effectiveness: A multivariate approach. Aviat. Space Environ. Med. 65: 904-909.

94. Rose, R.M., Fogg, L.F., Helmreich, R.L., and McFadden, T.J. 1994. Psychological predictors of astronaut effectiveness. Aviat. Space Environ. Med. 65: 910-915.

95. Ursin, H., Comet, B., and Soulez-Lariviere, C. 1992. An attempt to determine the ideal psychological profiles for crews of long-term space missions. Adv. Space Res. 12:301-314.

96. Santy, P.A., and Jones, D.R. 1994. An overview of international issues in astronaut psychological selection. Aviat. Space Environ. Med. 65: 900-903.

97. Santy, P.A., and Jones, D.R. 1994. An overview of international issues in astronaut psychological selection. Aviat. Space Environ. Med. 65: 900-903.

98. Fassbender, C., and Goeters, K.M. 1994. Psychological evaluation of European astronaut applications: Results from the 1991 selection campaign. Aviat. Space Environ. Med. 65: 925-929.

99. Thagard, N., Lessons from Mir, presentation to the Panel on Human Behavior, Committee on Space Biology and Medicine, May 2, 1997, National Research Council, Washington, D.C.

100. Kanas, N. 1985. Psychological factors affecting simulated and actual space missions. Aviat. Space Environ. Med. 56: 806-811.

101. Kanas, N., and Feddersen, W.E. 1971. Behavioral, Psychiatric, and Sociological Problems of Long-Duration Space Missions. NASA-TM-X-58067. NASA Johnson Space Center, Houston, Tex.

102. Lebedev, V. 1988. Diary of a Cosmonaut: 211 Days in Space. Phytoresource Research Information Service, College Station, Tex.

103. Doll, R.E., and Gunderson, E.K.E. 1971. Group size, occupational status, and psychological symptomatology in an extreme environment. J. Clin. Psychol. 27: 196-198.

104. Harrison, A.A., Clearwater, Y.A., and McKay, C.P., eds. 1991. From Antarctica to Outer Space: Life in Isolation and Confinement. Springer-Verlag, New York.

105. Lebedev, V. 1988. Diary of a Cosmonaut: 211 Days in Space. Phytoresource Research Information Service, College Station, Tex.

106. Oberg, J.E. 1981. Red Star in Orbit. Random House, New York.

107. Connors, M.M., Harrison, A.A., and Akins, F.R. 1985. Living Aloft: Human Requirements for Extended Spaceflight. National Aeronautics and Space Administration, Washington, D.C.

108. Kelly, A.D., and Kanas, N. 1992. Crewmember communication in space: A survey of astronauts and cosmonauts. Aviat. Space Environ. Med. 63: 721-726.

109. Bluth, B.J. 1984. The benefits and dilemmas of an international space station. Acta Astronautica 11: 149-153.

110. Johnson, J.C., and Finney, B.R. 1988. Structural approaches to the study of groups in space: A look at two analogs. J. Soc. Behav. Personality 1: 325-347.

111. Natani, K., and Shurley, J.T. 1974. Sociopsychological aspects of a winter vigil at a South Pole station. Pp. 89-114 in Human Adaptability to Antarctic Conditions (E.K.E. Gunderson, ed.). American Geophysical Union, Washington, D.C.

112. Rohrer, J.H. 1961. Interpersonal relationships in isolated small groups. Symposium on Psychophysiological Aspects of Spaceflight (B.E. Flaherty, ed.). Columbia University Press, New York.

113. Lebedev, V. 1988. Diary of a Cosmonaut: 211 Days in Space. Phytoresource Research Information Service, College Station, Tex.

114. Grigoriev, A.I., Kozerenko, O.P., and Myasnikov, V.I. 1987. Selected problems of psychological support of prolonged spaceflights. Proceedings of the 38th Congress of the International Astronautical Federation (IAF). IAF, Paris.

115. Kanas, N. 1991. Psychological support for cosmonauts. Aviat. Space Environ. Med. 62: 353-355.

116. Palinkas, L.A. 1992. Going to extremes: The cultural context of stress, illness and coping in Antarctica. Soc. Sci. Med. 35: 651-664.

117. Kelly, A.D., and Kanas. N. 1992. Crewmember communication in space: A survey of astronauts and cosmonauts. Aviat. Space Environ. Med. 63: 721-726.

118. Hogg, M.A. 1992. The Social Psychology of Group Cohesiveness: From Attraction to Social Identity. Harvester Wheatsheaf, New York.

119. Lebedev, V. 1988. Diary of a Cosmonaut: 211 Days in Space. Phytoresource Research Information Service, College Station, Tex.

120. More back talk from space. 1983. San Francisco Chronicle, December 8.

121. Cooper, H.S.F., Jr. 1976. A House in Space. Holt, Rinehart & Winston, New York.

122. Kanas, N., Weiss, D.S., and Marmar, C.R. 1996. Crew member interactions during a Mir space station simulation. Aviat. Space Environ. Med. 67: 969-975.

123. Sandal, G.M., Vaernes, R., and Ursin, H. 1995. Interpersonal relations during simulated space missions. Aviat. Space Environ. Med. 66: 617-624.

124. Stuster, J. 1996. Bold Endeavors: Lessons from Polar and Space Exploration. Naval Institute Press, Annapolis, Md.

125. Sells, S.B. 1973. The taxonomy of man in enclosed space. Pp. 281-304 in Man in Isolation and Confinement (J.E. Rasmussen, ed.). Aldine, Chicago.

126. Palinkas, L.A. 1992. Going to extremes: The cultural context of stress, illness and coping in Antarctica. Soc. Sci. Med. 35: 651-664.

127. Kelly, A.D., and Kanas, N. 1994. Communication between space crews and ground personnel: A survey of astronauts and cosmonauts. Aviat. Space Environ. Med. 64: 721-726.

128. Nicholas, J.M., and Penwell, L.W. 1995. A proposed profile of the effective leader in human spaceflight based on findings from analog environments. Aviat. Space Environ. Med. 66: 63-72.

129. Helmreich, R.L., Foushee, H.C., Benson, R., and Russini, W. 1986. Cockpit resources management: Exploring the attitude performance linkage. Aviat. Space Environ. Med. 57:1198-1200.

130. Nelson, P.D. 1962. Leadership in small isolated groups. Report No. 62-13. U.S. Navy Medical Neuropsychiatric and Research Unit, San Diego, Calif.

131. Nicholas, J.M., and Penwell, L.W. 1995. A proposed profile of the effective leader in human spaceflight based on findings from analog environments. Aviat. Space Environ. Med. 66: 63-72.

132. Weybrew, B. 1991. Three decades of nuclear submarine research: Implications for space and antarctic research. From Antarctica to Outer Space: Life in Isolation and Confinement (A.A. Harrison., Y.A. Clearwater, and C.P. McKay, eds.). Springer-Verlag, New York.

133. Gunderson, E.K.E., and Nelson, P.D. 1963. Adaptation of small groups to extreme environments. Aerosp. Med. 34: 1111-1115.

134. Nelson, P.D. 1964. Similarities and differences among leaders and followers. J. Soc. Psychol. 63: 161-167.

135. Kanas, N., Weiss, D.S., and Marmar, C.R. 1996. Crew member interactions during a Mir space station simulation. Aviat. Space Environ. Med. 67: 969-975.

136. Fiedler, F.E., and Garcia, J.E. 1987. New Approaches to Effective Leadership: Cognitive Resources and Organizational Performance. Wiley, New York.

137. Burns, J.M. 1992. Leadership. Harper and Row, New York.

138. Smith, P.M., and Fritz, A.S. 1987. A person-niche theory of depersonalization: Implications for leader selection, performance and evaluation. Review of Personality and Social Psychology: Group Processes, Vol. 8 (C. Hendrick, ed.). Sage, Newbury Park, Calif.

139. Holland, A., NASA Operational Program, presentation to the Panel on Human Behavior, Committee on Space Biology and Medicine, May 1, 1997, National Research Council, Washington, D.C.,

140. Ferguson, M.J. 1970. Use of the Ben Franklin Submersible as a Space Station Analog, Vol. II: Psychology and Physiology. OSR-70-5. Grumman, Bethpage, N.Y.

141. Ferguson, M.J. 1970. Use of the Ben Franklin Submersible as a Space Station Analog, Vol. II: Psychology and Physiology. OSR-70-5. Grumman, Bethpage, N.Y.

142. Radloff, R., and Helmreich, R. 1968. Groups Under Stress: Psychological Research in Sealab II. Appleton-Century-Crofts, New York.

143. Haythorn, W.W., and Altman, I. 1963. Alone Together. Research Task MR-022-01-03-1002. Navy Department Bureau of Medicine and Surgery, Washington, D.C.

144. Dunlap, R.D. 1965. The Selection and Training of Crewmen for an Isolation and Confinement Study in the Douglas Space Cabin Simulator. No. 3446. Douglas Aircraft Company, Santa Monica, Calif.

145. McDonnell Douglas Astronautics Company. 1968. 60-day Manned Test of a Regenerative Life Support System with Oxygen and Water Recovery, Part II: Aerospace Medicine and Man-Machine Test Results. NASA CR-98501. McDonnell Douglas Astronautics Company, Santa Monica, Calif.

146. Jackson, J.K., Wamsley, J.R., Bonura, M.S., and Seeman, J. 1972. Program Operational Summary: Operational 90-day Manned Test of a Regenerative Life Support System. NASA CR-1835. National Aeronautics and Space Administration., Washington, D.C.

147. Santy, P.A., Holland, A.W., Looper, L., and Marcondes-North, R. 1993. Multicultural factors in the space environment: Results of an international shuttle crew debrief. Aviat. Space Environ. Med. 64: 196-200.

148. Nicholas, J.M. 1987. Small groups in orbit: Group interaction and crew performance on space station. Aviat. Space Environ. Med. 58: 1009-1013.

149. Nicholas, J.M. 1989. Interpersonal and group-behavior skills training for crews on space station. Aviat. Space Environ. Med. 60: 603-608.

150. Dunlap, R.D. 1965. The Selection and Training of Crewmen for an Isolation and Confinement Study in the Douglas Space Cabin Simulator. No. 3446. Douglas Aircraft Co., Santa Monica, Calif.

151. McDonnell Douglas Astronautics Company. 1968. 60-day Manned Test of a Regenerative Life Support System with Oxygen and Water Recovery, Part II: Aerospace Medicine and Man-Machine Test Results. NASA-CR-98501. McDonnell Douglas Astronautics Company, Santa Monica, Calif.

152. Helmreich, R.L., Foushee, H.C., Benson, R., and Russini, W. 1986. Cockpit resource management: Exploring the attitude-performance linkage. Aviat. Space Environ. Med. 57: 1198-1200.

153. Helmreich, R.L., Wilhelm, J.A., Gregorich, S.E., and Chidester, T.R. 1990. Preliminary results from the evaluation of cockpit resource management training: Performance ratings of flightcrews. Aviat. Space Environ. Med. 61: 576-579.

154. Kanki, B.G., and Foushee, H.C. 1989. Communication as group process mediator of aircrew performance. Aviat. Space Environ. Med. 60: 402-410.

155. Jackson, J.K., Wamsley, J.R., Bonura, M.S., and Seeman, J. 1972. Program Operational Summary: Operational 90-day Manned Test of a Regenerative Life Support System. NASA-CR-1835. National Aeronautics and Space Administration, Washington, D.C.

156. Belew, L.F. 1977. Skylab, Our First Space Station. NASA SP-400. National Aeronautics and Space Administration, Washington, D.C.

157. Kanas, N. 1991. Psychological support for cosmonauts. Aviat. Space Environ. Med. 62: 353-355.

158. Kanas, N. 1991. Psychological support for cosmonauts. Aviat. Space Environ. Med. 62: 353-355.

159. Thagard, N., presentation to the Panel on Human Behavior, Committee on Space Biology and Medicine, May 2, 1997, National Research Council, Washington, D.C.

160. Stuster, J. 1996. Bold Endeavors: Lessons from Polar and Space Exploration. Naval Institute Press, Annapolis, Md.

161. Sandal, G.M., Vaernes, R., Bergan, T., Warncke, M., and Ursin, H. 1996. Psychological reactions during polar expeditions and isolation in hyperbaric chambers. Aviat. Space Environ. Med. 67: 227-234.

162. Kelly, A.D., and Kanas, N. 1992. Crewmember communication in space: A survey of astronauts and cosmonauts. Aviat. Space Environ. Med. 63: 721-726.

163. Santy, P.A. 1994. Choosing the Right Stuff: The Psychological Selection of Astronauts and Cosmonauts. Praeger Scientific, Westport, Conn.

164. Rohrer, J.H. 1961. Interpersonal relationships in isolated small groups. Psychophysiological Aspects of Spaceflight (B.E. Flaherty, ed.). Columbia University Press, New York.

165. Bechtel, R.B., and Berning, A. 1991. The third quarter phenomenon: Do people experience discomfort after stress has passed? Pp. 261-266 in From Antarctica to Outer Space: Life in Isolation and Confinement (A.A. Harrison, Y.A. Clearwater, and C.P. McKay, eds.). Springer-Verlag, New York.

166. Palinkas, L.A., Suedfeld, P., and Steel, G.D. 1995. Psychological functioning among members of a small polar expedition. Aviat. Space Environ. Med. 50: 1591-1596.

167. Sandal, G.M., Vaernes, R., Bergan, T., Warncke, M., and Ursin, H. 1996. Psychological reactions during polar expeditions and isolation in hyperbaric chambers. Aviat. Space Environ. Med. 67: 227-234.

168. Kanas, N., Weiss, D.S., and Marmar, C.R. 1996. Crew member interactions during a Mir space station simulation. Aviat. Space Environ. Med. 67: 969-975.

169. Stuster, J. 1996. Bold Endeavors: Lessons from Polar and Space Exploration. Naval Institute Press, Annapolis, Md.

170. Gunderson, E.K.E. 1974. Psychological studies in Antarctica. Human Adaptability to Antarctic Conditions (E.K.E. Gunderson, ed.). American Geophysical Union, Washington, D.C.

171. Christensen, J.M., and Talbot, J.M. 1986. A review of the psychological aspects of spaceflight. Aviat. Space Environ. Med. 57: 203-212.

172. Connors, M.M., Harrison, A.A., and Akins, F.R. 1985. Living Aloft: Human Requirements for Extended Spaceflight. National Aeronautics and Space Administration, Washington, D.C.

PART IV

Research Priorities and Programmatic Issues

13

Setting Priorities in Research

The preceding chapters present recommendations for research and research priorities in a wide range of disciplinary and interdisciplinary fields relevant to space biology and medicine. This chapter considers the question of overall priorities for research supported by the National Aeronautics and Space Administration (NASA) in the next decade, taking into account budgetary realities and the need for clearly focused programs.

The highest priority for NASA support should be given to research meeting the following criteria:

1. Research aimed at understanding and ameliorating problems that may limit astronauts' ability to survive and/or function during prolonged spaceflight. Such studies include basic as well as applied research and ground-based investigations as well as flight experiments. NASA programs should focus on aspects of research in which NASA has unique capabilities or that are underemphasized by other agencies; and

2. Fundamental biological processes in which gravity is known to play a direct role. As above, programmatic focus should emphasize NASA's capabilities and take into account the funding patterns of other agencies.

A lower priority should be assigned to areas of basic and applied research that are relevant to fields of high priority to NASA but are extensively funded by other agencies, and in which NASA has no obvious unique capability or special niche.

In the near term, until the research facilities of the International Space Station come online or an additional Spacelab mission is provided, NASA-supported research will necessarily be largely directed toward ground-based investigations designed to answer fundamental questions and frame critical hypotheses that can later be tested in space. Indeed, as the preceding chapters have emphasized, understanding the basic mechanisms underlying biological and behavioral responses to spaceflight is essential for designing effective countermeasures and protecting astronaut health and safety both in space and upon return to Earth. For these reasons, the following recommendations for high-priority areas of research over the entire life sciences program place greater emphasis on ground-based studies.

PHYSIOLOGICAL AND PSYCHOLOGICAL EFFECTS OF SPACEFLIGHT

The committee considers the following areas of research to be the most important for ensuring astronaut health, safety, and performance during and after long-duration spaceflight. The specific order of research priorities among these areas is likely to shift depending on the nature of the planned missions; for this reason the topics are not presented in an order of priority.

Loss of Weight-bearing Bone and Muscle

Bone and muscle deterioration is one of the best-documented deleterious effects caused by spaceflight in humans and animals. The reduction in bone mass has been shown to exceed 1 percent per month in weight-bearing bones, even when an in-flight exercise regime was followed, making this one of the major barriers to long-term human space exploration. Dramatic losses in strength and changes in functional properties of weight-bearing muscles have also been observed even after short-duration flights.   Exercise has been only partially successful in preventing bone loss and muscle weakness. Development of effective countermeasures requires advances in several areas of basic research.

Recommendations

• *Research should emphasize studies that provide mechanistic insights into the development of effective countermeasures for preventing bone and muscle deterioration during and after spaceflight.*
• *Ground-based model systems, such as hindlimb unloading in rodents, should be used to investigate the mechanisms of changes that reproduce in-flight and postflight effects.*
• *A database on the course of microgravity-related bone loss and its reversibility in humans should be established in preflight, in-flight, and postflight recording of bone mineral density.*
• *Hormone profiles should be obtained on humans before, during, and after spaceflight.*
• *The relationship between exercise activity levels and protein energy balance in flight should be investigated.*

Vestibular Function, the Vestibular Ocular Reflex, and Sensorimotor Integration

Over the past 10 years, extensive experimental research has been conducted on humans to better understand how the space environment affects the control of posture and movement in astronauts. Because of this, considerable information is now available regarding spatial orientation, postural control, the vestibular ocular reflex (VOR), and space motion sickness in microgravity. In future work, it will be important to extend these findings from human studies to mechanistic studies in suitable animal models. This should provide a better understanding of the basic mechanisms operating at the cellular and molecular levels in the control of posture and movement in microgravity.

We know that compensatory mechanisms function effectively in the vestibulomotor pathways on Earth and that compensatory mechanisms also occur in space.

Recommendations

• *Experiments to determine the basis for the compensation on Earth and in space, and to evaluate whether the mechanisms are the same, should receive the highest priority, since these compensatory mechanisms operate in astronauts entering and returning from space and may have a profound effect on their performance in space and their postflight recovery on Earth.*

• *In-flight recordings of signal processing following otolith afferent stimulation should be made by a trained physiologist serving as a payload specialist, to determine how exposure to microgravity affects central and peripheral vestibular function and development.*

• *From ground-based experiments, we know that the vestibulo-oculomotor system is capable of learning new motor patterns in response to sensory perturbations; therefore, future investigation should focus on determining whether and how these mechanisms are affected by exposure to microgravity.*

Orthostatic Intolerance Upon Return to Earth Gravity

Significant progress in cardiovascular research occurred during the 1990s on a series of Spacelab missions, but orthostatic hypotension, present since the earliest human spaceflights, still affects a high percentage of astronauts returning from flights of relatively short duration. It is an even greater problem for shuttle pilots, who must perform complex reentry maneuvers in an upright, seated position. The incidence and magnitude of orthostatic hypotension will increase with longer-duration flights planned for the space station and both lunar or Mars missions. The problem remains despite the use of extensive antiorthostatic countermeasures by both U.S. and Russian space programs. The committee recommends several areas of research.

Recommendations

• *Current knowledge of the magnitude, time course, and mechanisms of cardiovascular adjustments should be extended to include long-duration exposure to microgravity.*

• *The specific mechanisms underlying inadequate total peripheral resistance observed during postflight orthostatic stress should be determined.*

• *Current antiorthostatic countermeasures should be reevaluated to refine those that offer protection and eliminate those that do not. Studies should avoid confounding effects of multiple, simultaneous interventions unless data support these combinations. Priority should be given to interventions that may provide simultaneous bone and/or muscle protection.*

• *Appropriate methods for referencing intrathoracic vascular pressures to systemic pressures in microgravity should be identified and validated, given the observed changes in cardiac and pulmonary volume and compliance.*

Radiation Hazards

The biological effects of exposure to radiation in space pose potentially serious health effects for crew members that must be controlled or mitigated before initiation of long-term missions beyond low Earth orbit. High priority is given to the following.

Recommendations

• *Determine the carcinogenic risks following irradiation by protons and high-atomic-number, high-energy (HZE) particles.*

• *Determine if exposure to heavy ions at the level that would occur during deep-space missions of long duration poses a risk to the integrity and function of the central nervous system.*

• *Determine how the selection and design of the space vehicle affect the radiation environment in which the crew has to exist.*

• *Determine whether combined effects of radiation and stress on the immune system in spaceflight could produce additive or synergistic effects on host defenses.*

Physiological Effects of Stress

The immune system has close interactions with the neuroendocrine system. Results of these studies indicate a close association between alterations in status of the immune system and the state of the neuroendocrine system of the host.

Recommendation

Interactions between the hypothalamic-pituitary-adrenal (HPA) axis and the immune system during spaceflight should be analyzed to determine the role that the host response to stressors plays in alterations in host defenses.

Psychological and Social Issues

Aspects of living and working in space that have been well-tolerated by astronauts during short-duration missions are likely to have significant impacts on health, well-being, and performance during long-duration missions. Mechanisms of response to physiological and psychosocial stressors encountered in spaceflight must be better understood in order to ensure crew safety, health, and productivity during prolonged residence in space. These mechanisms require an interdisciplinary approach since many of the physiological changes (e.g., endocrine, immune, cardiovascular, neurovestibular) likely to occur during prolonged exposure to microgravity will have important implications for behavior and performance. Likewise, many of the characteristics of the psychosocial environment of long-duration missions—such as interpersonal conflicts, restrictions on privacy and territoriality, social monotony, and prolonged isolation from family and friends—have important implications for these physiological systems by virtue of their influences on the HPA axis.

Research is recommended in two areas, both of which will require the development of noninvasive techniques for the ongoing assessment of behavior and performance.

Recommendations

1. Research should be conducted on the neurobiological (circadian, endocrine) and psychosocial (individual, group, organizational) mechanisms underlying the effects of physical (microgravity, hazards) and psychosocial (isolation, confinement) environmental stressors on cognitive, affective, and psychophysiological measures of behavior and performance. Such research should be interdisciplinary and conducted in ground-based analogue settings as well as in flight.

2. The efficacy of existing countermeasures (screening and selection, training, monitoring, support) should be determined.

• *Studies of the use of psychophysiological measures in the implementation of these counter-*

measures and the effects of microgravity on the kinetics and efficacy of psychopharmacological medications should be interdisciplinary in nature.

    • *In those instances where existing countermeasures are found to be ineffective, new countermeasures should be developed that effectively contribute to optimal levels of crew performance, individual well-being, and mission success.*

FUNDAMENTAL GRAVITATIONAL BIOLOGY

Mechanisms of Graviperception and Gravitropism in Plants

Multicellular plants respond to changes in the direction of the gravitational vector by altering the direction of growth of roots and stems. The gravitropic response requires (1) perception of the gravitational vector by gravisensing cells, (2) intracellular transduction of this information, (3) translocation of the resulting signal to the sites of reaction (i.e., sites of differential growth), and (4) reaction to the signal by the responding cells (i.e., initiation of differential growth). In some systems, the gravity-perceiving cell is also the site of reaction (e.g., in the *Chara* rhizoid).

Recommendations

    • *Studies of graviperception should concentrate on three problems:*
    1. The identity of the cells that perceive gravity in multicellular plants;
    2. The intracellular mechanisms by which the direction of the gravity vector is perceived; and
    3. The threshold value for graviperception—this will require a spaceflight experiment.

    • *Studies of gravitropic transduction should focus on the nature of the cellular asymmetry that is set up in a cell that perceives the direction of the gravity vector.*
    • *Studies on the translocation step should concentrate on the nature and mechanism of the translocation of the signals that pass from the site of perception to the site of reaction.*
    • *Studies on the reaction step should focus on the mechanism(s) by which gravitropic signals cause unequal rates of cell elongation, and on the possible effects of gravity on the sensitivity of these cells to the signals.*

Mechanisms of Graviperception in Animals

Work on space research is concerned with whether those parts of the vestibular system that are gravity sensitive (otolith organs) can develop and function adequately in microgravity. In addition, it is important to determine whether gravity influences the sensory systems that depend for their development and function on vestibular input. This includes the other sensory systems that interact directly with the vestibular system, the multiple brain regions containing neural space maps, and finally those areas in the brain capable of responding to alterations in their activity by neuroplastic changes.

Recommendations

    • *Space-based experiments are needed to test the role of gravity on the embryonic development and maintenance of the vertebrate vestibular system. Prior to this, ground-based studies are needed to identify the critical periods in vestibular neuron develpment. In both Earth and space-based studies, it is important to characterize vestibular development at several different times to determine the sequence of intermediate events leading up to the final outcome. In space studies, controls for the effects of nongravitational stresses, including loud noise and vibration, must be performed on the ground so that the space experiments are designed to isolate the effects of microgravity from the effects due to other stresses.*

• Pre- and postflight functional magnetic resonance imaging (fMRI) studies should be conducted with astronauts to determine the effects of microgravity on neural space maps.

Effects of Spaceflight on Reproduction and Development

To determine whether there are developmental processes that are critically dependent upon gravity, organisms should be grown through at least two full generations in space.

Recommendations

• Key model animals should be grown through two life cycles; highest priority should be given to vertebrate models (e.g., fish, birds, and small mammals such as mice or rats). If significant developmental effects are detected, control experiments must be performed (including the use of a space-based 1-g centrifuge) to determine whether gravity or some other element of the space environment induces these developmental abnormalities.

• An analogous experiment should be carried out with the model plant Arabidopsis thaliana *to confirm results obtained on Mir with a preliminary experiment using* Brassica rapa. *The ideal experiment will require the development of a suitable plant growth unit and of* Arabidopsis *plants containing stress-indicator genes and/or mutations conferring insensitivity to environmental stresses.*

14

Programmatic and Policy Issues

Programmatic and policy issues raised in the 1987 Goldberg report and its 1991 follow-up centered largely on (1) strategic planning for life sciences research, including planning for facilities and research on the projected space station; (2) accessibility and timely publication of research data, especially results of flight experiments; and (3) the need for improved cooperation with the National Institutes of Health (NIH) and National Science Foundation (NSF) as well as international space life sciences agencies and investigators.[1,2] The response of the National Aeronautics and Space Administration (NASA) to a number of these concerns and recommendations was prompt and effective. Research management and strategic planning were significantly improved by reorganization of NASA's advisory structure. Creation of a universal peer review process has settled long-standing concerns of the Committee on Space Biology and Medicine (CSBM) and the external life sciences community. Initiation of the NASA Specialized Centers of Research and Training (NSCORT) program fostered the development of specific, interdisciplinary research foci and increased the scope of interaction of the agency with the academic community. Effective programmatic interactions with NIH and, to a lesser degree, NSF were put into place and have prospered. International cooperation has been a focus for the International Space Station (ISS) as well as for shuttle-based research, and the cooperative programs with Russia, most notably the residence of U.S. astronauts on Mir, have developed to an extent that could not have been predicted at earlier times.

However, significant concerns in the program and policy arena remain unresolved. These focus principally on issues relating to various aspects of strategic planning and conduct of space-based research, utilization of ISS for life sciences research, mechanisms for promoting integrated and interdisciplinary research, collection of and access to human flight data specifically, and publication of and access to space life sciences research in general. The following sections summarize the committee's concerns and provide recommendations for NASA's consideration.

SPACE-BASED RESEARCH

Criteria for Space Research

Flight opportunities for the life sciences will continue to be precious for the next decade, and facilities—especially expert crew time—will be extremely limited, at least until assembly of the International Space Station is complete. It is therefore crucial that priority for these limited opportunities be given to experiments of the highest possible quality. In the past, a number of factors, including pressures on funding and limitations in crew time and on-board facilities and resources, have led to compromises in experimental rigor that have resulted in diminished quality of study results. A renewed emphasis is now required on experimental best practices, as defined in the following guidelines, which should be clearly disseminated to the proposing research community.

1. *Justification*. A thorough prior justification should be made, based on solid experimental evidence and/or theoretical considerations, demonstrating or clearly predicting that a measurable and significant change due to effects of microgravity on the phenomenon under investigation can reasonably be expected during spaceflight. Flight experiments should be carried out only when ground-based experiments have demonstrated a clear and critical need for space-based data. The experiments chosen for support should be those that promise the greatest reward in terms of science and address the major questions. They should not be selected because they fit some preexisting or planned facility. If a facility such as the ISS is not the appropriate place for a particular, important experiment, it should not be carried out there.

2. *Hypothesis*. A clear-cut hypothesis should be presented that can be tested in a convincing manner under the conditions of the flight experiment.

3. *System*. A limited number of model systems should be selected and studied in detail in order to build a comprehensive database of flight results. Model systems should be chosen that have been well characterized in ground-based studies and/or have been demonstrated to have some unique advantage for the space experiment under consideration. Where possible, the use of organisms or cells with well-characterized genetics and genetic manipulability will be advantageous. Conditions should be designed to avoid, or at least adequately account for, confounding variables of the space environment (see item 5).

4. *Hardware*. Flight hardware must be tested extensively prior to flight in ground-based simulation experiments in order to demonstrate the capacity of the apparatus to maintain normal function and to demonstrate any differences induced by the hardware compared with standard conditions. Hardware design should take into consideration the effects of microgravity on fluid dynamics and gas exchange. If necessary, equipment modifications must be made and the apparatus retested sequentially until a satisfactory baseline is attained. Any equipment that cannot meet a minimal standard should not be used for experiments.

5. *Controls*. For in-flight experiments, it is crucial to distinguish among the following:
   • Effects resulting directly from microgravity per se;
   • Indirect effects of microgravity, such as the lack of convective flow and the profound attendant effects on solute and gas exchange; this may be a particularly debilitating problem for cell culture experiments (although it is presumably minimized in flow systems such as the Bioreactor), and consideration must be given to the possibility of decreased rates of renewal of nutrients and oxygen and/or buildup of waste products, maintenance of bubbles, surface tension effects, and so forth;
   • Other flight- and space-related perturbations of the environment, including changes in gravity (acceleration and deceleration, including those due to crew movements, causing changes in space

platform momentum during the course of the experiment), vibration, and temperature and atmospheric fluctuations. As noted above, these environmental stresses induce stress responses in biological systems and may result in significant perturbations of physiology. Since such variables are difficult to reproduce on the ground and render unambiguous interpretations of in-flight results essentially impossible, it will be critically important to minimize these variables in the construction of hardware and ISS facilities. An important control for nongravitational effects of the space environment is the use of ground- and space-based variable force centrifuges; the variable force centrifuge is properly considered an essential facility for research carried out on ISS. However, it is important to emphasize that the centrifuge does not allow an investigator to distinguish between direct effects of microgravity on the biological system and indirect effects resulting from changes in fluid dynamics in microgravity. For experiments employing cell and tissue cultures, plants, or mice, it may be possible to detect the presence and extent of effects of environmental stress using cells or transgenic organisms containing indicator genes coupled to specific stress-inducible promoters (see Chapter 4).

6. *Data collection and interpretation.* Data collection should be made in such a way as to allow valid statistical analyses, and the number of experimental subjects should be large enough to permit this. Conclusions drawn from data obtained during spaceflight must be made in the context of the controls summarized in item 5. Efforts should be made to utilize the growing capacity for the application of dedicated microprocessors for process control, data storage, and rapid communication in real time with ground-based teams (see next section).

7. *Repetition of experiments.* Care should be taken to ensure that repetition of experiments is possible in order to directly compare independent sets of data and make solid conclusions about experimental consistency and statistical significance.

8. *Publication.* Every effort must be made to bring data and conclusions to a timely formal written presentation for stringent peer review and publication in a readily accessible scientific journal. For investigators with previous NASA funding, publication of the earlier results in peer-reviewed journals should be an important criterion for continued support.

Recommendation

NASA should emphasize the criteria described above in NASA Research Announcements (NRAs) and scientific peer review for flight experiments, in funding decisions, and in the development of flight protocols.

Development of Advanced Instrumentation and Methodologies

All future life science flight experiments, and especially the utilization of the International Space Station to best advantage, will depend on the availability of advanced instrumentation capable of carrying out the sophisticated measurements and analyses required by the research questions and experimental approaches described in the preceding chapters. In many cases, adaptation and/or miniaturization of existing technologies will meet investigators' needs. Examples from space physiology might include dual-beam x-ray absorptiometry for measurement of bone density; automated, submicroassay methods for assay of hormones and other blood chemicals; and mass spectrometers and ultrasonic flow meters for respiratory gas measurements. Similarly, for cell biological experiments, the adaptation of newly emerging technologies, such as automated screening for expression of reporter genes or molecular force measurements on single cells, may ultimately be needed. However, the development of entirely novel flight-certified instrumentation will also be necessary if the scientific objectives described in this report are to be achieved. NASA should work with the broad life sciences community to identify those capabilities that will be essential within the next decade, and with industry to catalyze their development. In developing advanced instrumentation, NASA must recognize that although the apparatus must be as adaptable as possible, generic equipment may not be adequate or appropriate to meet the needs of all experiments. False economies in equipment choices may result in unacceptable costs in failed experiments. Several experiments on the International Microgravity Laboratory (IML) missions, for example, were seriously compromised by the requirement to use generic equipment that was not actually adequate for the experiment.

NASA should take every opportunity to make use of advanced instrumentation developed in other countries. There is no justification for money to be spent developing a piece of equipment in this country when a suitable item has already been developed elsewhere. Likewise, NASA must coordinate its own efforts so that it does not support multiple projects for development of the same instrument. For example, NASA is currently supporting four competing projects for the advanced plant growth facility.

In addition, the importance of facile data and information transfer between space- and ground-based investigators cannot be overestimated. The capability for direct, real-time communication between the station-based experimenter and ground-based principal investigator (PI) at the investigator's own laboratory is vital for the efficient conduct of experiments and is considered a high priority by astronaut-experimenters. Rapid responses to unanticipated experimental difficulties or mid-experiment questions about protocols or methods are generally impossible when communication must go through Mission Control, and it is clearly unrealistic to expect ground-based PIs to travel to the Johnson Space Center (JSC) for the duration of their experiments. Filtering question and response through intermediate layers of Mission Control is slow, inevitably degrades the quality of the information transfer, wastes crew time, and is a significant impediment to appropriate conduct of the experiment. This will be even more important for the longer-term, more complex experiments that will be possible on ISS. Current plans for ISS do not appear to be responsive to this need.

On-board data storage and analysis capacity and the capability for fast, real-time down- and uplinks are of the first importance. The ability to uplink a recorded video of new experimental procedures would increase crew flexibility and allow greater iteration in experimental protocols, while on-board cameras would allow the PI to observe and evaluate both experimental procedures and experimental samples.

The centrifuge facility at Ames Research Center will continue to be valuable for future ground-based experiments employing hypergravity to probe mechanisms of response to changes in the gravitational force. For example, it is likely that the response of the musculoskeletal system to gravitational force represents a continuum from microgravity through normal gravity to hypergravity. In such cases, mechanistic studies carried out in a relatively cost-effective way in hypergravity would be valuable in predicting responses and formulating hypotheses for critical testing under microgravity conditions.

Recommendations

• *NASA should work with the broad life sciences community to identify and catalyze the development of advanced instrumentation and methodologies that will be required for sophisticated space-based research in the coming decade.*

• *NASA should take advantage of advanced instrumentation developed in other countries.*

• *The capability for direct, real-time communication between space-based experimenters and principal investigators at their home laboratories should be a high-priority objective for the International Space Station.*

Utilization of the International Space Station for Life Sciences Research

Issues relating to design and utilization of the ISS are a major concern for the committee. Repeated changes in design of the ISS and the diversion of funds intended for scientific facilities and equipment into construction budgets have provoked alarm among the life sciences user community. Construction of the variable force centrifuge, essential for controlled life sciences experiments, has been substantially delayed, and there is concern that increases and overruns in construction costs may force additional delays in the design of hardware and the availability of other equipment or research facilities, and/or downgrading of their specifications. Involvement of the user community in the design of hardware and software remains problematic. If ISS is to meet its objectives as an advanced facility for the conduct of space life sciences research, the user community must continue to be brought into the actual planning phases. Although there have been multiple meetings and committees convened to define the scientific requirements for ISS, implementation of the resulting recommendations into the actual decision-making process concerned with ISS design seems to be deficient. If there is to be a viable, effective community of life sciences investigators to make use of ISS, researchers must be more directly involved in planning and design decisions made now. Issues relating to the adequacy of power, data transmission to and from Earth, and availability of crew time for research are also matters of significant concern.

In addition, delays in the schedule for station construction have greatly constrained flight opportunities for the life sciences over the next 5 years because of the shift of shuttle missions from science to construction. This hiatus poses a real and serious threat to the integrity of the academic space life sciences communities. Although limited, small-scale shuttle flight opportunities may be programmed in the intervening period, it is entirely unclear whether these will be adequate to maintain the momentum or interest of existing NASA-supported life scientists, much less attract the new investigators who will be important for future new directions in space-based research. Given the delays in bringing ISS to effective utilization, NASA should seriously consider reinstituting at least one additional Spacelab flight. The facility is already built and available, and the previous Spacelab flights have been highly successful in terms of science return. The cost to the progress of space life sciences (including perhaps research important for the success of a crewed Mars mission) of a prolonged period without any ability to conduct major flight experiments is likely to be high.

It is not completely clear that, even after the ISS is available, it will necessarily be the most cost-effective platform for all flight experiments. There will continue to be high-priority experiments that require relatively short times (1 to 2 weeks) in space and do not have to be conducted on ISS, especially in its early years when available crew time will be extremely limited. NASA should investigate the possibility that continued use of Shuttle missions for such purposes might be of economic and scientific advantage.

Recommendations

- *To better ensure that the ISS will adequately meet the needs of space life sciences researchers, NASA should continue to bring the external user community as well as NASA scientists into the planning and design phases of facility construction.*

- *NASA should make every effort to mount at least one Spacelab life sciences flight in the period between Neurolab and the completion of ISS facilities.*

- *NASA should determine whether continuation of shuttle missions for short-term flight experiments after the opening of ISS would be economically and scientifically sound.*

SCIENCE POLICY ISSUES

Peer Review

The Division of Life Sciences initiated a system of peer review in 1994 for all NASA-supported investigators, in which investigator-initiated proposals from NASA scientists as well as from the external community are subjected to the same rigorous, NIH-style, initial scientific peer review by expert panels drawn from the total scientific community. The new process has answered long-standing concerns of the academic community and CSBM, and has the committee's strong support. The committee also strongly endorses the current policy that places responsibility in the NASA Headquarters Division of Life Sciences for establishing peer review panels and for funding decisions. The impact of recent changes in NASA organization—specifically, the transfer of the Office of Life and Microgravity Sciences and Applications to the Human Exploration and Development of Space enterprise, and the transfer of program management responsibilities from the Headquarters Division of Life Sciences to Johnson Space Center—is not yet clear. Even though peer review remains a function of NASA Headquarters, the committee is concerned that decentralization may make coherent oversight and strategic planning difficult to maintain.

The new universal peer review system has been in place only 3 years at the time of this writing. Analyses of the results of the competition have been carried out annually, and plans for periodic review of the process are appropriate. It will be especially important to evaluate regularly the composition of scientific review panels as it relates to review of space-based research where hardware, environmental factors, and the availability of crew time and expertise may place limits on the feasibility of experiments that are otherwise scientifically worthy of support.

Recommendations

- *Responsibility for the establishment of peer review panels and for funding decisions should remain a function of the Headquarters Division of Life Sciences.*
- *NASA should regularly evaluate the composition of scientific review panels to ensure that the feasibility of proposed flight experiments receives appropriate expert evaluation.*

Integration of Research Activities

Early Development of Integrated Teams for Planning of Flight Experiments

Flight experiments, with their inherent complexity and high cost, are dependent on a degree of teamwork far greater than that required for most ground-based research, which allows for ongoing modification and improvement of instruments and facilities. This teamwork must begin with a close interaction and exchange of ideas between the investigators and NASA managers and engineers. There must be agreement at the outset on the objectives of the experiment and the requirements for instrumentation and crew time to meet these objectives. Because of the long lead times, high costs, and inflexibility of space-based research, it is crucial that investigators be brought together with design engineers at the beginning of the project planning process as an integrated team, responsible for operating specifications, design and testing of necessary hardware, and working together through all phases of the project. No instruments or experimental facilities should be flown that do not adequately meet the investigators' needs as stated and approved.

Funding and Guiding of Interdisciplinary Research

This report has as a major theme the need for multilevel, multi- and interdisciplinary approaches to address problems in all areas of ground-based research in space biology and medicine. Two mechanisms for funding and oversight of interdisciplinary research programs are currently in operation within the life sciences program: NASA Specialized Centers of Research and Training (NSCORT), first established in 1991, and the new (1997) National Space Biomedical Research Institute. The NSCORT program, which focuses on designated priority areas of disciplinary research in universities, represents a "classical" mechanism for fostering and support of interdisciplinary research, and the potential advantages and problems of center-type programs are relatively well understood, given the extensive experience of funding agencies such as NIH and NSF, as well as NASA itself. However, NASA should carefully assess its relative funding priorities for individual versus multi-investigator and center grants, and should consider whether NSCORTs are, in fact, the most productive way to foster interdisciplinary research and increase the value of the research program in life sciences. The life sciences program is small, and few, if any, universities have more than a handful of investigators conducting space-related life sciences research. This raises the potential concern that any one institution may not have a sufficient number of high-quality researchers to create a critical mass in the area of the NSCORT program. On the other hand, creation of multisite NSCORTs, like the recently established New Jersey Center of Research and Training, complicates the development of mechanisms ensuring the close interaction among investigators that is a prime objective of the centers.

The very recent establishment of the National Space Biomedical Research Institute as a multi-university, multi-investigator consortium with close ties to Johnson Space Center marks a major change in the conduct of NASA life sciences research. The institute concept has promise as a mechanism to foster and facilitate multidisciplinary research and to bring highly qualified biomedical investigators into space physiology and related areas of research. However, establishment and maintenance of true collaborations and interdisciplinary activities are likely to be complicated by the geographical dispersion of institute members, whose research continues to be conducted in their own laboratories on their home campuses. Thus, the development of effective mechanisms for fostering close communications and interactions among the component laboratories will be crucial to success, as will the development of effective procedures for oversight and review of the institute's progress and performance. It will also be essential that potential impacts of the very substantial institute funding on the life sciences research budget and overall program over the coming years be adequately monitored and evaluated. The committee is concerned that in a continuing era of tight budgets, funding of the overall life sciences research program might fall into a "rob-Peter-to-pay-Paul" mode, in which areas of research and biomedical investigators external to the institute are frozen out.

Recommendations

- *Principal investigators of projected flight experiments should be brought together with NASA managers and design engineers at the beginning of the planning process to function as an integrated team responsible for all phases of planning, design, and testing. This integration of scientists and engineers should continue throughout the life of the project.*
- *NASA should regularly review and evaluate the NASA Specialized Centers of Research and Training (NSCORT) program to determine whether this mechanism provides the best way to foster interdisciplinary research and increase the scientific value of the life sciences research program.*

• *NASA should regularly review and evaluate the performance of the National Space Biomedical Research Institute and the impact of its funding on the overall life sciences research program and budget.*

Human Flight Data: Collection and Access

Collection of Baseline Data

The disciplinary chapters of this report have repeatedly stressed the need for improved, systematic collection of baseline data on astronauts, preflight, in space, and postflight. In the past, in-flight tests and sample collection were often done at single, seemingly arbitrary times and at random and undefined points in the subjects' diurnal circadian cycle. In many cases, experimental depth and rigor have fallen victim to an unrealistic and overambitious scheduling of mission projects. Similarly, postflight testing has often suffered from variable timing and insufficient long-term follow-up. The resulting data are incomplete and even potentially misleading. In order to understand the changes induced by spaceflight and their significance for astronaut health and safety, as well as the effectiveness of countermeasures, it is essential that testing and sample collection be done on a well-considered and rigorous schedule, with a sufficient number of time points in flight and postflight to define adequately the time course of changes in flight, and of recovery postflight. The opportunity for sophisticated systematic monitoring of the effects of spaceflight over longer periods on the ISS is particularly important in thinking ahead to a future that includes crewed interplanetary missions.

Because the number of subjects will necessarily continue to be small, and individual variations in response are often large, each astronaut should serve as his or her own preflight control. Successful completion of sample and data collection will depend on the availability of advanced, miniaturized instrumentation for carrying out sophisticated physiological tests in flight, on the utilization of micro- and submicro-methods for analysis of small samples, and on the availability of appropriate cryostorage equipment in flight. Success will also be critically dependent on the cooperation of the astronaut subjects, who must understand and accept the rationale for the requisite testing and its long-term role in improving their health and ability to function for prolonged periods in space and in optimizing their return to Earth-normal physiology after return. Every effort should be made to encourage the astronauts to "buy into" the necessary clinical research—for example, by bringing investigators and astronauts together, early in the planning phase, for full and frank discussion of the experimental rationale, methodology, and long-term significance. Selection of astronaut crews should take into account the need to participate in experiments and to make the resulting data accessible to the relevant scientific community.

Access to Astronaut Data

According to the current NASA policy on protection and confidentiality of human subjects, astronauts may withhold permission for publication or dissemination of data obtained from their participation as subjects in clinical studies. Given the small number of participants in any given study, successful blinding of data to prevent identification of individual subjects is indeed difficult, and the issue of confidentiality is an important and complicated one. In large-scale clinical studies with statistically significant numbers of subjects, data can be presented without loss of significance in an aggregated form that essentially precludes the identification of individuals. This is not the case with studies involving a small number of subjects (e.g., a spaceflight crew) where inclusion of clinically significant physical characteristics and past histories may make the participant immediately identifiable. On the other hand, exclusion of these data and, certainly, total withdrawal of study results for members of a very small study population necessarily bias the results and risk their correct interpretation, with a waste of precious resources as the end result. The risk of drawing misleading conclusions is especially acute because individual variations in response tend to be large and lack of access to outlier data may mask significant problems.

Up to the present, only a small fraction of astronaut data has actually been released for use by qualified investigators, and the quality of the affected studies and the validity of the conclusions remain impossible to assess with confidence. The issue will become even more critical as ISS comes online, and longer-term human studies will be a high priority. The problem is particularly difficult because it sets two positive goods—confidentiality and accessibility of complete data—in potential conflict. However, the committee believes that the current policy and practice are counterproductive with respect to the cost-effectiveness of the affected—and in the worst case, unusable—research and already impede the search for ways to improve astronaut health and safety in space. NASA should seriously consider modifications of policy and practice that would better ensure full astronaut cooperation and compliance with the need for complete access to clinical data by qualified investigators. Results from medical experiments should not be used administratively to restrict subsequent flight opportunities for the participant astronauts.

Recommendations

- *NASA should initiate an ISS-based program to collect detailed physiological and psychological data on astronauts before, during, and after flight.*
- *NASA should make every effort to promote mechanisms for making complete data obtained from studies on astronauts accessible to qualified investigators in a timely manner. Consideration should be given to possible modifications of current policies and practices relating to the confidentiality of human subjects that would ethically ensure astronaut cooperation in a more effective manner.*

Publication and Outreach

An essential outcome of scientific research is publication—dissemination of results to the scientific community at large—and scientific peer review of findings and conclusions before publication is rightly considered a crucial component of the publication process. If any aspect of NASA's life sciences program has been deficient, it is its publication record, most especially in spaceflight experiments. There are several reasons for this poor publication record. A significant factor is that NASA has sometimes failed to provide funds (Research and Analysis funds) for data analysis and publication of flight experiments. Funding for such experiments has often been terminated before the principal investigator has had sufficient time to access the data and carry out the necessary analysis after the flight, leaving the investigator to somehow find other funds for data analysis and publication. NASA should provide a flexible mechanism whereby additional funding can be made available for these purposes for a reasonable period of time after completion of the flight to facilitate data analysis and publication of the results.

A second factor is that, in the past, NASA has not insisted on timely publication of results in peer-reviewed journals, for example, by including the publication record of NASA-funded investigators in peer-reviewed journals as an important criterion for continuation of research support. Too often, the results, if available at all, are found in NASA technical bulletins, which are not readily available to the scientific community and may not contain sufficient methodological information to allow adequate evaluation of the data. As an alternative to publication of individual experiments in separate journals, NASA might consider publishing a timely, detailed, single-volume, peer-reviewed compendium of the results of each flight, either in hard copy or online.

The Spaceline Archive, currently available online through the National Library of Medicine, was developed to make data from flight experiments more readily accessible to the general scientific community, and a second online archive has recently been established by JSC. However, entry of data from past flights is not yet complete at either site, access is not yet transparent, and much of the human spaceflight physiological data remains sequestered because of the current policy regarding release of data from human subjects, discussed above. The archive will be of limited value to the scientific community until data are complete and immediately accessible. The costs to the investigator of preparing data for archiving are real and an intrinsic part of the publication costs of the project; such costs should be included in NASA funding of flight experiments. Again, evaluation of the quality of the available data will be compromised unless attention is paid to including adequate detail on experimental conditions.

Recommendations

• *NASA should provide funding for data analysis and publication of flight experiments for a sufficient period to ensure analysis of the data and publication of the results.*

• *NASA should insist on timely publication and dissemination of the results of space life sciences research in peer-reviewed journals. For investigators with previous NASA support, the publication record should be an important criterion for subsequent funding.*

• *NASA should take as a high priority the completion of data entry into the Spaceline Archive and should ensure that access to the archive is simple and transparent. Funds for the preparation of data for archiving should be considered part of the direct costs of research projects.*

Professional Education

NASA should make every effort to ensure the professional training of graduate students and postdoctoral fellows in space and gravitational biology and medicine. Training programs should include components to enhance the retention of trainees in research areas of importance to NASA. These needs would be well met by a small, highly competitive program separate from the existing National Research Council fellowship program, designed to award individual postdoctoral fellowships to highly qualified candidates for training in laboratories of NASA-supported investigators outside NASA centers. Such a program could achieve the high visibility necessary to attract and catalyze the development of potential future leaders in the fields of space biology and medicine.

Recommendation

NASA should make every effort to support a small, highly competitive program of individual postdoctoral fellowships for training in laboratories of NASA-supported investigators in academic and research institutions external to NASA centers.

REFERENCES

1. Space Science Board, National Research Council. 1967. A Strategy for Space Biology and Medical Science for the 1980s and 1990s. National Academy Press, Washington, D.C.
2. Space Studies Board, National Research Council. 1991. Assessment of Programs in Space Biology and Medicine 1991. National Academy Press, Washington, D.C.

APPENDIXES

Appendix A

Acronyms and Abbreviations

ACTH	Adrenocorticotropic hormone
ADH	Antidiuretic hormone
ALS	Advanced Life Support (program)
ANS	Autonomic nervous system
AT	Ataxia telangiectasia
BMC	Bone mineral content
BMD	Bone mineral density
BMP	Bone morphogenetic proteins
CAM	Cell adhesion molecule
CELSS	Closed Ecological Life Support System program (NASA)
CRM	Crew resource management
CSA	Canadian Space Agency
CSBM	Committee on Space Biology and Medicine
CSST	Crew Status and Support Tracker
CVP	Central venous pressure
DEZ	Distal elongation zone (of a plant root)
DNA	Deoxyribonucleic acid
EEG	Electroencephalograph
EGG	Electrogastrogram
EKG	Electrocardiograph
ELISA	Enzyme-linked immunosorbent assay
ES	Embryonic stem (cell)
ESA	European Space Agency
EVA	Extravehicular activity
FGF	Fibroblast growth factor
fMRI	Functional magnetic resonance imaging
GCR	Galactic cosmic radiation

GH   Growth hormone
GM-CFU   Granulocyte macrophage colony forming unit
HPA   Hypothalamic-pituitary-adrenal axis
HU   Hindlimb unloading
HZE   High atomic number (Z), high energy
ICE   Isolated and confined environments
ICRP   International Commission on Radiological Protection
Ig   Immunoglobulin
IGF   Insulin-like growth factor
IL   Interleukin
IML   International Microgravity Laboratory
ISS   International Space Station
JSC   Johnson Space Center
LBNP   Lower body negative pressure
LEO   Low Earth orbit
LET   Linear energy transfer
LMS   Life and Microgravity Sciences (1996 shuttle mission)
LOFT   Line-oriented flight training
MRI   Magnetic resonance imaging
NAS   National Academy of Sciences
NASA   National Aeronautics and Space Administration
NASDA   National Space Development Agency of Japan
NCRP   National Council on Radiation Protection and Measurements
NIH   National Institutes of Health
NRA   NASA Research Announcement
NRC   National Research Council
NSCORT   NASA Specialized Centers of Research and Training
NSF   National Science Foundation
PCR   Polymerase chain reaction
PI   Principal investigator
PTH   Parathyroid hormone
RBC   Red blood cell
RBE   Relative biological effectiveness
RNA   Ribonucleic acid
RSA   Russian Space Agency
SAM   Sympathetic-adrenal-medullary system
SL-J   Space Laboratory-Japan
SLS   Spacelab for Life Sciences
SMS   Space motion sickness
SMSP   Shuttle-Mir Space Program
SPE   Solar particle event
TGF   Tumor growth factor
TNF   Tumor necrosis factor
VOR   Vestibular ocular reflex, vestibulo-ocular reflex

Appendix B

Glossary

Absorbed dose: Mean energy imparted by ionizing radiation to an object per unit mass (units: gray [Gy], rad).

Acquired immunity: Specific immune response to antigens, pathogens, tumors, and other foreign agents. It requires previous exposure to an agent, and the response is specific to the agent.

Acute effects of radiation: Effects that occur shortly after exposure to radiation, usually within a week. They result from exposure to radiation at relatively high doses, generally more than 1 Gy, and are usually due to the killing of cells in critical tissues in the body.

Acute phase proteins: Proteins secreted by the liver into the bloodstream to help the body respond to injury and stress.

Advanced Life Support System: Current NASA program in bioregenerative and nonbiological life support systems; supersedes the CELSS program.

Affect: Experience of emotion or feeling.

Affiliative need: Need to belong to a social group or to be in the presence of others.

Aldosterone: Hormone secreted by the renal cortex of the kidney and involved in the regulation of fluid and electrolyte balance.

Alveolus: Smallest unit of the lung, an air sac in which gas exchange occurs.

Amyloplast: Plant organelle that contains starch; because of its density, it sediments in response to the direction of the gravity vector.

Anabolic: Denoting a process by which living organisms convert simple substances into the complex materials of living tissue.

Anaerobiosis: Life in the absence of oxygen.

Anlage: Early stage of tissue development.

Anterograde amnesia: Loss of memory of events following trauma or shock.

Anteroposterior axis: Imaginary line running from head to tail in an animal.

Anthropometrics: Measurement of the dimensions of the human body (e.g., height, weight, length of trunk and limbs, and head and neck diameter).

Antibody: Member of a class of proteins called immunoglobulins, found in blood and other secretions and produced by lymphocytes in response to exposure to specific antigens with which they react.

Antigen: Substance recognized as foreign by the host immune system and that induces an immune response.

Arabidopsis thaliana: Flowering plant of the mustard family used for genetic and developmental studies of plants because of its small size, short generation time, and ease of laboratory culture.

Arthropod: Animal of the phylum that includes insects and crustaceans.

Ascinus: Ventilatory unit in the lung that includes terminal bronchioles and alveoli.

Asthenia: Lack or loss of strength; weakness or debility.

Ataxia telangiectasia: Disorder inherited as a recessive trait, characterized by neurological changes, immunological deficiency, increased susceptibility to cancer, and increased cellular radiosensitivity.

Autocrine: Denoting a mode of action in which a hormone is secreted from one part of a cell to act on another site within the same cell.

Autonomic nervous system: Complex nerve network that connects the central nervous system with the glands and smooth muscles of the body.

Autonomic vasomotor: Regulation of vascular peripheral resistance through parasympathetic and sympathetic neural stimulation of arteriole dilation and constriction.

Auxin: Plant hormone thought to be involved in control of gravitropism in plants; auxins regulate the rate of cell elongation, as well as a number of other developmental processes, in plants.

Basal energy expenditure: Energy expenditure by the body at rest.

B cells: Lymphocytes that play a major role in antibody production.

Biomechanics: Study of biological principles underlying motion in living organisms.

Bion: Russian space capsule that can support animals (e.g., monkeys, rats) and insects in orbit for up to 3 weeks.

Blastocyst: Early stage of a placental mammalian embryo.

Blastogenesis: Response of lymphocytes to exposure to an antigen or mitogen that involves their rapid and extensive replication, an indication of the host's ability to carry out an immune response.

Bone mineral content: Measure of bone mass based on absorption of x rays by the calcium in the bone.

Bone mineral density: Bone mineral content corrected for bone size (bone mineral content/ cross-sectional area).

Bone morphogenetic proteins: Large family of related molecules with proven role in development, discovered because of their ability to induce bone formation when injected into muscle or skin.

Brachiation: Use of upper extremity for grasping and ambulation.

Brassica rapa: Rapid cycling flowering plant used for developmental studies because of a short generation time, small stature, and hardy growth habit.

Caenorhabditis elegans: Nematode worm used in laboratory studies of genetics and development because of its short life cycle and simple structure.

Caloric stimulation: Infusion of cool or warm water (or air) into the external auditory canal to stimulate a horizontal semicircular canal.

Cancellous: Spongy, honeycomb-like interior of most bones.

Cell cycle: Complete generation of a cell, whose progress is tightly regulated at numerous steps.

Cell-killing effect: Cessation of cell division and/or metabolism. Sufficient doses of radiation can kill cells in the body, and this cell death is responsible for most of the acute effects of radiation.

Cell-mediated immunity: Specific immunity induced by previous exposure to an antigen or pathogen that can be protective to the host.

Cell transformation: Process by which cells in vitro, which have a limited ability to divide, are altered by radiation or chemicals so as to have unlimited potential for division.

Chrondrocyte: Cartilage cell.

Circadian rhythms: Regular biological cycles of sleep and activity characteristic of each species that synchronize an organism's internal environment with daily events in its surroundings.

***Cis*-acting factor or element:** DNA that binds to DNA to regulate gene expression.

Clinostat: Apparatus for rotating an object around its longitudinal axis. When this axis is horizontal (horizontal clinostat), the effect is to cancel out the direction of the gravity vector. Although this may mimic weightlessness, it is not identical to it.

Clone: Group of identical cells or organisms related by descent from a common ancestor.

Closed Ecological Life Support System: NASA program directed toward developing a plant-based system for spaceflight or extraterrestrial bases, to produce food and recycle waste products such as carbon dioxide. (Now part of the ALS program.)

Coleoptile: Sheath that covers the first leaf of a germinating monocot seedling (e.g., corn, oats) and protects the leaf from soil abrasion. Its growth is due primarily to rapid, auxin-regulated cell elongation.

Colony stimulating factor: Cytokine that regulates the production of white blood cells from bone marrow.

Conserved gene: A gene whose sequence is closely related in widely different types of organisms.

Coriolis acceleration: Inertial acceleration that arises when an object moves linearly with respect to a rotating reference frame.

Costamere: Protein complex consisting of cytoskeleton, transmembrane glycoproteins, and extracellular matrix that is at the level of Z bands involved in transferring tension from contractile elements to connective tissue and may serve as a mechanosensor for signal transduction.

Cupula: Dome-like structure that is part of the sensory receptor apparatus of a semicircular canal.

Cyclooxygenase 2: Enzyme that synthesizes prostaglandins and related substances.

Cytogenetic: Denoting the association of genes with particular locations on chromosomes.

Cytokine: Class of hormones that mediate immune (and other) responses.

Cytotoxic T cells: Lymphocytes responsible for specific destruction of pathogens and tumors after previous exposure.

Danio rerio: Freshwater fish (zebrafish) used in the study of developmental genetics because of its small size, transparency, rapid growth, and ease of culture.

Delayed-type hypersensitivity: Immune reaction observed in skin, characterized by swelling, reddening, and hardness.

Deterministic effects: Formerly known as nonstochastic effects, these may appear early or late after irradiation of an organism. Most deterministic effects involve cell killing.

Diastolic: The minimal blood pressure that occurs between beats while the heart is relaxed (e.g., for a blood pressure of 120/80, the diastolic pressure is 80).

Differential cDNA libraries: Collections of molecularly cloned DNA fragments representing the RNA molecules found in a specified subset of cell types.

Differential display: Method for identifying different RNA molecules found in a collection of cells.

Differentiation: Change of a cell from a generalized to a specialized type.

Distal elongation zone: Region of a root just behind the meristem, in which cells produced in the meristem first begin to elongate as well as increase in width.

Dominant negative: Mutant protein that prevents the action of its normal counterpart.

Dose: See *absorbed dose*.

Dose-effect (dose-response) model: Mathematical formulation used to predict the magnitude of an effect produced by a given dose of radiation.

Dose equivalent: See *equivalent dose*.

Dose rate: Quantity of absorbed dose delivered per unit of time.

Dystrophin: Cytoskeletal protein whose absence causes Duchenne muscular dystrophy.

Electrogastrogram: Recording of the electrical activity of the gastrointestinal system made with surface electrodes.

Electromyographic: Physiological recording of muscle action potentials generated during muscle contractile activity.

Electron volt: Unit of energy (equal to 1.6×10^{-19} J) (1 eV is equivalent to the amount of energy gained by an electron passing through a potential difference of 1 V).

Embryonic stem cells: Cultured early embryonic mouse cells that, after injection into developing mouse embryos, can be used to produce clones of differentiated cells in a mature mouse, including germ cells.

Endochondral: Bone development on a cartilage scaffold.

Endocrine: Pertaining to secretions from one organ that have a specific effect on another organ.

Endolymph: Fluid within the membranous labyrinth of the inner ear.

Endothelial cells: Cells that line blood vessels.

Endothelin: Regulatory peptide produced by endothelial cells.

Enkephalin: Neuropeptide related to pain control.

Enzyme-linked immunosorbent assay: Sensitive and rapid measurement based on the selective binding of antibodies.

Epinephrine: Catecholamine hormone produced by adrenal in response to stress.

Epiphyses: Extremities of long bones where growth and development occur.

Epitope: Portion of an antigen that determines its capacity to combine with its corresponding antibody in an antigen-antibody interaction.

Equivalent dose: Absorbed dose averaged over an organ or tissue and weighted for the radiation quality and type of radiation involved.

Ethylene: Gaseous hormone produced by plants that adversely affects plant growth.

Excess cancers: Number of individuals in a population who develop cancer over and above the number that would be expected to do so normally.

Expansins: Family of cell-wall localized plant proteins involved in loosening cell walls and permitting plant cells to expand.

Extensor digitorum longus: Anterior leg muscle that straightens the toes and lifts them and the foot off the ground.

Extravasation: Escape of blood from a vessel into the connective tissue.

Fast-twitch fibers: Skeletal muscle fibers that contract rapidly and are maximally activated for each muscle fiber action potential.

Fibroblast growth factor: Family of regulatory peptides that stimulate cell proliferation.

Fibronectin: Large extracellular glycoprotein with binding domains for heparin, cell, and collagen.

Flexor: Muscle that acts to decrease the angle of a joint.

Flight simulation: Ground-based training or experimental condition designed to simulate specific conditions of long-duration missions, including microgravity, isolation, and confinement.

Focal adhesion kinase: Enzyme activated by cell adhesion.

Forebrain: Top part of the brain mass, including the cerebrum, corpus callosum, thalamus, and hypothalamus.

Fractionation: Delivery of a given total dose of a radiation as several smaller doses, separated by intervals of time.

Functional magnetic resonance imaging: Technique for measuring the structural and functional organization of the brain.

Galactic cosmic rays: Ionizing radiation originating from the galaxy.

Gamma rays: Short-wavelength electromagnetic radiation of nuclear origin with an energy of about 10 keV to 9 MeV.

Genetic effects of radiation: Effects that arise from damage to genes in the germ cells of a parent which do not appear in the parent but may be passed on to offspring.

Genetic screen: Search through the progeny of a mutagenized population for particular types of mutants.

Genome adaptation syndrome: Model of the effects of stress on physical and psychological mechanisms, developed by Hans Selye.

Glucocorticoids: Steroid hormones produced by the adrenal cortex (cortisol, hydrocortisone, corticosterone) during stress.

Gluconeogenesis: Synthesis of glucose from noncarbohydrate precursors.

Golgi tendon organ: Sensory receptor within a muscle tendon that responds to tension.

Granulopoiesis: Formation of neutrophils and other white blood cells.

Gravitaxis: Swimming of an organism in a particular direction in response to the direction of the gravity vector.

Gravitropism: Orientation of plant stems and roots in response to the force of gravity.

Gray (Gy): SI unit of absorbed dose, equal to the energy transferred by ionizing radiation to a mass of matter corresponding to 1 joule/kg (equals 100 rads).

Growth factor: A protein that signals cells to divide or differentiate.

G-threshold: Minimum value of gravity that elicits a response in an organism.

Haptic: Relating to or based on sense of touch (i.e., arising from grasping, manipulating, or touching with the hand, or from contact with the body surface).

Helix-loop-helix superfamily: Family of proteins that dimerize with each other in order to bind to DNA and activate gene expression.

Helper T cells: Lymphocytes responsible for assisting B cells in making specific antibodies and helping cytotoxic T cells to kill their targets.

Hemopoiesis: Generation of blood cells in the bone marrow.

Hindlimb immobilization: Reduction of mechanical load and function in the hind limb of an experimental animal model.

Hindlimb unloading: Simulation of microgravity unloading of muscles in an animal model by employing various harnessing strategies to elevate the animal and prevent its hindlimbs from weight bearing on the floor.

Hippocampus: Part of the brain that is essential to the transfer of information from short-term to long-term memory.

Histogenesis: Process of formation of a tissue, with its characteristic cell types, from a progenitor cell population.

Histomorphometry: Computer-aided quantitative histology.

Homeobox: Domain found in a number of proteins that bind DNA and thereby regulate different genes; many of these proteins are necessary for critical events in animal development.

Homologous recombination: Replacement of a gene or segment of DNA with the corresponding gene or segment from another source.

H-reflex: Hoffman's reflex; reflexive contraction of a muscle elicited by electrical activation of muscle afferent fibers.

Humoral factors: Hormones and other substances delivered via the circulation.

Humoral immunity: Specific antibody-mediated immunity that is most effective against extracellular pathogens.

Hypercortisolinemia: Excess of cortisol.

Hypermetabolism: Abnormally increased utilization of material by the body.

Hypocotyl: Portion of a plant stem between the root and the first pair of leaves (i.e., the cotyledons).

Hypophysectomized: Describing an animal whose pituitary has been removed.

Hypothalamic-pituitary-adrenal axis: Neuroendocrine axis that interacts with the immune response. Mediates response to stress and other factors which could have a profound effect on immune responses and resistance.

HZE particles: Heavy (high-atomic-number), high-energy particles (e.g., carbon or iron nuclei) in cosmic rays, with an energy range of about 10^2 to 10^3 MeV per nucleon.

Immunoglobulin: Serum glycoprotein that can bind specific molecules.

Innate immunity: Nonspecific immunity that is always present and constitutes the first defense against pathogens.

Insulin-like growth factor: Proteins related to insulin that influence the survival, metabolism, and proliferation of cells.

Integrins: Dimeric cell surface receptors that mediate cell-cell and cell-matrix interaction and are involved in attachment, migration, and signaling.

Interferons: Cytokines that play a fundamental role in regulation of immune response.

Interleukins: Cytokines that play a major regulatory role in immune response.

Interstitium: Extracellular connective tissue between cells.

Intronic enhancer: DNA sequence that activates transcription from a promoter by serving as a specific binding site for gene regulatory proteins.

Ionizing radiation: Radiation that can penetrate and deposit its energy at random within cells and tissues by ejecting electrons from atoms, thereby ionizing them.

Ischemic necrosis: Cell and tissue death resulting from severe reduction of blood flow.

Knockout mice: Mice from which certain genes have been removed.

Knockout mutation: Deletion or mutation that results in the inability to manufacture the protein product of the mutant gene.

Labyrinthectomy: Destruction of the labyrinth of the inner ear.

Laminin-2 (merosin): Member of the family of basement membrane proteins present in striated muscle.

Latent period: Time between exposure to an agent and expression of a disease.

Latent viral epitope: Inactive virus to which carriers will demonstrate an immune response.

Linear energy transfer: Average amount of energy lost per unit of particle track length. Low-LET radiation is characterized by light charged particles such as electrons. High-LET radiation is characterized by heavy charged particles such as alpha particles and heavy nuclei.

Lymphocyte: Type of white blood cell responsible for immunological function.

Macrophage: White blood cell that can nonspecifically engulf pathogens and destroy them.

Mechanostat: Hypothetical mechanism (receptor) that senses mechanical strain and maintains bone mass accordingly.

Mitogen: Substance that induces cell division.

Molecular cascade: Group of proteins that act sequentially in the same process.

Muscle spindle: Sensory receptor within a muscle that is sensitive to its length and rate of change in length; spindle sensitivity is controlled by gamma motoneurons of the spinal cord innervating striated muscle fibers.

Musculovenous pumping: Compressive squeezing of blood from intramuscular veins by contracting skeletal muscle fibers.

Myocyte: Differentiated skeletal, cardiac, or smooth muscle cell.

MyoD family: Closely related myogenic proteins that can activate the expression of muscle-specific genes.

Myopathy: Degenerative disease of muscle thought to originate from a primary defect in muscle tissue.

Natural killer cells: Cells found in the body under normal conditions that can destroy target virus-infected and tumor cells by an as yet unknown recognition mechanism.

Neoplastically transformed cells: Tissue culture cells changed in vitro from growing in an orderly pattern and exhibiting contact inhibition to growing in a pattern more like that of cancer cells.

Neural imaging: Term used to refer to the application of dyes or radioactive compounds that allow the visualization of neuron morphology (e.g., computer tomography, magnetic resonance imaging).

Neurolab: Shuttle mission (1998) dedicated to studies in the neurosciences.

Neuroplasticity: Long-term changes in structure or function exhibited by neurons after changes in their activity.

Neutrophil: Type of white blood cell.

Nonstochastic effect: See *deterministic effects*.

Nuclear matrix: Fibrous material within the cell nucleus.

Nystagmus: Rhythmic oscillation of the eyeballs.

Oculomotor: Related to the control of eye movement.

Os calcis: Heel bone.

Osteoblast: Bone-forming cell.

Osteocalcin: Protein selectively expressed by bone-forming cells and present in bone matrix.

Osteoclast: Multinucleated, bone-resorbing cell related to macrophages.

Osteogenic: Forming bone.

Otoconia: Calcium carbonate crystals embedded in a gelatinous membrane of the inner ear, whose motion depends on acceleration.

Otolith: Component of the inner ear sensitive to linear acceleration.

Oxidative burst: Sign of neutrophil function indicating oxygen-based destruction of a foreign invader.

Paracrine: Referring to a mode of action in which hormones act on neighboring cells.

Parasympathetic: Pertaining to the division of the autonomic nervous system that controls most of the basic metabolic functions essential for life.

Parathyroid hormone: Hormone produced by the parathyroid gland in response to low plasma calcium.

Patch clamp: Method for measuring voltage and current across cell membranes by using glass microelectrodes.

Periosteum: Exterior layer of bone composed of several cell layers and fibrous tissue.

Plantar flexion: Movement at the ankle joint that results in the ball of the foot moving toward the floor.

Plasmodesmata: Complex cytoplasmic tubes that connect the cytoplasm of two plant cells and allow small molecules to pass freely between the cells.

Polar auxin transport: Cell-to-cell movement of auxin in plants.

Prolactin: Hormone responsible for control of lactation in females.

Proprioception: Conscious awareness of the positions of various parts of the body in space provided by joint and muscle sensory inputs; also, sensory stimulation arising from receptors within muscles.

Prostaglandin: Regulatory molecule derived from lipids that act locally.

Protonema: Filament of cells that grows from a fern spore.

Proto-oncogene: Gene that codes for proteins involved in the signaling pathways stimulated by growth factor; when mutated, this gene causes cancer and is called an oncogene.

Protoplasts: Plant cells from which the cell wall has been removed.

Quality factor: LET-dependent factor by which absorbed doses are multiplied to obtain the equivalent dose for radiation protection purposes.

Radiosensitivity: Relative susceptibility of cells, tissues, organs, and organisms to the injurious action of radiation.

Recall antigen: Standard testing antigen to which most individuals have been exposed and will respond by exhibiting delayed hypersensitivity.

Relative biological effectiveness: Biological potency of one type of radiation compared with another that produces the same biological end point.

Reporter genes: Genes that, when fused to the gene of interest, yield a hybrid protein that can be readily detected or visualized in intact cells and tissues.

Reticulocyte: Young red blood cell.

Retrograde amnesia: Loss of memory for events during a circumscribed period prior to a traumatic event, brain injury, or damage.

Rhizoid: Root-like organ of some algae, mosses, and ferns that consists of a single cell or filament of cells.

Root cap: A protective cap of cells at the distal end of a root that covers its meristem and senses the direction of gravity.

Saccadic: Pertaining to rapid movement of the eyes from one position to another.

Saccule: Smaller chamber of the membranous labyrinth of the inner ear.

Sarcolemma: Connective tissue sheath surrounding a muscle fiber that is composed of the muscle plasma membrane, the basement membrane, and the adjoining   connective tissue or endomysium.

Sarcomere: Fundamental unit of contraction arranged in repeating series to form myofibrils within a striated muscle cell.

Saturation mutagenesis: Treatment of a sufficient population of some organism with a mutagen, so that descendants have a high probability of exhibiting mutations in all genes.

Sensory deprivation: Term referring to the deficits that a neuron undergoes after losing input from part of its afferent fibers due to nerve lesion, exposure to drugs, or decreased exposure to adequate sensory stimulus.

Sievert: SI unit of radiation equivalent dose, equal to dose in grays times a quality factor, times other modifying factors (1 sievert = 100 rem).

Signal transduction: Process by which a cell responds to a stimulus.

Skylab: Prototype U.S. space station that flew three missions of 28, 56, and 84 days' duration in the early 1970s.

Sled: Rail-mounted device for delivering controlled linear acceleration to a subject positioned on it.

Slow-twitch fibers: Skeletal muscle fibers that contract slowly and are maximally activated for each muscle fiber action potential.

Solar maximum: Period of maximum probability of emission of solar event radiation (e.g., protons, alpha radiation, electromagnetic energy).

Solar minimum: Period of minimum probability of emission of solar event radiation.

Solar particle event: Flux of energetic ions and/or electrons of solar origin.

Soleus: Slowly contracting muscle in the leg that produces plantar flexion of the foot to raise the body up against gravity.

Somatic effects of radiation: Effects arising from damage produced in various tissues of an irradiated individual's body.

Somatosensory: Pertaining to sensory stimulation of tactile receptors of the body surface.

Somites: Paired blocks of mesodermally derived cells organized segmentally along the developing spinal cord to generate bone (vertebrae) and skeletal muscle cells.

Sopite syndrome: Form of motion sickness associated with prolonged exposure to unusual motion or gravity conditions whose primary features are drowsiness, fatigue, lack of initiative, apathy, and irritability.

Spacelab: Module specially constructed for the Shuttle payload bay for use in scientific experiments; a self-contained laboratory that can be loaded onto shuttle missions as needed.

Space map: Organization of groups of neurons into specific arrangements in the central nervous system that reflect the structure and/or function of the spatial environment representing the sensory or motor systems.

Spemann's organizer: Region of developing amphibian embryo in which cells play a critical role in signaling positional information to nearby cells.

Statolith: Dense body within a plant cell, whose sedimentation provides information to the cell about the direction of the gravity vector.

Stochastic effects: Effects resulting from random events; their probability of occurrence in an exposed population of cells or individuals is a direct function of dose; these effects are commonly regarded as having no threshold.

Stromal cells: Resident cells in the bone marrow that do not develop into blood cells but may support this process; they can differentiate into bone or fat cells.

Subtractive hybridization: Method by which genes expressed in a tissue-specific manner can be enriched for cloning.

Sympathetic: Pertaining to the division of the autonomic nervous system that is active in emergency conditions of extreme cold, violent effort, and emotions.

Systolic: Peak blood pressure that occurs at the end of heart contraction (e.g., for a blood pressure of 120/80, the systolic pressure is 120).

Telomere: Special DNA structure at the ends of chromosomes.

Th1 cytokine profile: Profile of cytokines produced by helper T cells that promote the development of a cell-mediated immune response.

Th2 cytokine profile: Profile of cytokines produced by helper T cells that promote the development of an antibody-mediated immune response.

Third-quarter phenomenon: Marked decline in individual and group performance believed to occur between the midpoint and the beginning of the last quarter of a specified period of isolation and confinement.

Thrombin: Protease that stimulates clotting and other cellular processes via a specific receptor.

Tibialis anterior: Predominantly fast muscle that dorsiflexes the ball of the foot off the floor during walking.

Tonus: Muscle tension that is present independent of voluntary innervation.

Topographic: Orderly representation reflecting regional anatomy and/or function.

***Trans*-acting factor:** Protein that binds to DNA and regulates gene expression.

Transformed cell lines: See *cell transformation.*

Transgenic mice: Mice into which genetic material from another organism has been experimentally inserted.

Transport code: Computer program that calculates the particle distributions and energy behind a specific shield, derived from the basic nuclear cross sections for interactions and fragmentation in shielding.

Trochanter: Anatomical region in the hip bone.

Tropic response: Change in direction of growth by a plant organ in response to an environmental factor such as the direction of gravity.

Tumor growth factor-β: Regulatory peptide that stimulates bone formation when injected into bone.

Tumor necrosis factor: Peptide cytokine involved in inflammation; produced by macrophages.

Tumor necrosis factor-α: Cytokine that destroys tumor cells and regulates immune responses.

Tumor suppressor: Gene or gene product that acts to regulate cell replication.

T-wave: Brain wave having a frequency of only 5 to 7 cycles per second.

Ubiquitination: Covalent linkage of a small protein, ubiquitin, to the lysine amino groups of proteins, which targets them for proteolysis.

Vestibular: Pertaining to the sensory system concerned with maintenance of posture and balance by perceiving gravity (linear acceleration) and rotary movements of the head.

Vestibular hair cells: Cells in the vestibular system of the inner ear.

Vestibular ocular reflex: Passive eye movements elicited by activation of receptors in the vestibular apparatus.

Visual field: Entire expanse of space visible at a given instant without moving the eyes.

Water use efficiency: Ratio of CO_2 assimilated into carbohydrates during photosynthesis to H_2O lost from leaves by transpiration; a high ratio is desirable for maximum efficiency in plants.

Xeroderma pigmentosum: Inherited disease in which individuals are highly susceptible to sun-induced cancer.

Z-axis: Longitudinal axis of the body.

Zeitgeber: Periodic feature in the social and physical environment that entrains the circadian rhythms of an organism.

Appendix C

Workshops

CELL BIOLOGY WORKSHOP

February 12-15, 1996
Houston, Texas

Sessions

Gene Expression
Cytoskeleton
Hormones and Intracellular Signaling
Cell Membranes
Cell-Matrix and Cell-Cell Interaction

Participants

Elliot Meyerowitz,* California Institute of Technology, *Co-chair*
Steven E. Pfeiffer,* University of Connecticut Health Center, *Co-chair*
Daniel D. Bikle, Veterans Administration Medical Center, San Francisco, Calif.
Janet Braam, Rice University
William R. Brinkley, Baylor College of Medicine
Elisabeth Burger, ACTA, Free University, Amsterdam, The Netherlands
W. Zacheus Cande, University of California at Berkeley
Robert E. Cleland,* University of Washington

Don W. Cleveland, University of California, San Diego
David R. Colman, Mt. Sinai School of Medicine
Mary F. Dallman,* School of Medicine, University of California, San Francisco
E.A. Dawidowicz, Marine Biological Laboratory, Woods Hole, Mass.
Charles Emerson, School of Medicine, University of Pennsylvania
John Frangos, University of California, San Diego
Clara Franzini-Armstrong, University of Pennsylvania
Robert D. Goldman, Northwestern University Medical School
Alan D. Grinnell, School of Medicine, University of California at Los Angeles
Magnus Höök, The Texas Medical Center
Donald E. Ingber, Boston Children's Hospital
Steve Kay, University of Virginia
Ching Kung, University of Wisconsin, Madison
Ira Mellman, Yale University School of Medicine
Marian L. Moore-Lewis, University of Alabama, Huntsville
Richard I. Morimoto, Northwestern University
John E. Mullet,* Texas A&M University
Mary Jane Osborn, University of Connecticut Health Center
Barbara Pickard, Washington University
Howard Rasmussen, Medical College of Georgia
Urs Rutishauser, School of Medicine, Case Western Reserve University
Edward Schultz, University of Wisconsin Medical School
Tom K. Scott, Visiting Scientist, NASA Headquarters
Robert B. Silver, Marine Biology Laboratory, Woods Hole, Mass.
Kenneth Souza, NASA Ames Research Center
Herman Vanderburgh, The Miriam Hospital, Providence, R.I.
Pedro Verdugo, University of Washington

Program Committee

Charles J. Arntzen, Robert E. Cleland,* Mary F. Dallman,* Sandra J. Graham,* Anthony P. Mahowald,
    Mary Jane Osborn,* Steven E. Pfeiffer*

---

*Participated in development of workshop summary.

DEVELOPMENTAL BIOLOGY WORKSHOP

August 19-22, 1996
Irvine, California

Sessions

Gametogenesis, Early Development
Pattern Formation
Specialized Organogenesis
Neural Development
Axis Formation and Cellular Determination

Participants

Anthony P. Mahowald,* University of Chicago, *Chair*
Jeffrey R. Alberts, Indiana University
Norma M. Allewell,* University of Minnesota
Robert Baker, New York University Medical Center
Marianne Bronner-Fraser, California Institute of Technology
Catherine E. Carr, University of Maryland
Robert E. Cleland,* University of Washington
Edward M. De Robertis, University of California, Los Angeles
James H. Eberwine, University of Pennsylvania Medical Center
Lewis J. Feldman, University of California, Berkeley
Steven R. Heidemann, Michigan State University
Robert K. Ho, Princeton University
Richard Hyson, Florida State University
Abraham D. Krikorian, State University of New York, Stony Brook
Michael S. Levine, University of California, San Diego
Elliot Meyerowitz,* California Institute of Technology
Randall T. Moon, School of Medicine, University of Washington
D. Kent Morest,* University of Connecticut Health Center
Mary E. Musgrave, Louisiana State University
Lee Niswander, Memorial Sloan-Kettering Cancer Center
Eric N. Olson, University of Texas Southwestern Medical Center at Dallas
Mary Jane Osborn,* University of Connecticut Health Center
Kenna D. Peusner,* George Washington University Medical Center
Steven E. Pfeiffer,* University of Connecticut Medical School
Richard M. Schultz, University of Pennsylvania
Antoinette Steinacker, Marine Biological Laboratory, Woods Hole, Mass.
Paul W. Sternberg, California Institute of Technology
Frank M. Sulzman, NASA Headquarters

---

*Participated in development of workshop summary.

Charles Wade, NASA Ames Research Center
Debra J. Wolgemuth, Columbia University College of Physicians and Surgeons

Program Committee

Norma M. Allewell,* Robert E. Cleland,* Sandra J. Graham,* Anthony P. Mahowald,* Elliot Meyerowitz,* Mary Jane Osborn,* Kenna D. Peusner,* Steven E. Pfeiffer*

---

*Participated in development of workshop summary.

SYSTEMS PHYSIOLOGY WORKSHOP

May 28-31, 1997
Washington, D.C.

Sessions

Endocrine and Neuroendocrine Systems
Immunology
Sensorimotor Integration
Bone and Muscle Physiology
Cardiovascular and Cardiopulmonary Systems

Participants

James Lackner, Brandeis University, *Chair*
Clarence P. Alfrey, Baylor College of Medicine
Claude D. Arnaud,* University of California at San Francisco
C. Gunnar Blomqvist,* University of Texas Southwestern Medical Center
David Cardús, Baylor College of Medicine
Charles A. Czeisler, Brigham and Women's Hospital, Harvard Medical School
John P. Donoghue, Brown University
Donald S. Faber, Allegheny University of the Health Sciences
David Felten,* University of Rochester
Sherre Florence, Vanderbilt University
Mary Anne Frey, NASA Headquarters
Francis (Drew) Gaffney, Vanderbilt University Medical Center
Bernard P. Halloran, Veterans Administration Medical Center, San Francisco
Stephen M. Highstein, Washington University School of Medicine
Lynette A. Jones, Massachusetts Institute of Technology
Mary Jane Osborn,* University of Connecticut Health Center
Kenna D. Peusner, The George Washington University Medical Center
G. Kim Prisk, University of California at San Diego
Danny A. Riley,* Medical College of Wisconsin
Catherine L. Rivier, The Salk Institute for Biological Studies
Gideon A. Rodan,* Merck Research Laboratories
Victor S. Schneider, NASA Headquarters
Gerald Sonnenfeld,* Carolinas Medical Center
T. Peter Stein,* University of Medicine and Dentistry of New Jersey
Frank M. Sulzman, NASA Headquarters
Larry Suva, Beth Israel Deaconess Medical Center
Norman E. Thagard, Florida State University

---

*Participated in development of workshop summary.

Marc Tischler, University of Arizona
Joan Vernikos, NASA Headquarters
Charles E. Wade, NASA Ames Research Center
G. Donald Whedon, Consultant, Biomedical Research, Clearwater, Fla.
Ronald L. Wilder, National Institute of Arthritis and Musculoskeletal and Skin Diseases,
    National Institutes of Health

Program Committee

Francis (Drew) Gaffney, Sandra J. Graham,* James Lackner, Mary Jane Osborn,* Kenna D. Peusner,
    Gideon A. Rodan*

---

*Participated in development of workshop summary.

Appendix D

Committee Biographies

Mary J. Osborn is professor and head of microbiology at the University of Connecticut Health Center. Dr. Osborn's fields of specialization are biochemistry, microbiology, and molecular biology. Current research interests include biogenesis of bacterial membranes. Dr. Osborn has served on numerous distinguished committees, including the National Science Board (1980-1986), the President's Committee on the National Medal of Sciences (1981-1982), the Advisory Council of the National Institutes of Health's Division of Research Grants (1989-1994; chair, 1992-1994), the Advisory Council of the Max Planck Institute of Immunobiology (1974-1978), the Board of Scientific Advisors for the Roche Institute for Molecular Biology (1981-1985; chair, 1983-1985), and the Governing Board of the National Research Council (1990-1993). Memberships include the National Academy of Sciences, American Association for the Advancement of Science, American Society of Biochemistry and Molecular Biology (president, 1981-1982), American Chemical Society (chairman, Division Biological Chemistry, 1975-1976), the American Academy of Arts and Sciences (fellow; council, 1988-1992), Federation of American Societies for Experimental Biology (president, 1982-1983), American Society for Microbiology, and American Academy of Microbiology. Dr. Osborn received a B.A. from the University of California, Berkeley, and a Ph.D. (biochemistry) from the University of Washington.

Norma Allewell is a professor of biochemistry (and of chemistry and computational science, as well as vice provost for Research and Graduate/Professional Education, at the University of Minnesota. Dr. Allewell has expertise in the fields of molecular biophysics, structural biology, and biochemistry. Research interests include protein structure, function, and design; macromolecular interactions; and computer modeling. Memberships include the Biophysical Society (president, 1993-1994), American Association for the Advancement of Science (fellow), American Society for Biochemistry and Molecular Biology, and Sigma Xi. Dr. Allewell received a B.Sc. (honors) from McMaster University and a Ph.D. (molecular biophysics) from Yale University.

Robert Cleland is a professor of botany and director of the Biology Program at the University of Washington. His expertise is in physiology, plant hormones, and cell walls. The main focus of Dr. Cleland's research has been the mechanism by which plant cell elongation is controlled. He has served on the gravitropism panel and has had experience with National Aeronautics and Space Administration (NASA) programs while serving on internal NASA Advisory Committees. In addition to his experience in plant physiology, he has familiarity with biological issues associated with closed ecological life support systems (CELSS). This is important in the context of the Space Station and long-duration human exploration of space. Memberships include the American Society of Plant Physiologists (secretary, 1971-1973; president, 1974-1975), the American Association for the Advancement of Science (fellow), and the American Society for Gravitational and Space Biology. Dr. Cleland received a B.A. from Oberlin College and a Ph.D. (plant physiology) from the California Institute of Technology.

F. Andrew (Drew) Gaffney is professor of medicine and director of Clinical Cardiology at Vanderbilt University Medical Center. Dr. Gaffney's expertise is in the medical aspects of spaceflight, and his central interest is the elucidation of interactions between body fluid distribution and neurohumoral regulatory mechanisms of cardiovascular control. His recent studies have addressed mechanisms of orthostatic hypotension, especially those associated with exposure to the microgravity of space. Since 1984, Dr. Gaffney has been heavily involved in the space program, which has given him unique perspective and insight into NASA's life sciences research program. As a payload specialist on the Spacelab Life Sciences 1 Mission, Dr. Gaffney went into space with a central venous catheter in place, which provided the first critical invasive data on the cardiovascular effects of weightlessness. Dr. Gaffney received an M.D. from the University of New Mexico.

James R. Lackner, who is director of Brandeis University's Ashton Graybiel Spatial Orientation Laboratory and Riklis Professor of Physiology, is an expert on neurovestibular systems. Research interests concern human spatial orientation and motor control, and his research concentrates on adaptation to spaceflight and the etiology of space motion sickness. Memberships include the American Society for Gravitational and Space Biology, the Aerospace Medical Association, the Society for Neuroscience, the Psychonomics Society, and the International Brain Research Organization. He has been honored by election to the Bárány Society and the International Academy of Astronautics, and has been awarded the Arnold B. Tuttle Award of the Aerospace Medical Association for "Outstanding Contributions to Aviation Medicine." Dr. Lackner received a B.Sc. and a Ph.D. (psychology and brain sciences) from the Massachusetts Institute of Technology.

Anthony P. Mahowald is the Lewis Block Professor and chairman of the Department of Molecular Genetics and Cell Biology and the Committee on Developmental Biology at the University of Chicago. Dr. Mahowald specializes in developmental genetics and developmental biology. Research interests include developmental and molecular genetics of *Drosophila*; vitellogenesis; molecular analysis of maternal effect mutations affecting both germ cell formation and gastrulation; oogenesis; and sex determination of germ cells. Memberships include the National Academy of Sciences, American Academy of Arts and Sciences, Genetics Society of America (secretary, 1986-1988), American Society of Cell Biologists (council, 1996-1999), Society for Developmental Biology (president, 1989), and American Association for the Advancement of Science (fellow). Dr. Mahowald received a B.S. from Spring Hill College and a Ph.D. (biology) from Johns Hopkins University.

Elliot Meyerowitz is a professor of biology at the California Institute of Technology. Dr. Meyerowitz has expertise in molecular genetic methods for plant and animal research. Recent work of his laboratory has concentrated on three areas: the origin of developmental patterns in flowers, the control of cell division in meristems, and the mechanisms of plant hormone action. Memberships include the National Academy of Sciences, Genetics Society of America, American Association for the Advancement of Science (fellow), American Academy of Arts and Sciences, International Society for Plant Molecular Biology, American Society of Plant Physiologists, and Botanical Society of America. He has received the Pelton Award of the Botanical Society of America and the Conservation and Research Foundation, the Gibbs Medal of the American Society of Plant Physiologists, the Genetics Society of America Medal, the Mendel Medal of the U.K. Genetical Society, and the International Prize for Biology. Dr. Meyerowitz received an A.B. from Columbia University and an M.Phil. and a Ph.D. (biology) from Yale University; he did postdoctoral research at the Stanford University School of Medicine.

Lawrence A. Palinkas is director of research in the Division of Family Medicine; director of the Immigrant/Refugee Health Studies Program; and professor in the Department of Family and Preventive Medicine at the University of California at San Diego. He is a medical anthropologist and a leading expert on behavioral adaptation by groups to extreme and isolated environments, and his studies have included a focus on both the care and the prevention of adverse effects. He has expertise in medical ecological and applied anthropology; social, cultural, and psychiatric epidemiology; and environmental and health psychology. He has served on numerous NASA and U.S. Navy advisory groups on behavioral issues. He is a member of the American Public Health Association, American Anthropological Association (fellow), Society for Medical Anthropology, the Society for Psychological Anthropology, Society for Applied Anthropology (fellow), Society for Behavioral Medicine, and American Psychosomatic Society. Dr. Palinkas received a B.A. from the University of Chicago and an M.A. and a Ph.D. in anthropology from the University of California, San Diego.

Kenna D. Peusner is a professor of anatomy and cell biology at the George Washington University School of Medicine. Dr. Peusner is a neurobiologist specializing in intracellular electrophysiological and microscopic techniques to investigate neural structure and function. Her research is focused on characterizing synaptic transmission and ionic conductances and their role in the emergence of excitability in the developing and damaged central vestibular system. Memberships include the Neuroscience Society, Association for Research in Otolaryngology, New York Academy of Science, and American Association for the Advancement of Science. She received the Lindback Foundation award for distinguished teaching in the basic medical sciences, Jefferson Medical College. She is a grantee of the National Institute on Deafness and Other Communicative Disorders, National Institutes of Health. Dr. Peusner received a B.S. from Simmons College and a Ph.D. (anatomy) from Harvard University.

Steven E. Pfeiffer is a professor of microbiology at the University of Connecticut Health Center. Dr. Pfeiffer has expertise in molecular cell biology and neurobiology. His research interests are in molecular, cell, and developmental biology of the nervous system and myelinogenesis. He is the recipient of the Javitz Neuroscience Investigator Award from the National Institutes of Health. Memberships include the American Association of Cell Biologists; American Society for Neurochemistry; International Society for Neurochemistry, of which he is president; and Society for Neuroscience. Dr. Pfeiffer received a B.A. from Carleton College and a Ph.D. (molecular biology) from Washington University.

Danny A. Riley is a professor of cell biology, neurobiology, and anatomy at the Medical College of Wisconsin. Dr. Riley's expertise is in the mechanisms of muscle atrophy and nerve regeneration in animal models and humans, with emphasis on space biology. He was a recipient of the American Institute of Aviation and Astronautics Jeffries Medical Research Award in 1992 for outstanding contributions to the advancement of aerospace medical research and two NASA Group Achievement awards, for the Cosmos 2044 Biosatellite Team (1991) and the Spacelab Life Sciences-2 Team (1993). He is an elected member of the Board of Directors of the American Society for Gravitational and Space Biology (1989-1993, 1997-). Other memberships include the American Association of Anatomists, International Society of Electromyographic Kinesiology, Society for Neuroscience, American Society for Cell Biology, Aerospace Medical Association, and American Institute of Biological Sciences. Dr. Riley received a B.S. and a Ph.D. (anatomy) from the University of Wisconsin, Madison.

Gideon A. Rodan is the vice president for bone biology and osteoporosis research at the Merck Laboratories, in West Point, PA. Dr. Rodan is an expert in the field of hard tissues, and his research interests include cell biology of hard tissues, and the hormonal control of growth and differentiation in bone-derived cells. Dr. Rodan's research has focused on the relationship of osteoblasts and osteoclasts in bone resorption, bone formation, and osteoporosis. Dr. Rodan is a member of the American Society for Cell Biology, the Endocrine Society, the International Bone and Mineral Society, and has served as President of the American Society of Bone and Mineral Research. Dr. Rodan is the recipient of the Neuman Award, for pioneering research in the field of bone metabolism. He received his M.D. from Hebrew University, and his PhD from the Weizmann Institute of Science.

Richard Setlow is associate director for Life Sciences at Brookhaven National Laboratory. Dr. Setlow is an expert in the fields of radiation biophysics and molecular biology. He has served on numerous National Research Council committees. Research interests include far-ultraviolet spectroscopy; ionizing and nonionizing radiation; molecular biophysics; action of light on proteins, viruses, and cells; nucleic acids; repair mechanisms; and environmental carcinogenesis. Dr. Setlow received the Finsen Medal in 1980 for "Outstanding Contribution to Photobiology and Repair of Nucleic Acids" and the Enrico Fermi Award in 1989 from the U.S. Department of Energy for "pioneering and far-reaching contributions to the fields of radiation biophysics and molecular biology." Memberships include the National Academy of Sciences, American Association for the Advancement of Science, Biophysical Society, the American Society for Photobiology, Environmental Mutagen Society, and American Association for Cancer Research. Dr. Setlow received an A.B. from Swarthmore College and a Ph.D. (physics) from Yale University.

Gerald Sonnenfeld is director of research immunology and a senior scientist at the Carolinas Medical Center. He is also adjunct professor of diagnostic sciences at the University of North Carolina at Chapel Hill. His expertise in the field of immunology is interferon and cytokine research. Dr. Sonnenfeld has served on numerous peer review and advisory groups for immunology research programs of NASA and other agencies, and is currently program director of NASA's Space Biology Research Associates Program. Dr. Sonnenfeld is president of the American Society for Gravitational and Space Biology. Other memberships include the American Association of Immunologists, American Society for Microbiology, American Society for Virology, International Cytokine Society, International Society for Interferon Research (charter member), International Society for Antiviral Research, Sigma Xi, Society for Leukocyte Biology, and Tissue Engineering Society (founding board member). Dr. Sonnenfeld received a B.S. from the City College of New York and a Ph.D. (microbiology/immunology) from the University of Pittsburgh School of Medicine. He carried out a postdoctoral fellowship in immunology and infectious diseases at the Stanford University School of Medicine.

T. Peter Stein is a professor of surgery and nutrition at the University of Medicine and Dentistry of New Jersey. His expertise is in the areas of clinical nutrition and protein and energy metabolism during spacelift, lipid metabolism, clinical nutrition, nutritional assessment, and lung biochemistry. Dr. Stein was a co-winner of the American Institute of Aviation and Astronautics Jeffries Medical Research Award in 1992 for his work on Spacelab Life Sciences-1. Memberships include the American Association for the Advancement of Science, American Institute of Nutrition, American Society for Clinical Nutrition, American Physiological Society, Society for Parenteral and Enteral Nutrition, American Chemical Society, American College of Nutrition, and American Society for Gravitational Physiology. Dr. Stein received a B.Sc. from the Imperial College of Science and Technical University of London; an M.Sc. (biochemistry) from University College, University of London; and a Ph.D. (molecular biology/chemistry) from Cornell University.